ADAPTIVE DISASTER RISK ASSESSMENT
COMBINING MULTI-HAZARDS WITH SOCIOECONOMIC VULNERABILITY AND DYNAMIC EXPOSURE

Neiler de Jesús Medina Peña

Cover design

Camilo A. Triana C. / trianacartoon@gmail.com / trianacartoon.blogspot.com

ADAPTIVE DISASTER RISK ASSESSMENT
COMBINING MULTI-HAZARDS WITH SOCIOECONOMIC VULNERABILITY AND DYNAMIC EXPOSURE

DISSERTATION

Submitted in fulfillment of the requirements of
the Board for Doctorates of Delft University of Technology
and
of the Academic Board of the IHE Delft
Institute for Water Education
for
the Degree of DOCTOR
to be defended in public on
Monday, 21 June 2021 at 17:30 hours
in Delft, the Netherlands

by

Neiler de Jesús MEDINA PEÑA

Master of Science in Water Science and Engineering, UNESCO-IHE Institute for Water Education, the Netherlands

born in Medellín, Colombia

This dissertation has been approved by the
promotor: Prof. dr. D. Brdjanovic and
copromotor: Dr. Z. Vojinovic

Composition of the Doctoral Committee:

Rector Magnificus TU Delft	Chairman
Rector IHE Delft	Vice-Chairman
Prof.dr. D. Brdjanovic	TU Delft/IHE Delft, promotor
Dr. Z. Vojinovic	IHE Delft, copromotor

Independent members:

Prof.dr.ir. M. Kok	TU Delft
Prof.dr.ir. A.E. Mynett	TU Delft / IHE Delft
Prof.dr. J.P. O'Kane	University College Cork, Ireland
Prof.dr.-Ing. P. Fröhle	Hamburg University of Technology, Germany
Prof.dr.mr.ir. N. Doorn	TU Delft, reserve member

This research was conducted under the auspices of the Graduate School for Socio-Economic and Natural Sciences of the Environment (SENSE)

CRC Press/Balkema is an imprint of the Taylor & Francis Group, an informa business

Published by:
CRC Press/Balkema
enquiries@taylorandfrancis.com
www.crcpress.com – www.taylorandfrancis.com
ISBN 978-1-032-11617-4

ACKNOWLEDGMENTS

The end of another stage of my life; one that took more time than initially intended, but by far is the most memorable time I had lived, a period full of learning experiences in every aspect of my life. This journey would not have been possible or as enjoyable without the support of many people that I have the privilege to cross during the development of this dissertation.

I want to acknowledge my supervisory team, who guide me through the academic or personal challenges I faced during these years. To my promotor, Damir Brdjanovic, thank you for your support, your great patience and for that final push that encouraged me to finish the writing part of this dissertation. To my co-promotor, Zoran Vojinovic, thank you for convincing me to return to the Netherlands and trust me with this PhD. Thank you for all the ideas and excellent discussions we had throughout all these years and for the financial support. To Igor Nikolic, for the inspirational meetings and for always challenging me, I am thankful to you for always making me look to a bigger picture. To Arlex Sanchez, thank you for all, you did the extra mile with the support. More than a supervisor, you are a very good friend; you always made time for me even if you were extra busy; thank you for every discussion we had; they always helped me improve my ideas, results and conclusions, also thank you for the emotional support and thank you for inviting me to join your family, the BBQs, the Christmas dinners and for share a good beer or rum now and then.

I want to acknowledge the funding sources which made this research possible. The project PEARL (Preparing for Extreme And Rare events in coastal regions), an FP7 EU-Programme (Grant agreement no 603663), and RECONECT (Regenerating ECOsystems with Nature-based solutions for hydro-meteorological risk rEduCTion), from the Horizon 2020 EU-Programme (Grant Agreement no 776866). To the Colombian government through the Administrative Department of Science, Technology and Innovation COLCIENCIAS. (Departamento Administrativo de Ciencia, Tecnología e Innovación) For providing extra funding for my Ph.D. To the people I have the pleasure to work on the PEARL project, especially Linda Sorg, Daniel Feldmeyer, Arabella Fraser, Jaume Amoros and Chiara Cosco for the joint work during different deliverables. And to other members with whom I share a dinner, a drink or a conversation during the different project meetings. Special thanks to our partners in Sint Maarten that provides us with data, interviews and logistics, mainly in VROMI, the statistics department and the disaster management team. I also extend my gratitude to Jolanda Boots and Anique Karsten, for their help regarding administrative issues.

What is life without the constant support and company of friends? I am sure the list is incomplete, my apologies to those that I do not mention by name. Mauri, Jessi, Kun, Mario, Miguel, Pin, Fer, Vero, Alex Kaune, Aki, Pablo, Barreto, Juancho, Jairo, Diego and family, Irene, Stefan, Claudini, Mohanned, Berend, Adele, Thaine, Kelly, Zaki, Laura, Janis, Diva, Claire N, Mary Luz, Milk and Alex, you all made my life more amusing. To Nata y Jeffrey, thanks for the friendship and making my Iceland dream come true. To my sister Angie, we started the NL dream together, and you have been by my side unconditionally ever since; gracias totales. Can parcero, thanks for the food, beers, biking and the Dutch translation. To Mile and Till mi familia europea, gracias por recibirme y hacerme sentir parte de su hogar en cada visita. A mis amigos en Colombia, sin importar la distancia, los sentí siempre allí John, Camilo, Chato, Pou, James, Eli, Betty, Yonnatan, Cata Negra, Caro Arias, a los Sanitarios y a los INEM-itas.

To Pato, an unconditional friend, you helped me in so many ways; I will always be grateful to you; thank you for saving me by "forcing" me to start running. To Alida, thanks for always be there, advising me, sharing your delicious food, and for being my espantapajaros. To Yared, I definitively would not be here without you, thank you for your friendship, for improving the quality of my papers, (sorry for my English); I wish all the best to you, to Juliette and Eliana. Juan "el Pollo" Chacon, words won't be enough to thank your role in my life during all these years; you became my brother and make every day enjoyable at little Colombia, thank you for sharing your views, laughs, thoughts, weirds videos, cooking skills...

To the students I met during these lecturer years, thanks for the academic discussions and for challenging me; for the events and places you invited me Minh, Nikos, Vittorio, Nhilce, Marianne, Feroz. Fabia, Rachelle, and many more.

Para mi familia, una disculpa por estar ausente tantos años y perderme tantos momentos importantes para nuestra familia, los extraño y amo demasiado. A mis padres, muchas gracias por todos los sacrificios realizados a lo largo de mi vida que me han convertido en quien soy y ha donde he llegado. A Diana, John y Astrid gracias por su constante apoyo, ustedes y sus familias son una fuente constante de alegría, inspiración y admiración. Maria C, gracias por tu visita y compañía, eres una mujer muy especial y llegaras muy alto, Sofi, tú ya sabes lo que significas en mi vida, gracias por ese amor que me tienes y que no merezco, siempre estaré allí para ti; A Tomas, Gaby y Vicky, perdón por no estar cuando crecían, les deseo un futuro brillante y pueden contar con este tío pa'las que sea.

Camila... " la prima!!! ", my partner in crime, this achievement is also yours; no words can express my gratitude for all you did for me during the last years of my PhD; thank you for the time and adventures we had shared, for your constant and unconditional support, for your patience, and for believing in me even when I doubted myself. I hope I can pay you back what you did for me.

Summary

Climate change, combined with the rapid and often unplanned urbanisation trends, is associated with a rising trend in the frequency and severity of disasters triggered by natural hazards. Among the weather-related disasters, floods and storms (i.e. hurricanes) account for the costliest and deadliest in the last decades. The situation is of particular importance in Small Islands Developing States (SIDS) because their relative higher vulnerability to the impacts of climate change, due to their location, fragile economies, limited resources, and more vulnerable habitats. Therefore, SIDS must implement adaptation measures to face the impacts of climate change and those of the urbanisation growth; for which is necessary to have an appropriate Disaster Risk Assessment (DRA), which should include the hazard itself, the intrinsic socio-economic vulnerability of the system and the exposure of infrastructure and humans to the hazard.

Traditional DRA approaches for disaster risk reduction (DRR) have focused mainly on the natural and technical roots of risk, this is the modelling of the hazard and implementation of physical and structural defences, for which the hazard component is the centre. Traditional DRA methods pay no or little attention to the other dimensions of disaster risk, and do not often investigate the spatial and temporal relationships between the hazard, the vulnerability and the exposure components. A better alternative when dealing with DRA is a holistic risk assessment, which looks at risk as a whole, looking into the components and seeking to understand the interactions, interrelatedness and interdependences between different processes and parts of the whole.

Hence, DRA could be more successful if it considered the adaptive nature of vulnerability and exposure components in their frameworks. This dissertation's main objective is to develop and test a disaster risk modelling framework that incorporates socioeconomic vulnerability and the adaptive nature of exposure associated with human behaviour in extreme hydro-meteorological events in the context of SIDS. To accomplish the main objective, we developed a methodology to incorporate the adaptive nature of risk into traditional DRA. The so-called ADRA method incorporates elements of socioeconomic vulnerability that account for local characteristics of a particular case study and the dynamic nature of exposure to account for household protective behaviours (i.e. evacuation, in-situ preparation).

We test our modelling framework in a case study using the Caribbean island Sint Maarten (the Dutch side of Saint Martin) and using as hazard the most recent disaster caused by Hurricane Irma in September 2017. We use the findings of a fact-finding mission in the island in the aftermath of the hurricane that included a face-to-face and a web-based survey to collect key elements of vulnerability, exposure, evacuation and risk. The mission's findings allow us to propose a framework to assess socioeconomic vulnerability in the context of SIDS in a post-disaster context using an index-based approach. The method called PeVI has a modular and hierarchical structure with three components: susceptibility, lack of coping capacities, and lack of adaptation.

Furthermore, to assess the current levels of exposure, we use two approaches. First, we use the survey results to evaluate the actual evacuation rates observed during Hurricane Irma using logistic regression models. The regression models results allow us to identify some factors that can act as predictors of evacuation behaviour, and we extrapolate the results for the whole Sint Maarten. The results, shown as probabilistic evacuation maps, aim to measure at neighbourhood scale the likelihood (or not) to evacuate and lower (or not) the exposure levels. The second method is based on an agent-based model (ABM). The ABM is used to assess exposure to water-related natural hazards dynamically by modelling the flow of information from several sources during Hurricane Irma in Sint Maarten, and how the different sources and level of trust may influence a particular household to undertake protective actions at the household level. Using the ABM, we also provide probabilistic maps of protective behaviour; we model evacuation and in-situ protection as measures to reduce exposure levels in a household.

We end the dissertation by presenting a practical web-application for disaster risk management (DRM) and evacuation purposes on the island of Sint Maarten. The web application was conceptualised based on the main drivers of evacuation based on the finding of this research.

This research contributes to DRA using a new methodology that considers disaster risk not as a static attribute of the system, but as one in a constant adaptation by including the dynamic of the system due to households' behaviour. Incorporating behavioural adaptation into DRA frameworks may lead to a different representation of risk. Hence, the usability of the outputs for DRM policy and strategies may increase by offering a more holistic view of how vulnerability and exposure may evolve. Our methodology is a holistic assessment of risk, ADRA, assess disaster risk using an adaptive approach, in which the exposure component is explicitly quantified and mapped. ADRA is a people-centred approach and can be used to quantify which protective measures can be more useful to lower risk (to life) and show where those measures will have a more significant impact. In addition, the findings of this dissertation offer practical recommendations for disaster risk managers and policymakers in Sint Maarten to reduce the risk to natural hazards in the island.

SAMENVATTING

Klimaatverandering, in combinatie met snelle en vaak ongeplande trends in verstedelijking, wordt in verband gebracht met de toenemende frequentie en intensiteit van natuurgevaren. Van deze weer-gerelateerde rampen zijn overstromingen en stormen (oftewel orkanen) verantwoordelijk voor de duurste en dodelijkste rampen van de laatste decennia. De situatie is in het bijzonder belangrijk voor Kleine Eilandstaten in Ontwikkeling (SIDS) vanwege hun relatief hoge kwetsbaarheid voor klimaatverandering, als gevolg van hun locatie, kwetsbare economie, beperkte grondstoffen en kwetsbare huisvesting. Daarom moeten SIDS aanpassingsmethoden implementeren om de impact van klimaatverandering en verstedelijking tegen te gaan. Dit vereist een Rampenrisico-beoordeling (DRA), wat bestaat uit onder andere de natuurramp zelf, de intrinsieke socio-economische kwetsbaarheid van het systeem en de blootstelling van mens en infrastructuur aan de ramp.

Traditionele DRA methodes voor Rampenrisico-beperking (DRR) leggen de focus op voornamelijk de natuurlijke en technische oorzaken van het risico, oftewel het modelleren van de ramp en de implementatie van fysieke en structurele verdediging, waarvoor het gevaarcomponent centraal staat. Traditionele DRA methodes schenken weinig tot geen aandacht aan de ruimtelijke en tijdelijke verbanden tussen de ramp, kwetsbaarheid en mate van blootstelling. Een beter alternatief om met DRA om te gaan is een holistische risicobeoordeling, wat kijkt naar het risico als geheel, waarbij ieder component wordt bekeken en waarbij men tracht om de onderlinge interacties, verwevenheid en afhankelijkheid tussen de verschillende processen en het geheel te begrijpen.

Om die reden kunnen DRA succesvoller zijn als er rekening gehouden wordt met de adaptieve aard van kwetsbaarheids- en blootstellingscomponenten in de raamwerken. Het hoofddoel van dit proefschrift is om een raamwerk voor rampenrisico modellering te ontwikkelen en testen, wat gebruik maakt van menselijk handelen in extreme hydro-meteorologische gebeurtenissen in de context van SIDS. Om dit hoofddoel te bewerkstelligen hebben we een methodiek ontwikkeld om de adaptieve aard van risico in traditionele DRA op te nemen. Deze zogenaamde ADRA-methode omvat elementen van socio-economische kwetsbaarheid wat lokale karakteristieken verklaart in een bepaalde case study en de dynamische aard van blootstelling om huishoudelijk beschermend gedrag (oftewel evacuatie, in-situ voorbereiding) te verklaren.

We testen ons modelleerraamwerk in een case study op het Caraïbische eiland Sint-Maarten met als ramp de recente orkaan Irma van September 2017. We gebruiken de bevindingen van een fact-finding-mission in de nasleep van de orkaan met een face-to-face en virtuele enquête om de belangrijkste elementen van kwetsbaarheid, blootstelling, evacuatie en risico te verzamelen. De bevindingen van de missie hebben ons in staat gesteld om een raamwerk voor te stellen waarmee socio-economische kwetsbaarheid beoordeeld kan worden in de context van SIDS in een post-natuurramp context met een geïndexeerde aanpak. Deze zogenaamde PeVI methode heeft een modulair en hiërarchische structuur met drie componenten: vatbaarheid, gebrek aan zelfredzaamheid, en gebrek aan aanpassing.

Verder gebruiken we twee aanpakken om de huidige blootstellingsniveaus te beoordelen. Ten eerste gebruiken we enquêteresultaten om de actuele evacuatiegraden die geobserveerd zijn tijdens orkaan Irma te evalueren met behulp van logistieke regressiemodellen. De resultaten van de regressiemodellen laten ons enkele factoren identificeren die als voorspellers voor evacuatiegedrag gebruikt kunnen worden voor heel Sint-Maarten. De resultaten, getoond als probabilistische evacuatiekaarten, hebben als doel om op wijkniveau de waarschijnlijkheid op evacuatie te meten en om de blootstellingsniveaus te verlagen. De tweede methode is gebaseerd op een agent-based models (ABM). Het ABM wordt gebruikt om de dynamische blootstelling aan water-gerelateerde natuurrampen te beoordelen door de informatiestroom van meerdere bronnen tijdens orkaan Irma op Sint-Maarten te modelleren, en hoe de verschillende bronnen en het vertrouwensniveau een bepaald huishouden kan beïnvloeden om beschermende acties te ondernemen op huiselijk niveau. Met behulp van het ABM genereren we probabilistische kaarten van beschermend gedrag; we modelleren evacuatie en in-situ protectie als middelen om blootstellingsniveaus in een huishouden te reduceren.

We eindigen het proefschrift door een praktische webapplicatie te presenteren met als doel Rampenrisico-management (DRM) en evacuatie op het eiland van Sint-Maarten. De webapplicatie was geconceptualiseerd op basis van de belangrijkste drijfveren voor evacuatie gebaseerd op de bevindingen van dit onderzoek.

Dit onderzoek draagt bij aan DRA door gebruik te maken van een nieuwe methodiek die rampenrisico niet als statisch attribuut van het systeem beschouwt, maar als een attribuut in constante adaptatie door de dynamiek van het systeem ten gevolge van huishoudelijk gedrag mee te nemen. Gedrag meenemen in DRA raamwerken kan leiden tot een andere representatie van risico. Daardoor kan de bruikbaarheid van de uitkomsten voor DRM-beleid en strategie toenemen door een meer holistisch beeld te tonen van hoe kwetsbaarheid en blootstelling kunnen evolueren. Onze methodiek is een holistische beoordeling van risico, ADRA, en beoordeelt rampenrisico met een adaptieve aanpak, in welke de blootstellingscomponenten expliciet gekwantificeerd en in kaart gebracht is. ADRA is een mensgericht aanpak en kan gebruikt worden om te kwantificeren welke middelen meer effect hebben op risicoverlaging en tonen waar deze middelen een significantere impact zullen hebben. Daarnaast biedt dit proefschrift praktische aanbevelingen voor ramp-risico managers en beleidsmakers in Sint-Maarten om het risico van natuurrampen op het eiland te verlagen.

CONTENTS

1

Introduction

1.1 MOTIVATION

In a changing climate, disasters triggered by natural hazard events such as hurricanes, storm surges, and flash floods are projected to increase in severity and in frequency (Hoeppe, 2016; IPCC, 2014). Moreover, in addition to a changing climate, the rapid and often unplanned expansion of urban areas and in particular those located close to coastal regions is also exposing more people and economic assets to disasters triggered by natural hazards, and it is projected that more disasters associated with the expansion of urban coastal cities will continue in the near future (Harrison and Williams, 2016; Kundzewicz et al., 2013; Sterzel et al., 2020). This combination of urbanisation trends, increased numbers of natural hazard events and demographic growth are creating the perfect scenario to have more frequent and more severe disasters.

Changes in climate are of special importance in the context of Small Island Developing States (SIDS), because SIDS are especially vulnerable to the associated impacts of climate change due to their location, fragile economies with limited diversification, restricted resources, and more vulnerable habitats (CRED-UNISDR, 2015; Robinson, 2017; Turvey, 2007). Impacts of climate change on SIDS can turn into significant loss of life and damage to property and infrastructure, and an easily damage the entire economy of a small country (UNFCCC, 2005).

According with the Insurance Information Institute (2019), there is a rising trend regarding global weather-related disasters and their associated losses (Figure 1.1). In addition, as reported in (CRED-UNISDR, 2015), floods and storm have been the main accountable source of economic impact (Figure 1.2 (a)), and also amongst the most catastrophic in terms of loss of life (Figure 1.2 (b)). Hence, adaptation and mitigation of the effects of climate change in coastal urban areas and of SIDS is necessary for the sustainability of these regions and to minimise the losses associated with disasters.

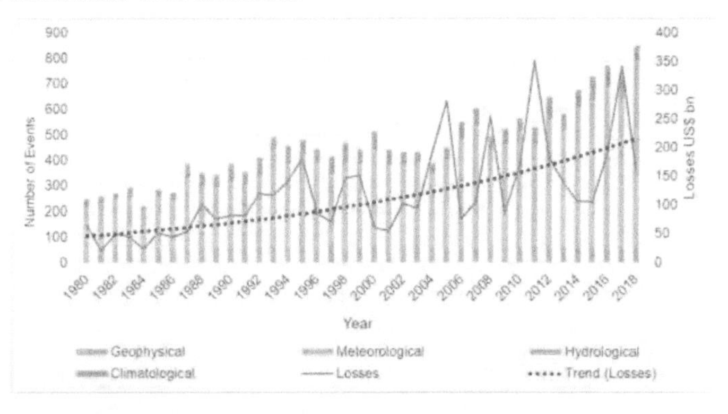

Figure 1.1 Global weather-related natural catastrophes by disaster type and associated losses (1980-2018). Based on data from the Insurance Information Institute (2019).

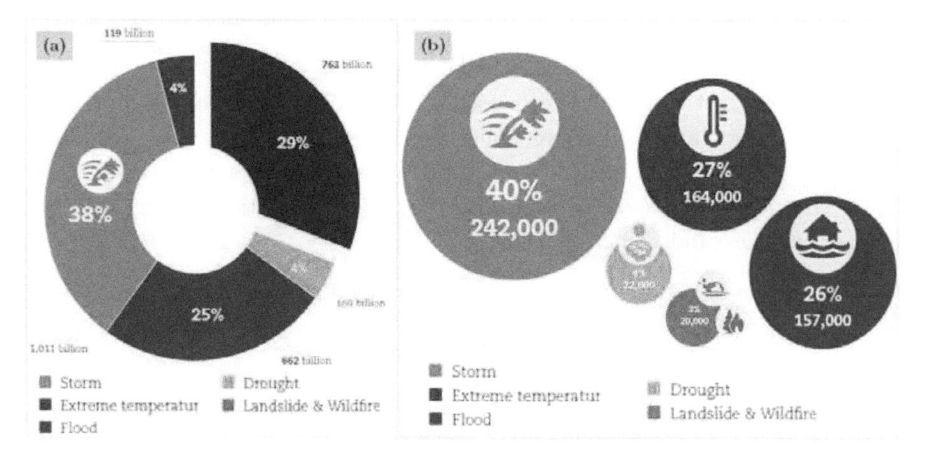

Figure 1.2 Impacts of disasters triggered by natural hazards (1995-2015). (a) Recorded economic damage by disaster type (USD). (b) Number of deaths by disaster type. Source: (CRED-UNISDR, 2015).

Implementation of climate-change adaptation measures should include planning programs, such as community-based development strategies, disaster risk assessment, assessment of the critical physical, social, economic, and environmental issues in combination with raising awareness, and communicating existing and future risks to local communities (Nurse et al., 2014; Robinson, 2017). In that regard, disaster risk management (DRM) has been the pillar to address or mitigate the impacts of weather-related disasters. The next section describes current approaches and defines DRM as well as identifying some of the gaps in this field.

1.2 Disaster Risk Management - DRM

1.2.1 Definitions

Disaster risk management is defined as a method to identify, assess and reduce risk through a series of strategies, policies and measures that aim to promote improvement in disaster preparedness, response and recovery (IPCC, 2012b). It is widely accepted by researchers and policy makers that the first step towards a sustainable DRM strategy is the proper assessment of the disaster risk (Samuels et al., 2009). In this thesis disaster risk assessment (DRA) is considered the first and an essential step in DRM, hence both terms DRM and DRA will be used interchangeably throughout this document.

According to UNDRR (2017), DRA is defined as *"A qualitative or quantitative approach to determine the nature and extent of disaster risk by analysing potential hazards and evaluating existing conditions of exposure and vulnerability that together could harm people, property, services, livelihoods and the environment on which they depend."*

From the definition presented above, it is necessary to adopt and tailor some definitions to the scope of this thesis. According to IPCC (2012a), disaster risk can be defined as the potential disruption of the normal functioning of a society or community with possible consequences for loss of life, injury, or destruction or damage of infrastructure, which can occur to a society in a specific period of time. In a technical sense, disaster risk is defined as the combination of three elements: hazard, exposure and vulnerability (Figure 1.3).

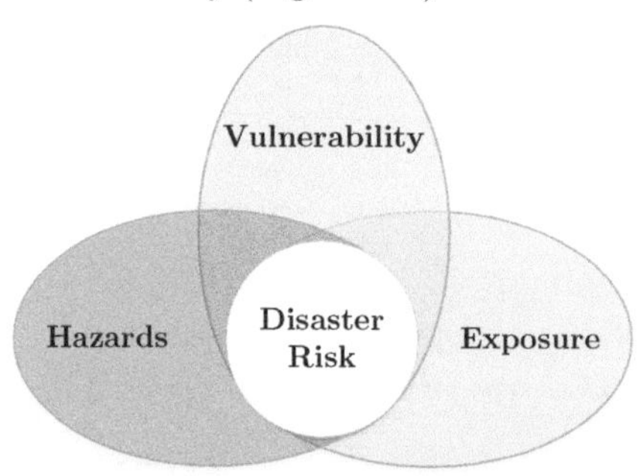

*Figure 1.3. Elements of risk. *Adapted from (IPCC, 2012b).*

In the context of this research, the term **hazards** refers to the possible future occurrence of natural or human-induced physical events that may have adverse effects on vulnerable and exposed elements (Birkmann, 2006); **exposure** refers to the inventory of elements (people and/or goods) in an area in which the hazardous events may occur (Cardona et al., 2012); and **vulnerability** refers to the propensity of exposed elements such as human beings, their livelihoods, and assets to suffer adverse effects when being exposed to and impacted by hazard events (Birkmann, 2006).

1.2.2 DRM Approaches

Due to the very nature of disasters and the elements that compose disaster risk, DRA is not a fixed science; instead, it is a method that is in constant evolution based on new concepts and understanding of the different elements that constitute it (Cardona et al., 2012). Traditional methods of disaster risk assessment in water-related events have mainly focused on the natural and technical roots of risk, the focus has been on reducing the likelihood of the hazard to cause an impact through physical and structural defences, for which the hazard component is the centre (Sayers et al., 2013), and as such these methods are limited in comparison to current theories. Traditional approaches evaluate the system by looking at the parts and linearly combining these or simply assessing the individual components without any real integration (Vojinović, 2015). These conventional methods are known as Integrated Flood Risk Management approaches (PEARL, 2016; Vojinović and Abbott, 2012).

However, there are also other roots that need to be taken into account, such as social, economic and technological roots that are better suited to measuring the vulnerability and exposure components. A better alternative when dealing with DRA is a holistic risk assessment, which looks at risk as a whole, looking into the components but also seeking to understand the interactions, interrelatedness and interdependences between different kinds of processes or parts of the whole (Aerts et al., 2018; Cardona et al., 2012; Vojinović and Abbott, 2012).

1.2.3 DRM Gaps and Requirements

Traditional DRM approaches have failed to address risk analysis from a holistic point of view. Of the three elements that compose risk, hazard modelling can be considered to be the technical component; it is also the most studied one, as it is relatively easier to undertake in comparison with the other elements of disaster risk (Birkmann et al., 2013). In contrast, current vulnerability and exposure assessments require a more holistic approach (Cardona et al., 2012).

Vulnerability Assessment

There are a huge number of vulnerability assessments to natural hazards in the literature. Nguyen et al. (2016) present an extensive review of 50 studies on the use of vulnerability indices associated with the impacts of climate change on coastal areas across a range of hazards. However, Nguyen et al. (2016) concluded that there is a lack of standardisation of concepts and methods to assess vulnerability, making them difficult to compare for different areas; they call for the adoption of a consistent and standard methodology and justify pursuing indicator-based vulnerability assessments. The call for the use of indexes to have a consistent set of metrics to assess vulnerability is not new; similar recommendations are also presented in Comfort et al. (1999) and Cutter et al. (2003).

In addition, some existing methods to assess vulnerability lack the adaptability required to look holistically into the drivers of vulnerability. Current methods are not flexible or easily adaptable to reflect the local characteristics of a particular case study (Turner et al., 2003; Vojinović, 2015), rather offering a generic picture of vulnerability based on standard parameters, which are normally extracted from census data. For this thesis, it was also necessary to have a vulnerability assessment that could incorporate in the analysis the special characteristics of small island states. SIDS are categorised as the most vulnerable nations in the world, given their higher and continual exposure to the effects of climate change and because of their relative geographical isolation (Scandurra et al., 2018).

Furthermore, holistic vulnerability assessments benefit from having a new method that allows the effects of a recent disaster to be captured and by using field data collection rather than a desk study.

Exposure Assessment

Regarding exposure, traditional methods have used the term exposure as part of the vulnerability component of disaster risk. Exposure is a necessary, but not sufficient, determinant of risk. It is possible to be exposed but not vulnerable, and the opposite can also be true, this is to say, it is possible to be vulnerable but not exposed to a particular hazard (Cardona et al., 2012). Traditionally, the exposure component has been expressed as a physical vulnerability (i.e. land use, existence of buildings) (Balica et al., 2012), but while this assumption may be true to assess risk to infrastructure or to assess economic impacts, it may not be the case to account for an individual's or household's exposure. Traditional DRA methods fail to incorporate the actions that individuals or households may undertake to reduce their exposure: for example, evacuating from risky areas or taking proactive and precautionary measures against the impact of the hazard, such as elevating their house or installing hurricane windows or shutters before or after being impacted.

DRA Needs

As shown above, current DRM practices rarely integrate the effects of local characteristics, and do not often investigate the spatial and temporal relationships that exist between the hazard, the vulnerability and the exposure. Hence, DRM could be more successful if it considered the adaptive nature of vulnerability and exposure components in their frameworks. Disaster risk should not be assessed as a static attribute of the system, but as one in constant adaptation by including the impact on the system due to the behaviour of individuals, businesses and governments (Aerts et al., 2018). Incorporating behavioural adaptation into DRA frameworks may lead to the better representation of risks. Hence, the usability of the outputs for DRM policy and strategies may increase by offering a more holistic view of how vulnerability and exposure may evolve. However, approaches that are able to incorporate explicitly such adaptivity are currently underdeveloped (Cardona et al., 2012).

It seems clear that new theories and methods are needed in order to assess risk from a holistic point of view. In terms of the vulnerability component, assessments based on an index-based approach can serve for this purpose. Index-based approaches can be used to incorporate these elements that make the vulnerability of SIDS unique, as well as to incorporate elements that can be changed after a disaster, such as risk perception and awareness, among others.

In terms of the exposure component, the latest knowledge of Complex Adaptive System (CAS) theory and agent-based models (ABM) can be used to undertake the challenge of assessing this component holistically. CAS is a suitable framework because it allows the complexity of risk to be captured, how risk may evolve from actions, and interactions within and between human systems and the natural environment. Using Agent-Based Models (ABM), it is possible to simulate these interactions and determine the exposure component using an adaptive method rather than the static method that current methodologies use.

1.3 Research Objectives

The aim of this thesis is to develop and test a disaster risk modelling framework that incorporates: socioeconomic vulnerability and the adaptive nature of human behaviour in extreme hydro-meteorological events in the context of a small island developing state. The specific objectives are:

1. To propose an adaptive disaster risk assessment framework that incorporates elements of socioeconomic vulnerability and human behaviour.
2. To develop a protective action model based on an ABM to evaluate the exposure component of disaster risk under extreme hydro-meteorological events.
3. To evaluate the potential benefits of an Adaptive Disaster Risk Assessment (ADRA) framework in comparison with traditional approaches for DRA.

1.4 Research Questions

Based on the objectives of this research, this thesis explores and addresses the following questions:

1. What elements of socioeconomic vulnerability are important in an adaptive risk framework in the context of a SIDS?
2. What are the main predictors of adaptive behaviour to reduce exposure in a SIDS?
3. How beneficial is ADRA over traditional DRA?

1.5 Research Approach

1.5.1 Scope

This PhD was carried out within the European Commission's Seventh Framework Program Preparing for Extreme And Rare events in coastaL regions (PEARL) project. Due to project objectives and requirements, this research aimed to develop adaptive risk management strategies for coastal communities against extreme hydro-meteorological events, minimising social, economic and environmental impacts and increasing the resilience of Coastal Regions. Furthermore, the case study site was selected based on the needs of the PEARL project.

1.5.2 Workflow

In order to address the main objective of this research, we present a comprehensive framework for disaster risk assessment that integrates the adaptability of disaster risk. The proposed framework builds on the steps for the disaster risk assessment approach proposed by the United Nations Development Programme (UNDP, 2010). In Figure 1.4 we present the steps adapted to include the novelties of the proposed methodology in this dissertation. It is worth noting that steps 2 and 3 can be assessed in parallel or in reverse order based on the objectives of the study.

Step one: corresponds to the understanding of the current situation, the needs and gaps for the objectives, and the case study. In this step we identified the data needs as well as data collection. This thesis uses data collected during a field campaign and complemented with information and data from previous research or provided by other researchers from the PEARL project in which this PhD took place.

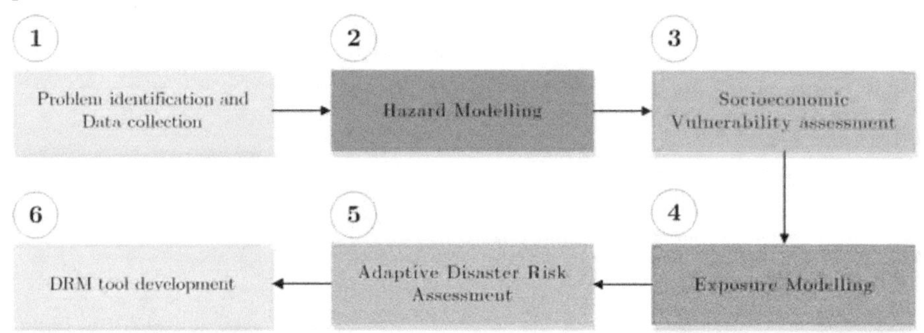

*Figure 1.4. Adaptive Disaster Risk Assessment steps. *adapted from (UNDP, 2010).*

Step 2: consists of the modelling of the potential hazards; this is the identification of possible physical threats, their location, intensity and likelihood of occurrence. In this dissertation, the type of hazards to be modelled corresponds to floods and hurricane winds.

Step 3: the vulnerability assessment is performed, which aimed to capture the multifaceted phenomena of socioeconomic vulnerability. Vulnerability was carried out in this dissertation using an index-based approach.

Step 4: exposure assessment consist in determining the likelihood of individuals or infrastructure to be exposed to a specified hazard. In this dissertation exposure was evaluated in two steps. First, an analysis of the observed evacuation behaviour during Hurricane Irma was performed. And, second, using the results of the observed evacuation patters we implemented an ABM for modelling protective actions done at household level to account for the adaptive nature of the exposure component.

Step 5: consists of the comparison and assessment of the results of traditional DRA methodologies against our results using the proposed ADRA framework. ADRA was computed by combining the results of the hazard modelling with those of the socioeconomic vulnerability and exposure ABM model.

Step 6: Based on the findings of the most relevant elements of socioeconomic vulnerability and the main predictor of evacuation in Sint Maarten, we have developed a tool aimed to be used for disaster risk management, in particular for evacuation.

1.5.3 Case Study

Study Area

The island of *Saint Martin* is located in the Leeward Islands in the northeast Caribbean Sea. The island is divided into two administrative units (Figure 1.5): the northern part *Saint-Martin* with an area of 53 km^2 is an overseas collectivite of France, and the southern part *Sint Maarten* with an area of 34 km^2 is one of the constituent countries of the Kingdom of the Netherlands (Vojinović and Van Teeffelen, 2007).

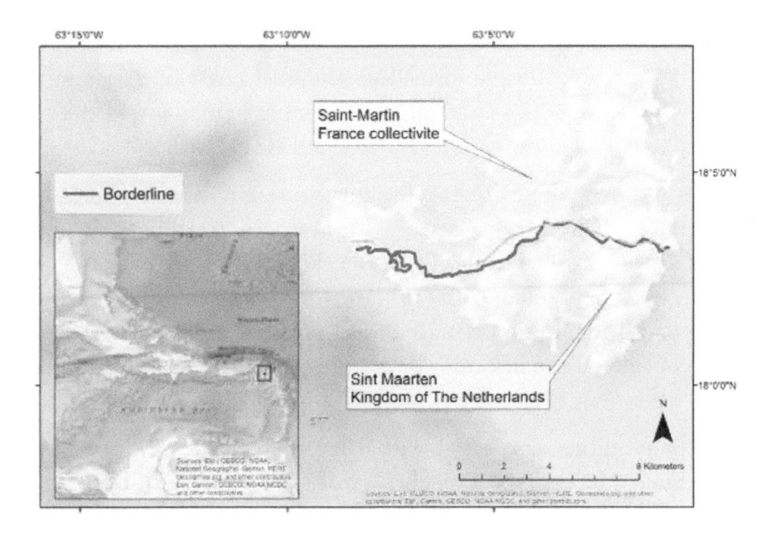

Figure 1.5. Location of Sint Maarten in the Caribbean Sea.

The magnitude and path of Hurricane Irma exposed the entire population of the island. However, this research focuses only on the Dutch part of the island. The official population on the Dutch side was 40,535 in 2017 (STAT, 2017). However, the numbers may not include all the undocumented immigrants, whose increase in numbers is considered one of the most significant social issues and drivers of vulnerability on the island (Bosch, 2017). According to non-official sources and during the interviews conducted during the fieldwork after Hurricane

Irma, the research team estimates that around 10,000 illegal immigrants might have been living on the Dutch part of the island before Hurricane Irma struck in September 2017. Previous figures put the number of undocumented immigrants close to 20,000 people (Geerds and de With, 2011).

The geographic location of the island, in the Atlantic Hurricane belt, exposes Sint Maarten to numerous hazards; the most noticeable are hurricanes which can cause (a combination of) strong winds, storm surges, pluvial flooding and mudslides. Since records began in 1851, a total of 20 major hurricanes (Category 3 or higher on the Saffir-Simpson Hurricane Wind Scale) have hit Sint Maarten. The most notable major hurricanes that have affected the island include Hurricane Donna in 1960, Hurricane Luis in 1995, Hurricane Lenny in 1999 and more recently Hurricane Irma in 2017 (the most catastrophic on record to date) (MDC, 2015).

Hurricane Irma Synopsis in Sint Maarten

The 2017 Atlantic hurricane season was one of the most active on record since records began in 1851 (NOOA, 2017). The 2017 season produced 17 named storms; 10 became hurricanes of which six were categorised as major hurricanes, i.e. Category 3 or higher on the Saffir-Simpson Scale (MDS, 2018). Hurricane Irma was the ninth named hurricane of the 2017 hurricane season, which originated from a tropical wave formed on the west coast of Africa around Cabo Verde on 27 August and dissipated on 13 September in mainland USA (Cangialosi et al., 2018), causing widespread destruction across its path.

During its lifetime, this catastrophic hurricane made seven landfalls, four of which occurred as Category 5 across the northern Caribbean Islands. Irma's second landfall was on the Small Island Developing State (SIDS) of the island of Sint Maarten on 6 September 2017 (Figure 1.6), with maximum recorded winds of 295 km/h and a minimum pressure of 914 mb (Cangialosi et al., 2018). At the time, it was considered the most powerful hurricane on record in the open Atlantic basin.

In terms of fatalities associated with Hurricane Irma, 11 direct deaths were reported on Saint-Martin (the French part of the island) and 4 on Sint Maarten (the Dutch part). Also, one indirect death was reported on Sint Maarten (Cangialosi et al., 2018). It is important to mention that during our fieldwork it transpired that the community of Sint Maarten believe that the reported number does not reflect the real number of casualties on the island associated with Irma. Their beliefs are based on the level of destruction and the gossip that circulated the island in the aftermath of the hurricane. During the survey, we heard that the real death toll ranges from 200 up to 1,000 deaths, with Irma primarily affecting the undocumented immigrant population. Also, the reports of injured people are estimated at around 250 to 300 people caused by Hurricane Irma on the island (ECLAC, 2017).

Figure 1.6. Satellite image of Hurricane Irma on 6 September 2017. Sint Maarten can be seen through the eye of the hurricane1.

Hurricane Irma also caused significant economic damage by destroying homes, schools, public buildings, businesses, and infrastructure. It is estimated that over 90% of housing had some damage, with 50% of these suffering from average damage or worse. It is also estimated that around one-third of the buildings were destroyed entirely (Netherlands Red Cross, 2017). The direct physical damage on the island was estimated at around US$1 billion (ECLAC, 2017).

1.6 Thesis Outline

Given the research motivation, questions, objectives and scope already presented, this thesis is structured in ten chapters (Figure 1.7), including the introductory section (Chapter 1) and the outlook and reflections from this dissertation (Chapter 10). The research chapters are structured as follows:

Chapter 2 contains the description of the fact-finding mission that was carried out in the case study in the aftermath of Hurricane Irma. It includes the methodological design of the tools used to collect data in the post-disaster environment as well as the main finding of the fieldwork with regard to elements of the socio-economic vulnerability, exposure and risk on the island.

[1] Source:https://www.weerplaza.nl/weerinhetnieuws/live-blog/orkaan-irma-raast-over-bovenwindse-eilanden/3567/

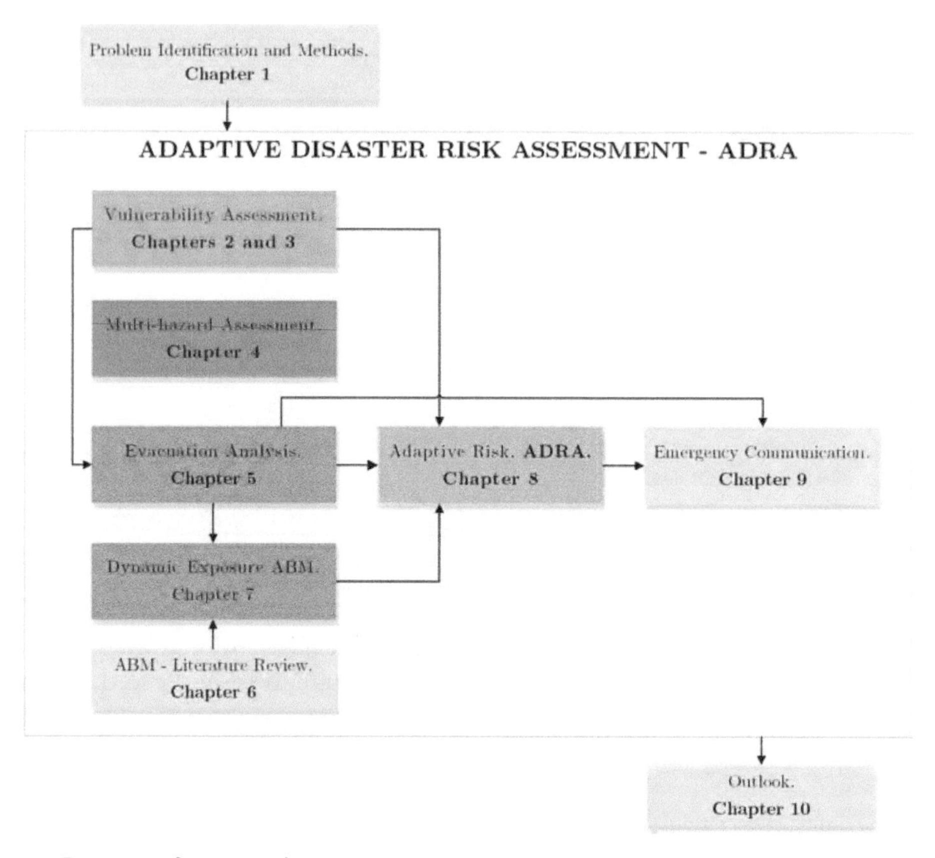

Figure 1.7 Overview of the methodological approach of the thesis.

In **Chapter 3,** we present a framework to assess socio-economic vulnerability in SIDS in a post-disaster context. We assess the vulnerability by computing a vulnerability index in combination with a principal component analysis. The index-based vulnerability-assessment approach, called PeVI, has a modular and hierarchical structure with three components: susceptibility, lack of coping capacities, and lack of adaptation, which are further composed of factors and variables. With the data collected in the aftermath of Hurricane Irma, we could incorporate into PeVI elements that can change after a disaster (e.g. risk awareness, risk perception, and access to information).

Chapter 4 presents the results of a multi-hazard assessment for Sint marten using Hurricane Irma as the base hazard. We took into account wind hazard from a maximum gust wind model and flood hazard to account for potential pluvial flood and storm surge in the study area using a synthetic, but plausible scenario, corresponding to a 100-year recurrence interval and a storm surge of 0.5 m.

Chapter 5 details the evacuation behaviour analysis observed in the case study during Hurricane Irma in September 2017. Based on a review of previous evacuation behavioural studies we have examined several factors to assess the validity as predictors (or not) of observed evacuation behaviour using logistic regression models and we selected those proven to be statistically significant to build a regression model of evacuation behaviour that is used in the setup of behavioural rules of the ABM in Chapter 7, as well as to map the evacuation behaviour on the Dutch part of the island.

Chapter 6 corresponds to a state-of-the-art literature review in the use of ABM in water-related disasters. The extensive literature review on this topic is presented because human behaviour is the central concept of the adaptive disaster risk framework of this dissertation, and ABM was chosen as the modelling tool to capture variability in the exposure component of risk. In the review, among other topics, we present some identified knowledge gaps, methodological issues and suggestions to enhance ABM applications as a novel tool in DRM, and we offer some recommendations and future directions, some of which were taken into account during the model setup of the ABM in Chapter 7. An important comment is that the literature review for the other concepts of disaster risk used in this dissertation is presented in a separate section in each corresponding chapter.

Chapter 7 contains the ABM model used to assess exposure to water-related natural hazards dynamically. The ABM model the flow of information during Hurricane Irma in Sint Maarten, and make predictions regarding protective actions at the household level. The ABM model uses the findings of Chapter 5 to setup the evacuation rules of the agents.

Chapter 8 contains our adaptive disaster risk assessment framework (ADRA), and its application in the case study of Sint Maarten. We use Hurricane Irma as the hazard component (Chapter 4), the vulnerability assessment carried out in Chapter 3, and the exposure component was incorporated using the results of the ABM model presented in Chapter 7. Then we compare our results against traditional DRA methodologies, and conclusions are drawn.

Chapter 9 presents a web application that was developed to be used as a tool for disaster risk management (DRM) and evacuation purposes on the island of Sint Maarten. The web application was conceptualised based on the main drivers of evacuation found in Chapter 5 as well as users' needs identified during the fieldwork presented in Chapter 2.

2

Capturing Elements of Vulnerability, Exposure and Risk

This chapter presents a household survey and the main findings related to vulnerability, exposure and risk to extreme weather events in the aftermath of the category 5 Hurricane Irma in Sint Maarten. The post-disaster context posed challenges in relation to data collection, determination of sample size and timing of the fieldwork. The survey was conducted using a combination of face-to-face interviews and web-administered questionnaires. This method proved useful in achieving a better coverage of the study area as well as obtaining a greater overall response rate. With regards to the timing of the survey, it was found that a period of six months after the hurricane for the field data campaign was adequate in terms of availability of resources and emotional distress of respondents. Data collected in the survey was categorised into general household information, hurricane preparedness and reaction, and risk perception/awareness. Survey findings show that the factors that increased vulnerability and risk on the island include a high tenancy rate, low insurance coverage, lack of house maintenance, low evacuation rate, not receiving a clear warning, and lack of preparation. The factors that reduce vulnerability include high hurricane awareness at a household level and high tendency of rebuilding houses with comparable quality to houses that can sustain hurricanes.

This chapter is based on:

Medina, N., Abebe, Y., Sanchez, A., Vojinović, Z., & Nikolic, I. (2019). Surveying After a Disaster. Capturing Elements of Vulnerability, Risk and Lessons Learned from a Household Survey in the Case Study of Hurricane Irma in Sint Maarten. Journal of Extreme Events, 6(2). Doi:10.1142/S2345737619500015.

2.1 INTRODUCTION

In the immediate aftermath of a hurricane concerns and efforts need to be focused on the relief effort, needs assessment, safety, health, and well-being of inhabitants (Alexander, 2015; Petak, 1985; Tan, 2013; Walle et al., 2013). However, when the immediate emergency has passed, reconstruction should commence and the bringing back of living standards to at least the pre-disaster status. As identified in previous work (Vojinović, 2015), an assessment of risk and vulnerability to hurricanes and floods is vital for reconstruction efforts and future planning activities.

In this chapter we present the main findings of a field data collection campaign, lessons we learnt while collecting data in a post disaster environment on Sint Maarten after the devastation caused by Hurricane Irma in September 2017, and some of the main finding regarding vulnerability, exposure and risk to disasters triggered by natural hazards in the island.

2.1.1 Need for post-disaster data

Due to the lack of data available that could be used in the computation of vulnerability and risk, it was evident that collecting data in Sint Maarten was important for the reconstruction efforts and future planning activities, especially data concerning vulnerability, exposure and risk of population and infrastructure to extreme weather events at household level as explained above. The World Bank report on Data Against disasters triggered by natural hazards (Amin and Goldstein, 2008) offers a good guide to understanding the different data needs during the different phases of a disaster. The information needs in a recovery phase vary from losses per household and economic and business losses to data needed on the availability of water, schools, and health facilities, levels of vulnerability to disasters triggered by natural hazards, the status of land ownership, among others.

Several authors have addressed the need for information at a household level in different phases of a disaster. The work of Birkmann et al. (2016) present some of the findings of a household survey in a highly exposed area to extreme weather events for the Megacity of Lagos in Nigeria. In their approach, the survey focused on different characteristics of vulnerability, resilience, and transformation as a critical element for any planning and decision-making in the context of climate change. Similarly, Shah et al. (2018) present an approach to elaborate household vulnerability and resilience assessment to flood disasters in two districts in Pakistan affected by the floods of 2010; The research was carried out using a dataset of 600 face-to-face household interviews, in which it was identified how the different components of vulnerability change from place to place and resulted in specific recommendations for each one of the study districts. Bird et al. (2011) highlight the importance of conducting post-disaster surveys to gain a better understanding of human behaviour and the significance of this information in improving community-based disaster risk mitigations.

In addition, post-disaster data have to be collected and managed to assess the risk and to support the tasks of various organisations such as government, the scientific community, financial institutions and nongovernmental organisations (NGOs) (Wirtz et al., 2012). The importance of data collection just after a disaster to understand and meet the needs of the affected population has also been recognised by the United Nations Office for the Coordination of Humanitarian Affairs (OCHA, 2016). This office, supported by the Ministry of Foreign Affairs of the Netherlands, established a global humanitarian data centre aiming to centralise, process, visualise and analyse humanitarian data.

2.1.2 Challenges to data collection after a disaster

In any data collection campaign, it is vital to have clearly defined objectives, which is especially true in post-disaster data collection since the acquisition of this data is more challenging for any researcher in many ways. First, researchers will probably conduct the data collection in an environment with restricted transportation, accommodation and food supplies among others (Benight and McFarlane, 2007; Henderson et al., 2009; Morton and Levy, 2011). Second, researchers must navigate through emotional distress and post-traumatic stress disorder (PTSD) in the community to be addressed in the post-disaster phase (Haney and Elliott, 2013; SAMHSA, 2016). Third, an evacuated and relocated population might be out of reach to be surveyed (Kessler et al., 2008).

Given all these possible limitations that can be faced during post-disaster data collection, it is crucial for a research team to keep in mind that conventional approaches in the design of surveys may not apply or may be misleading in times of disaster and need to be adapted to local circumstances (Lavin et al., 2012; Liang et al., 2012; Richardson et al., 2009). Accordingly, it is important to find a balance between keeping the research quality and maintaining sufficient sample size with the limitations that can be experienced in the field.

Sampling strategy and methods

Surveying in post-disaster circumstances should, therefore, consider among other aspects, the best possible sampling strategies and methods. Norris (2006) presents a summary of usage of different sampling strategies such as convenience sampling, census, purposive sampling, and random and quasi-random sampling. Norris's findings show that convenience and random sampling methods are used more often in studies after disasters triggered by natural hazards. Liang et al. (2012) also evaluate three sampling methods for selecting houses for post-hurricane damage assessment simple random sampling, equal spatial sampling, and route-based strategy, and conclude that the route-based sampling method showed an acceptable level of performance.

The method of post-disaster surveying is also an important factor. Since disasters usually result in infrastructure damage, conventional survey modes such as face-to-face and telephone surveys are proved to be challenging to conduct (Kessler et al., 2008). In the first case, the main challenges include damaged roads that impede the travel of interviewers, concerns of criminal hazards and disease, difficulty contacting people either because they have been evacuated or because the area is entirely devastated (Henderson et al., 2009; Kessler et al., 2008). In the case of telephone surveys, the main challenges are telephone service interruptions and lower participation or response rates, especially from low-income people (Henderson et al., 2009; Kessler et al., 2008). The final selection of a survey method should be done based on the characteristics of the type of disaster, the population size affected by the disaster and human idiosyncrasies (Henderson et al., 2009). But given the nature of disasters, several authors (see Kessler et al., 2008; Skinner and Rao, 1996) agree on multiple-frame sampling as the preferred method to minimise some of the most concurrent issues in post-disaster surveys, as it achieves a better representation of the population in the final sample selected compared to single frame methods.

Timeframe for data collection

Another critical aspect in post-disaster data collection is when to perform the fieldwork. The decision should equally consider: (i) sufficient time for the relief of the population and (ii) not waiting for too long after the event as the target population could forget critical aspects to be collected in the fieldwork (Henderson et al., 2009; Lavin et al., 2012). Kessler et al. (2008) also note that surveys need to be performed as soon as possible after a disaster so that the outcomes of the survey can be used for planning decisions. Although it was not possible to find a unified definition on the best time to perform a survey, based on the reviewed papers for this work, the "normal" time to start data collection campaigns vary from 4 to 6 months after a disaster occurred.

One of the most documented and studied disasters is Hurricane Katrina in 2005; authors have performed studies in the impacted area of that hurricane and studies regarding hurricane Katrina can be considered as a standard or reference case study. The role of professionals from the region of New Orleans conducting research after hurricane Katrina was studied and shows that 6 months elapsed before the collection of information was optimal. The reasoning behind this lag was to obtain funding, to allow enough time for the target population to return to a "normal" situation after the hurricane, to properly train students for the field mission and bureaucracy (Haney and Elliott, 2013). Also, in a research focus on lessons learned in survey methodologies on the impact of Hurricane Katrina to the population of New Orleans, Henderson et al. (2009) summarised 4 different projects carried out in this area, the starting timeframe of which vary from a minimum of 4 months up to 6 months.

Ethics in the data collection

It is common practice to deploy research teams into an area hit by a disaster. While many of these projects and research are conducted with the best of the intentions, some, unfortunately, are opportunistic (Sumathipala et al., 2010). As such, several authors have addressed some of the most common ethical issues while surveying in a post-disaster community and offer suggestions on how to avoid or minimise the risk of violating these issues and how to balance the critical need for research in and after the disaster, with the ethical responsibility to protect the research participants. (Hendriks et al., 2015), points out that all disaster research activity should be balanced with the ethical responsibility to protect vulnerable participants. It is suggested that the selection of participants should not be done solely by the research team but include local and relief organisations in the selection process.

Similarly, Sumathipala et al. (2010) presents some of the main ethical challenges that need to be accounted for while designing and performing post-disaster research, some of which are: targeting most vulnerable population such children and women, the mental health implications of the disaster on the target population, experiments and sample collection without ethical approval. Also, Sumathipala et al., stresses that research in a post-disaster community needs to be contextual and regionalised, the use of international guidelines is essential, but the local context should never be left out in the design and research work. Additionally, O'Mathúna (2010), points out that while doing research, the protection of participants and minimising harm are the researchers' highest ethical priorities. In this work, O'Mathúna also comments on the importance of including formal ethical approval, informed consent, balancing burdens and benefits, participant recruitment, coercion, the role of compensation, and conflicts of interest.

2.2 SURVEY DESIGN AFTER HURRICANE IRMA

2.2.1 Conceptual Design

Data preparation

For the data collection and survey campaign, having a good source of building locations improves the accuracy of the study through improving random sampling selection methods, increasing precision by saving time and valuable resources, and increasing the sample size (Kaiser et al., 2003; Roper and Mays, 1999). In Sint Maarten, the main challenge while preparing the survey was the lack of a consistent and up-to-date geographic database of residential buildings. The official source of information available to the team dated back to 2008, and it was provided as a shapefile by the Ministry of Public Housing, Spatial Planning, Environment and Infrastructure of the Government of Sint Maarten (Ministerie van Volkshuisvesting, Ruimtelijke Ordening, Milieu & Infrastructuur – VROMI). After visually assessing the quality of this file, it was concluded that the data was outdated and did not accurately represent the number and location of buildings on the island (see Figure 2.1-a and Figure 2.1-b). The nonexistence of a new shapefile at the moment of data preparation was confirmed by the ministry of infrastructure of the island.

The geographic dataset was updated following a similar approach of previous research studies with inadequate data where Open Street Maps (OSM) data was used as a source of information for disaster management. This follows the experiences by Latif et al. (2011), where the authors discuss the role of Open Street Maps for disaster management in Bangladesh. They describe the process of using Yahoo Aerial Imagery within OSM as a backdrop for map production. Michael (2014) reports on a project called 'missing maps' which is an initiative led by the Humanitarian OpenStreetMap team to create free, digital maps of the world. With the concept of using satellite images as background in OSM, volunteers all over the world can trace the outlines of buildings, roads, parks, and other urban elements to create free maps. Michael (2014) reported that the first big test case was Lubumbashi, a city in the Democratic Republic of Congo and explained the potentiality of this approach in Haiti after the 2010 earthquake. In a similar manner, Homberg et al. (2017) offer a relatively complete list of which datasets should be collected before a disaster strikes. In this work, Homberg presents a data sets preparedness index that is computed based on completeness, recency and accuracy, and reliability and is tested in a case study after typhoons in The Philippines and floods in Malawi. The geographic components in their computation use datasets available from OSM for these two case studies.

In the case of Sint Maarten, OSM data and official data (from VROMI) were merged and cleaned, creating an updated map of residential buildings. The merged shapefile was further improved by visual comparison with a high-resolution SPOT satellite image of the island taken on 16 February 2017 (i.e., prior to Hurricane Irma). The spatial resolution of the SPOT image is 20 cm (IGN, 2017). With this approach, it was possible to detect and include newly constructed buildings that were not represented in the shapefile (see Figure 2.1-(a)) and to remove buildings that no longer exist on the island (i.e., demolished structures) (see Figure 2.1-(b)).

Because the household survey target population was limited to residential buildings, it was necessary to check the land use or type of each building. The data cleansing was carried out in four phases (Figure 2.2): (i) Buildings listed as residential were selected from the original VROMI database. (ii) Building use was then verified with the OSM attribute of land use. (iii) By using the locations of commercial and industrial buildings in Google Maps, it was possible to remove some buildings mislabelled as residential, and (iv) by using the location of hotels and accommodation reported in booking.com and Airbnb web pages, it was possible to detect buildings that were incorrectly categorised as residential. Buildings with mixed use were not accounted for residential.

Figure 2.1. Sources of buildings data – OSM, VROMI and SPOT satellite image. (a) Shows the buildings shapefile extracted from OSM using the satellite image as a background. The blue polygons represent buildings existing in the database whereas the rest are new buildings missing in the OSM database. (b) Shows the VROMI shapefile in beige and red line over the SPOT image. It is observed that some buildings have been demolished.

The data cleansing results in the classification of buildings into eight different land use types: 1. Residential, 2. Hotel/Accommodations, 3. Commercial/Industrial/Services, 4. Sport/Leisure, 5. Medical/Health centres. 6. Religious, 7. Educational, 8. Correctional (Figure 2.2). A total of 11,128 units were of residential use, and those were the ones selected as the target population size for the household survey.

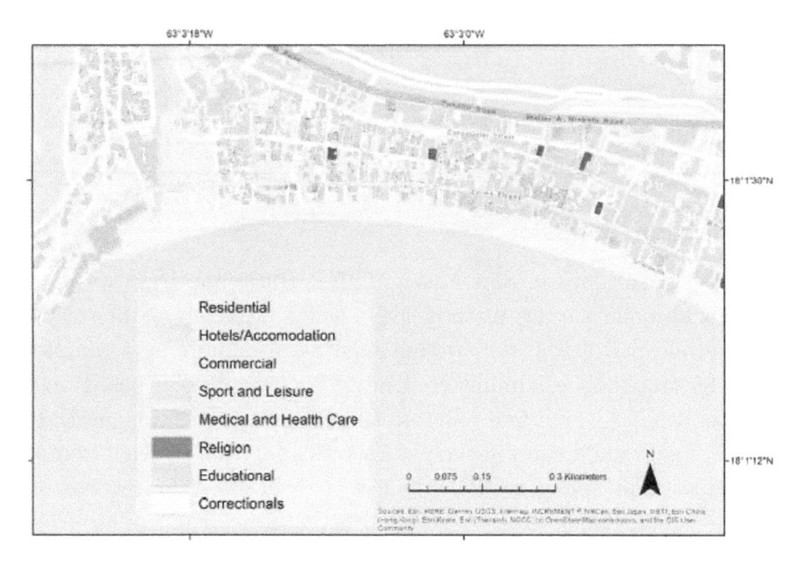

Figure 2.2. Categorisation of buildings by type of use.

Sample Size and Household Random Sampling

Using a random sampling method, for a target population of 11,128 residential buildings, the number of respondents needed to obtain statistically significant results was 267 houses. This number of houses would provide a confidence level of 95% and a 6% margin of error. Furthermore, assuming a 90% response rate to the survey, the sample size was adjusted to 296 residential houses and rounded up to 300 houses.

A random sampling technique was used across the entire geodatabase in order to select the residential buildings to be surveyed. For this, the *Sampling Design Tool for ArcGIS*[2] *was used*. It was important that the sample buildings were uniformly distributed among the eight major administrative districts of Sint Maarten since they have different socio-economic characteristics and therefore different capacity to respond and adapt to the consequences of major weather events. Table 2.1 shows the number of selected houses per district to be interviewed.

Table 2.1. Number of existing, targeted and selected households per district

District Name	Number of Houses	% Houses per district	# Houses per Percentage	# Houses Randomly Selected
Cole Bay	2071	18.6%	56	58
Cul de Sac	2524	22.7%	68	67
Little Bay	1079	9.7%	29	28
Lower Prince's Quarter	2389	21.5%	65	66
Lowlands	334	3.0%	9	5
Philipsburg	457	4.1%	12	11
Simpson Bay	367	3.3%	10	7
Upper Prince's Quarter	1907	17.1%	51	58

2.2.2 Collection Modes

The collection of data in Sint Maarten after Hurricane Irma was carried out using a mixed-mode survey methodology, using face-to-face interviews as the primary method and a web self-administered questionnaire as a complementary method. The face-to-face administered interview was chosen based on previous post-disaster studies which find out that respondents prefer this method because they feel listened to and can tell their stories, this serves as a relief from the post-disaster stress and increases the chances to obtain a better response rate (Henderson et al., 2009; Norris, 2006). Furthermore, the ability to help the

[2] https://coastalscience.noaa.gov/project/sampling-design-tool-arcgis/

respondent to understand specific questions when in doubt (Irvine et al., 2012) and the certainty of covering the whole extension of the questionnaire were important (Holbrook et al., 2003).

Also, performing a face-to-face interview allowed the team to get a general knowledge of the island and to collect extra information through a semi-structured interview at the end. This part of the interview gave the opportunity to focus on crucial topics and was often used to solicit information about issues that are considered to be sensitive (Lavrakas, 2008) such as government performance, income, immigration, riots, and looting, which were widely discussed by respondents in our survey.

One of the critical challenges in Sint Maarten during the data collection was the limited access to some specific areas to perform face-to-face interviews. The factors include difficulty to access gated houses without a doorbell, presence of dogs that made it difficult or impossible to reach the front door, houses currently uninhabited due to reconstruction works, non-permanent residents in houses used for summer holidays, no residents at the time and date when the visits were performed, and gated condominiums where security guards denied access to the team. These challenges were mainly but not limited to high-income areas, and led to low coverage in some districts.

As the overall response rate from the face-to-face interview was lower than that designed for the study, the team included a secondary method to collect data. The complementary collection method was a web self-administered version of the face-to-face survey. The secondary method was used to increase the response rate, due to its faster response rate at a lower cost (Duffy et al., 2010; Lefever et al., 2007). In addition to increasing the response rate, web surveys had been used previously in "hard to reach" populations to expand the geographical coverage of the study (Baltar and Brunet, 2012).

2.2.3 Survey preparation

The first step to prepare the questionnaire was the selection of the language to administer the interview. It is important to note that there are more than 25 registered nationalities on the island with Dutch and English being the official languages of the country and large French and Spanish speaking communities. Despite the multiple nationalities and languages present on the island, the survey was designed in English.

The selection was made on the basis that English is one of the official languages, is the official language spoken in the administration of the island and is also the preferred language within the population due to the high volume of North American tourists, which in 2016 accounted for 63 % of stayovers (STAT, 2017). However, in the implementation of the survey, the team found some areas of the island where only Spanish was spoken and given the presence of a native

Spanish speaker in the team, the survey was performed in their native language. The transcription into digital format was done in English.

The questionnaire for the face-to-face interviews consists of three parts (Figure 2.3-a). The topics covered are: (i) general household information, (ii) hurricane preparedness and reaction, and (iii) risk perception/awareness. The questionnaire has between 46 and 51 questions depending on evacuation decision taken by the household during Hurricane Irma. The online version of the survey questionnaire was created using Google Forms and also designed only in English.

The link to access the survey was shared through the Facebook pages of different organisations located on the island such as churches and aid/help groups created after Hurricane Irma and was sent to individuals that were identified as active participants in those groups and contacts created during the field mission. Participants were asked to share and distribute the survey among friends and relatives on the island. The online version kept all the original questions of the face-to-face questionnaire, but it was reorganised in such a way that it optimises the screen presentation and the navigation of respondents.

The final design consisted of 11 sections (Figure 2.3-b). A copy of the face-to-face interviews is presented in Appendix A. The copy of the online version can be accessed through (PEARL, 2018).

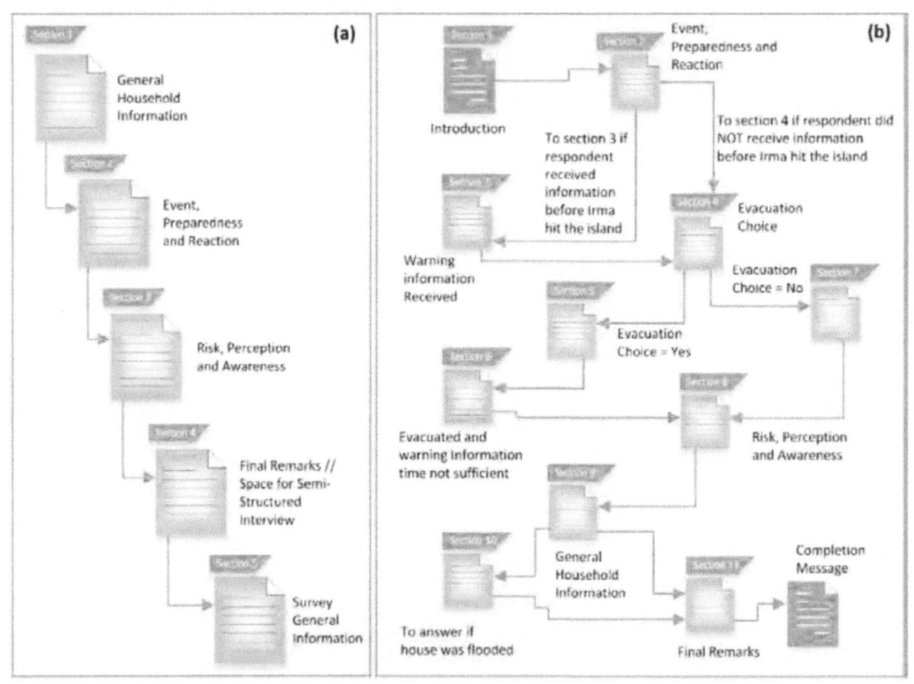

Figure 2.3. Survey questionnaire structure. (a) Face-to-face Survey. (b) Web Survey

2.2.4 Pre-Survey logistics and preparation

Given the limited resources on the island, even five months after Hurricane Irma, all the materials needed to perform the field surveys were prepared and printed before travelling to the island. These included 350 hard copies of the face-to-face questionnaires, pre-prepared maps showing the locations of the 300 randomly sampled houses, and named fieldwork papers following similar approaches that can be found in humanitarian projects carried out by the Red Cross (American Red Cross; Eros, 2018). In total, the study area was subdivided into 45 fieldwork papers, and all the targeted households for the survey were already marked with a unique identifier that was used both in the maps and in the hard copy of the questionnaire.

2.3 Survey

2.3.1 Data Collection

The face-to-face surveys were conducted by two of the authors of this study. The fieldwork was carried out from 12 February until 3 March of 2018. A pilot study was carried out on the first day to evaluate whether the initially chosen sampling framework and technique were adequate and useful for the characteristics of Sint Maarten. A total of 12 interviews were carried out in the pilot, and from the responsiveness of the inhabitants obtained, it was concluded that the face-to-face method was adequate for the data collection. Furthermore, the pilot was conducted by the two interviewers together, allowing the team to establish a common basis for performing the interviews.

For security and identification purpose, the vehicle used during the survey was properly marked as part of a post-Hurricane Irma impact assessment team with logos of the research institute leading the fieldwork (Figure 2.4). The interviewers also wore an institute t-shirt and hat which turned out to be crucial because some respondents only agreed to participate in the survey once they were entirely sure the team was not part of the government. This was mainly related to sharing sensitive private information such as immigration status, household size and income of respondents and the feeling that government reaction after Hurricane Irma was not good enough.

During the survey, if it was not possible to interview the initially randomly selected household, the interviewer proceeded to select a new one. The new selection was performed based on the closest house available to the original point that was willing to participate. It is worth mentioning that the initial survey protocol established to move to the right-hand side of the house, but this proved challenging on the field due to the limitations of a post-disaster area. Reasons that led to change the target house were: a) no presence of people in the house, b) no adult to respond at the time of the visit, c) the house was abandoned or under reconstruction and uninhabited after Hurricane Irma, or d) the person declined to participate. After proper identification and explanation of the purpose of the survey, the interviewer proceeded to ask the full questionnaire and the final semi-structured part as remarks.

Figure 2.4. Team identification on the field during interviews.

The survey protocol established that interviews would be carried out on Monday to Friday from 8:00 hours until 17:00 hours and during Saturdays from 8:00 hours until 15:00 hours, leaving Sundays as a preparation and resting day. The protocol was later adjusted to fit better the characteristics and daily behaviour of the residents of Sint Maarten. For example, Saturday and Sunday were fully scheduled for data collection. Adjusting the protocol had two main advantages on the overall performance of the fieldwork: (i) the respondent age group was normally distributed since, in weekdays, the team encountered a high number of retirees and (ii) it was possible to increase the chance of finding residents in some areas where potential respondents were not available during weekdays. In addition, working hours were also shifted to start at 9:00 am to avoid traffic jams in the early morning in the few main roads of the island. There was also a briefing of the main findings and challenges faced during the day in order to account for any change or adjustment to the survey protocol for the upcoming days. The protocol established that at the end of each week, a more extensive meeting would be held to adjust the survey based on the evidence of the work done until that moment. For example, based on assessments, a question related to the sources of warning information used by the respondent during Hurricane Irma was removed because it created confusion and respondents found it redundant to a previous question.

For the web version of the survey, the questionnaire was initially posted on 9 March 2018, and kept online for three weeks until 31 March to be consistent with the number of weeks that the face-to-face interview took place.

2.3.2 Response Rate, Confidence Level and Margin of error

The total number of respondents of the household survey using the mixed data collection mode was 260. A total of 207 responses correspond to the face-to-face interviews and 53 responses to the web survey. However, five of the web-survey respondents were living on the French side when Hurricane Irma hit the island and hence where discarded from the analysis. Therefore, the findings presented in the next section of this chapter is based on the analysis of the responses of 255 households. This corresponds to a response rate of 82.6% with a margin of error of the survey of 6.07% using a confidence interval of 95%.

To illustrate the geographical location of the respondents, Figure 2.5 shows the percentage of respondents according to the survey method used at neighbourhood level. As mentioned earlier, a face-to-face interview was conducted in most parts of Sint Maarten. However, in the less accessible, more prosperous, gated neighbourhoods of the island, the web-administered survey was best employed. Figure 2.5 illustrates that the web-administered survey was indeed complementary to the primary collection method.

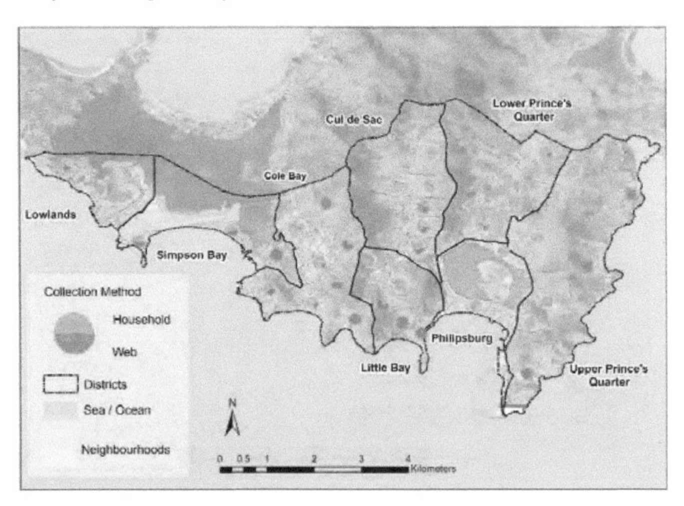

Figure 2.5. Survey collection methods per neighbourhood.

2.4 Survey – Results

2.4.1 Main findings

The following section presents the main findings of the household survey regarding vulnerability and risk assessment grouped into four categories: (1) Household and demographic parameters, (2) Information, awareness and experience of hurricanes, (3) Evacuation behaviour and (4) Risk perception. For the full report on the findings we refer to the full report (PEARL, 2018).

Household and demographic parameters

Census and demographic data are vital to compute vulnerability to extreme weather events (Cardona et al., 2012; Sorg et al., 2018; Vojinović et al., 2016). In Sint Maarten, this information was only available aggregated at country level. This data is too low-resolution to perform an in-depth analysis on the root causes of vulnerability, which requires block or neighbourhood level data. Hence, the household survey included questions collecting demographic information. One important demographic aspect is the country of origin as it can be directly related to some of the components of vulnerability such as language limitations, knowledge of the evacuation routes and experiences with past hurricanes Figure 2.6 shows that the comparison of respondents' countries of origin with those reported in the 2011 official census (STAT, 2017) are consistent. The main difference is that this research shows more respondents from The Netherlands and fewer respondents from Saint-Martin. However, in general, there is a good indication in terms of representation of the selected sample.

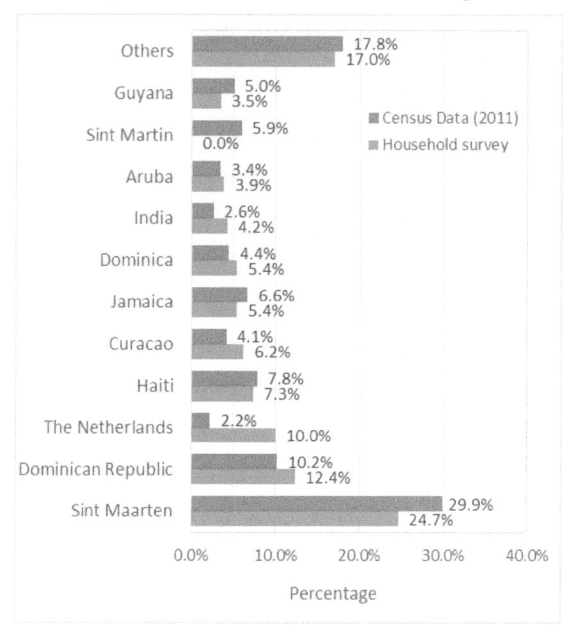

Figure 2.6. Country of Origin. Respondents of the household survey (blue bars) vs. census data 2011 (orange bars). This graph is intended to show that the sample size of the household survey adequately captures the multi-ethnicity of Sint Maarten when compared with the latest official census data available.

The number of years living in Sint Maarten is another indicator that can be used to assess vulnerability. Length of residence increases the knowledge of the associated risk of living in a hurricane-prone area, the sense of belonging and general knowledge of the island in case of evacuation, and increases the chances of having a social network. Figure 2.7 shows that 65% of the respondents have been living in Sint Maarten between 20 and 50 years. Hence, at least one member of the household has experienced at least one of the previous major hurricanes such as Hurricane Luis in 1995 or Hurricane Lenny in 1999. Hurricane Luis was repeatedly mentioned as the most significant hurricane the island had experienced before Hurricane Irma and was the reference hurricane in terms of the power of destruction.

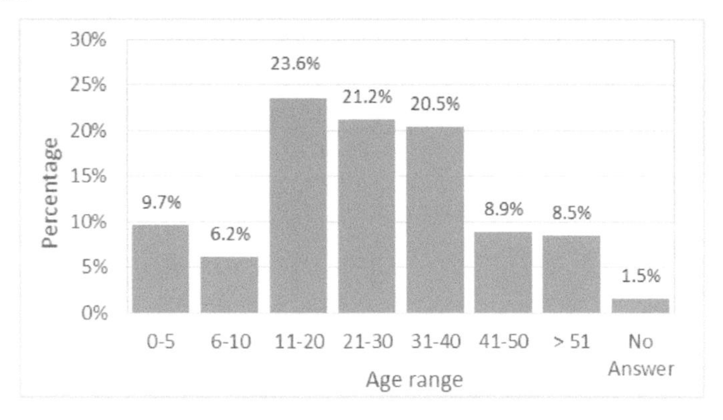

Figure 2.7. Number of years respondents have been living in Sint Maarten

House ownership and house material are also key demographic components that can be used to indicate the levels of vulnerability and risk. In Sint Maarten, 53% of the households live in rental houses, while the remaining 47 % live in their own houses (see Figure 2.8). Analysing household ownership by district, the proportion of rental houses is higher in the most deprived areas of the island. For example, in Lower Prince's Quarter and Philipsburg, 61% and 77.8% of the respondents are tenants.

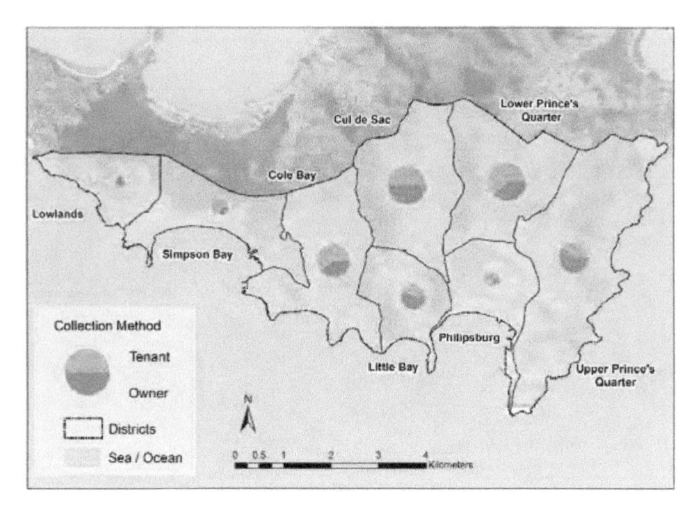

Figure 2.8. House ownership per district. It shows that except in the Lowlands districts, e.g. Upper Prince's Quarter and Little Bay districts, there are higher proportions of tenants in all the other districts.

Regarding decade of construction, approximately 60% of the houses in Sint Maarten were built after 1990, of which 40% were built between 1990 and 1999 (see Figure 2.9-(a)). That decade is worth mentioning since Hurricane Luis happened in 1995. Hurricane Luis was a tipping point in the history of Sint Maarten both in construction methods and awareness of hurricanes.

In Sint Maarten, 78% of the surveyed households live in concrete wall houses and approximately 20% live in wooden houses Regarding roof material, 69% of the respondents' houses are made of metallic sheets/Zinc (Figure 2.9-(b)). Using the data we collect it is not possible to correlate those house material with the ownership of the land.

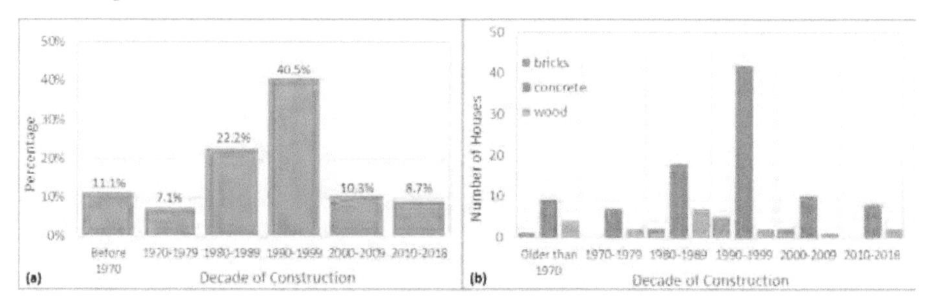

Figure 2.9. House construction characteristics. (a) shows the percentage of respondents' houses by decades of construction (b) actual numbers and materials (of walls) of construction by decades of construction

As illustrated in Figure 2.10, in Sint Maarten, 47% of respondents do not have house insurance while 26% of the respondents answer that they "do not know" if the house is insured. Evaluating this by district, regardless of the socioeconomic status, all of them, except the Lowlands, show a lower insurance coverage proportion.

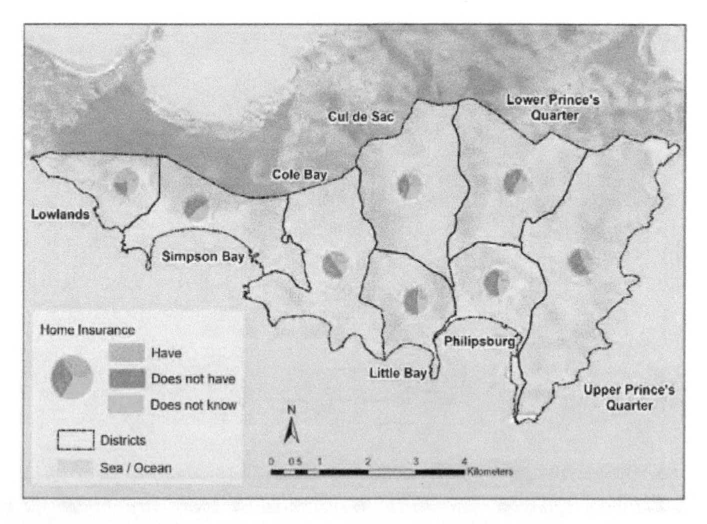

Figure 2.10. Home insurance coverage in Sint Maarten by district.

Information, Awareness, and Experience with Hurricanes and Storms

Prior experiences with hurricanes are of special importance, as they are associated with negative feelings such as worry and fear which have been reported to increase awareness and evacuation intentions (Demuth et al., 2016). In Sint Maarten, 80% of respondents reported they had experienced three or more hurricanes (Figure 2.11-(a)). Due to its strength and level of destruction, Hurricane Luis is the most remembered hurricane in Sint Maarten and constantly used as a reference for past experiences. Regarding the availability of information, respondents answered they know where to get up-to-date information on early warnings or evacuations when a hurricane is approaching the island. 77% of the respondents expressed that they know where to get information from a moderate to a great extent (Figure 2.11-(b)). Radio, Internet, and television, in that order, are the means to get the latest information during a hurricane.

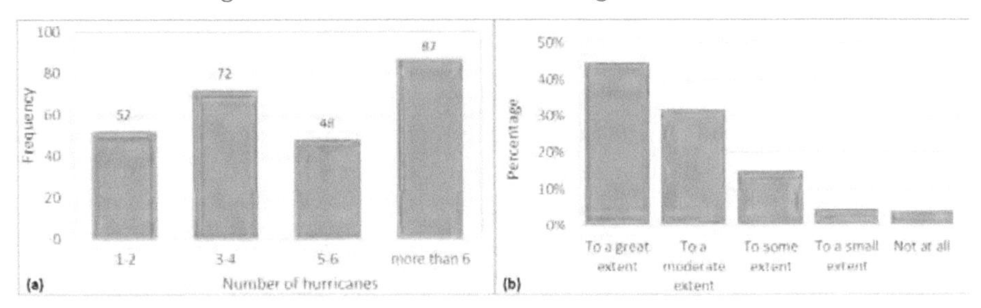

Figure 2.11. (a) Number of hurricanes remembered by the respondents and (b) Percentage to what degree people know where to get information about early warning and evacuation

Concerning information specifically about Hurricane Irma, the vast majority of people in Sint Maarten, 98%, were aware of the hurricane before it hit the island (Figure 2.12). Most people knew about Hurricane Irma well in advance to take measures to protect their lives and property. 55% of the respondents expressed that they were aware of Hurricane Irma in a range of 4 to 7 days and 19% from 8 to 14 days before Hurricane Irma's landfall (Figure 2.12-Series 1).

During the face-to-face survey, it was detected that there was a difference between the moment people were aware of the existence of Hurricane Irma and when they understood how severe the threat from this hurricane was to lives and properties. This signifies that respondents were not fully aware of the severity in advance. To capture the difference in the level of awareness of the hurricane by residents (i.e., hurricane existence and its severity), a new question was later included in the web survey regarding knowledge of the severity of Hurricane Irma. It was found out that about 80% of these respondents only considered Hurricane Irma as a real threat in a range of 0 to 3 days prior to landfall (Figure 2.12-Series 2). Despite the lower statistical significance of the web survey alone, the survey data shows that even if most residents knew about Irma four or more days before its landfall, they realized about its severity very close to the landfall.

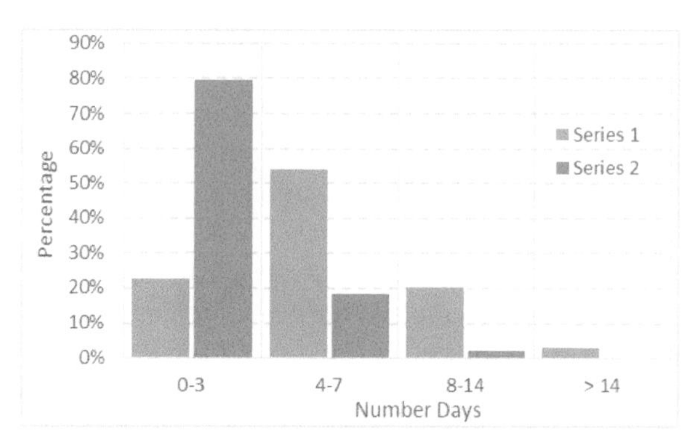

Figure 2.12. People's awareness of Hurricane Irma. (Series 1) How many days in advance they knew about Hurricane Irma and (Series 2) How many days in advance they knew about the severity of Hurricane Irma (only asked in the web survey).

Evacuation Behaviour

Concerning Hurricane Irma, 69% of the respondents did not evacuate (Figure 2.13). Of the 31% who evacuated, nearly 80% evacuated before the hurricane hit the island. But a remarkable 16% evacuated during the hurricane putting their lives in danger. The main reason to evacuate during the hurricane was that their houses collapsed during the hurricane and people had to run to neighbours houses looking for shelter. About 63% of those who chose to evacuate prior to the storm, sought shelter with friends' or relatives' whose houses they perceived were stronger than their own and to be with beloved ones during the hurricane. Hotels were the preferred evacuation locations for 21% of the evacuees, as they thought (incorrectly) that hotels were built to withstand a hurricane and that water and food would be available in the days after the hurricane. The number of people and assets exposed to certain hazards has previously been identified as one of the major drivers of vulnerability (Vojinović and Abbott, 2012).

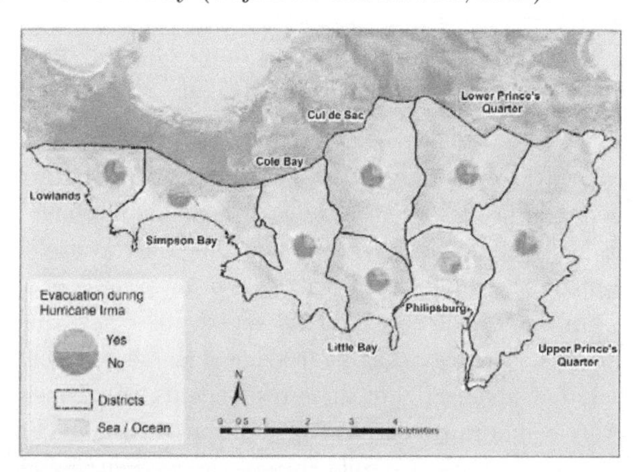

Figure 2.13. Evacuation percentage during Hurricane Irma by district. Philipsburg is the district with the highest proportion of evacuees.

On the other hand, only 3% evacuated to public shelters. In Sint Maarten, schools, churches, and community centres serve as official shelters and there are no buildings dedicated to serving only as such. The reasons given for this low percentage is that residents perceived their own houses (or chosen location to evacuate) as stronger structures than the official shelters. The perceived strength of houses to resist the hurricane was the most significant reason for not evacuating, with 86% of respondents giving this reason (Figure 2.14). Also, the belief that conditions of public shelters may not be adequate in terms of basic supplies (i.e., water, food, beds) was another reason not to choose to evacuate to them, with 21 % of respondents saying this is somewhat or more influential in their choice. Residents also remarked that staying at home has an extra advantage as they can protect their house against looting and repair window shutters and secure doors during the hurricane (especially when the hurricane's eye passes).

To those respondents who decided not to evacuate during Hurricane Irma, a set of questions was asked to gain a deeper insight into how influential different reasons were for staying at home. Among those were how strong they thought Hurricane Irma would be, how strong they thought their houses were, influences based on previous experiences that led them not to evacuate, not having enough information or not trusting the source from where they are getting it among other factors. The predominant reasons people gave were the perception of more wellbeing and safety when staying at home, the perception of having stronger houses compared to the public shelters and past experiences with hurricanes. We found that the third main reason not to evacuate was the perception that Hurricane Irma would not be a threat. Around 34% of respondents (see Figure 2.14) considered it from *to somewhat influential* to *extremely influential* in the 5-point Likert-type scale response used for these questions (Vagias, 2006).

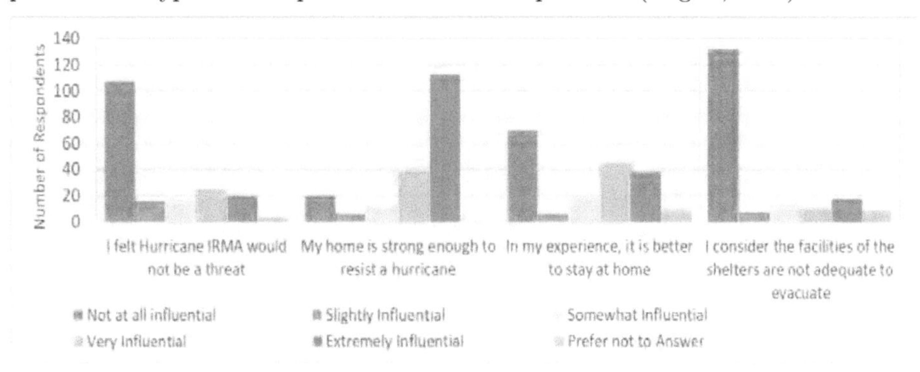

Figure 2.14. Main factors that influenced the decision not to evacuate during Hurricane Irma in Sint Maarten.

The factors that had non or little influence on the decision to not evacuate were unawareness of where to evacuate (89%), need of assistance to move to a shelter (96%), fear of looting (88%), not receiving any warning (91%) and no trust in institutions (94%).

The perception on the island is that the infrastructure for current officially designated shelters is not strong enough and does not provide enough resources. People do not see any added value to evacuate to those shelters. However, the perception about the location of the existing ones is appropriate in terms of geographical coverage for 40% of the respondents.

Risk perception

Most of the respondents consider the Government of Sint Maarten as the main body responsible for preventing losses during a hurricane. However, inhabitants recognize their role in the disaster management during an extreme event (second in frequency of responses as shown in Figure 2.15-(a)), and the Dutch government is listed as the third most common answer in frequency. It came as a surprise since the island became a constituent country in 2010 based on the 2000 status referendum. The referendum, among other reasons, was held to look for economic and administrative independence on the island.

More than 60% of the interviewees responded that the Government of Sint Maarten could have prepared better to reduce the direct impact of Hurricane Irma (Figure 2.15-(b)). The perception of the citizens is that the government of the island could have better regulations in place for the construction of buildings and the land use development on the island regarding the material, methods, and location of the houses. They also noted that inspection and enforcement of the regulations are necessary. In addition, the respondents mentioned that better warning information is needed from the authorities, with clear messages containing both arrival time, strength of the hurricane and clearly stating the need to evacuate (or not) from some areas.

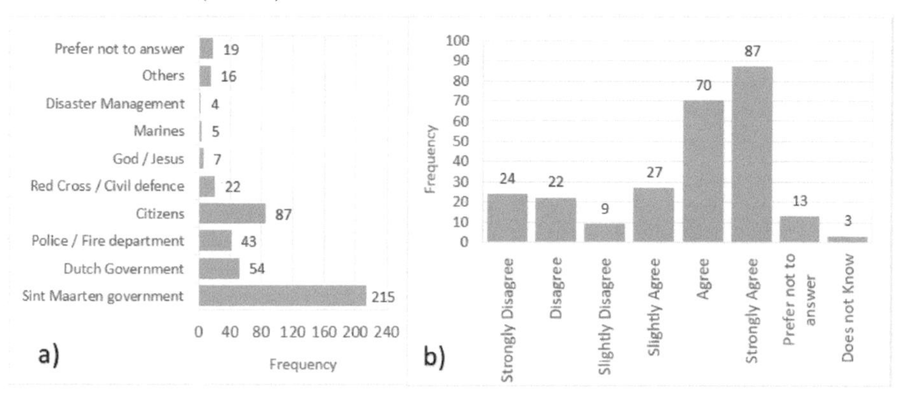

Figure 2.15. (a) Perception of respondents regarding who is responsible on the island during a disaster. (b) Evaluation of respondents on the performance of the government's preparedness and prevention of disasters.

To evaluate how Hurricane Irma may influence future evacuation behaviours on the island, the likelihood of evacuations with different intensities of storms or hurricanes was explored. This question was posed as a Likert-type chart with five categories: Definitely would not – Probably would not – About 50-50% - Probably would and Definitely would. Only the two extremes of this Likert question present a significant number of respondents as shown in Figure 2.16.

What is observed for the respondents is that with the increase in the magnitude of a forecasted hurricane, there is an associated increase in willingness (and perceived need) to evacuate and a decrease in those that definitely would not evacuate at almost the same rate. It is remarkable that in a future Category 5 hurricane, the number of people who expressed a definite willingness to evacuate is almost the same as those that definitely would not evacuate. What this means for Sint Maarten residents is that despite the recent disaster associated with Hurricane Irma, half of the respondents still prefer to seek shelter in his/her own house.

We recognise that self-accounted responses must be treated with special caution (Baker, 1991). Survey responses may be bias for the still fresh memory of the disaster caused by hurricane Irma, and that actual evacuation behaviour may be different form the answers we manage to collect. Despite the cautionary that we must observe with this data, we believe it is still is a good starting point to explore how the impact from Hurricane Irma may shape future evacuation in the island and serve as basis to establish shelter capacity and resources allocation, as well as coordination of stage or sectorial evacuation if needed.

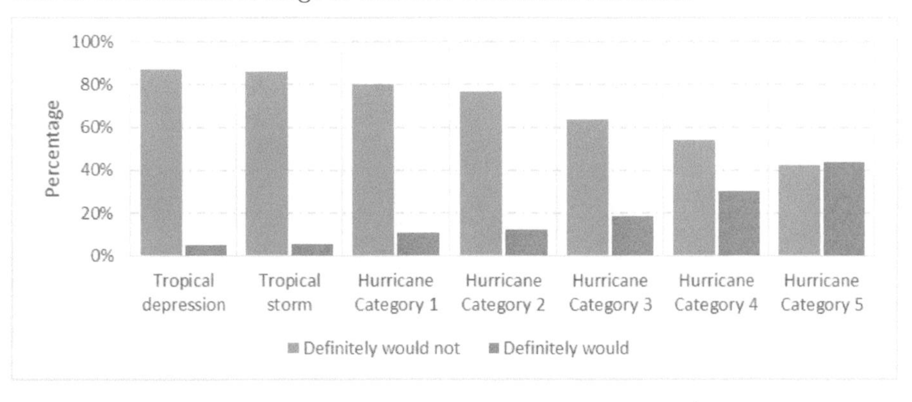

Figure 2.16. Future evacuation behaviour in relation to storm/hurricane categories (Saffir-Simpson Hurricane Scale)

2.4.2 Discussion and interpretation

Household and demographic parameters

After hurricane Luis in 1995, most of the residents of Sint Maarten built their homes to withstand a Category 4 hurricane. The subsequent hurricanes that hit the island did not cause significant damage to properties or lives. The perception of the research team is that the combination of these factors led to a point where residents felt prepared to withstand any hurricane based on their previous experiences, but the strength of a Category 5 hurricane like Hurricane Irma caught most of the island's residents unprepared. Hurricane Irma greatly surpassed Luis in terms of wind speed and barometric pressure. Additionally, residents did not take into account that even if the lesser hurricanes that occurred in the last two decades did not cause significant structural damage, they might have weakened the structure slightly each time.

From the fieldwork, the research team observed that people are rebuilding their damaged houses in concrete (i.e., concrete walls and roofs) (Figure 2.17-(a)). However, it was also observed that those who cannot afford or are not willing to build in concrete fall in two categories. The first is those using better reconstruction materials, such as longer or extra screws to fix metallic sheet roofs to the structure and the second category is those self-repairing with no or little guidance from the government using low-quality materials due to economic limitations and availability of materials on the island (Figure 2.17-(b)). The first category can be considered as a positive change and is a clear indication of adaptation to the disaster.

Figure 2.17. Observed reconstruction in the island. (a) With stronger materials (block, concrete). (b) With weak materials (wood).

House ownership affects the vulnerability of households, as tenants may not feel the same responsibility in strengthening the houses they live in and are less willing to spend money on a property they do not own. On the other hand, landlords may take a longer time to repair damaged houses under a rental contract. Furthermore, in most cases rented houses in Sint Maarten are not insured, and reconstruction of damaged rented houses may take longer since the local rent laws do not explicitly specify who is responsible for that task.

The low insurance coverage is contrary to the regulations, in Sint Maarten that it is mandatory to have home insurance when taking out mortgages. The reasons mentioned by respondents for the low house insurance coverage on the island include the high rate of premiums, low trust of the insurance companies, not getting paid what the house is worth, slow claim processing and poor client service. The low coverage rates of house insurance in conjunction with high tenancy is alarming for Sint Maarten considering how vulnerable some neighbourhoods are to disasters.

In terms of land development in Sint Maarten, large parcels of land are leased by private landlords. On the leased lands, the landlord does not allow the lessee to build concrete houses. Even if the building code specifies that the construction must be done using concrete or other solid materials, lessees may be obliged to build wooden houses. That further increases the vulnerability of the households. The reason behind the prohibition of building in concrete is that landlords are required to pay a more significant compensation from them if they want to terminate the lease. Some of these private lands do not have public infrastructure such as canals to collect the rainwater, or the installation of underground electricity due to the government lacking the authorization to build there.

Information, Awareness, and Experience with Hurricanes and Storms

Hurricane Luis had been repeatedly mentioned during the survey, and it is observed that Hurricane Luis has changed among others people's awareness to hurricanes. The behaviour of the inhabitants of the island has changed positively, especially immediately after the hurricane. As depicted in Figure 2.9-b, immediately after Luis, most inhabitants adapt, and hence, reconstruct or built new houses of materials such as concrete to withstand strongest hurricane wind forces.

However, that adaptive behaviour is lost over time as the social memory fades. Due to its strength and devastating effect, Hurricane Irma will be the reference and most remembered hurricane in the future, and it will boost the awareness of the population. Similar to the adaptive behaviour after Hurricane Luis, this study observed that inhabitants were rebuilding stronger residences after Irma. Such behaviour is not unique to Sint Maarten. For example, (Colten and Giancarlo, 2011) report that adaptive policies initiated in response to hurricanes in the U.S. Gulf Coast were gradually ignored as social memory fades. This effect had already been reported during the days previous to the hit of hurricane Isaac in 2012 in the affected areas from Hurricane Katrina in the United States, where residents showed a more proactive behaviour towards the hurricane warnings and preparation when compared with other areas that were hit by less severe hurricanes.

Regarding hurricane preparedness, one of the most common complaints of the respondents is that although they were aware of Hurricane Irma approaching the island, they did not have enough time to react and prepare to withhold the effects of the hurricane. It is important to mention that some schools were open until September 4 and some people had to work on September 5, a day before Irma's landfall. Other reasons respondents mentioned for the lack of preparedness included that they did not receive a clear warning on the severity of Hurricane Irma, the hurricane intensified and moved to the island rapidly, and there was a shortage of materials to strengthen houses or the price of materials escalated just before Hurricane Irma's landfall.

Hurricane severity and communities' experiences and memories also play an essential role in preparedness. For example, residents that were affected by Hurricane Katrina in the U.S. showed a more proactive behaviour and preparedness to Hurricane Isaac in 2012 compared to areas hit by less severe hurricanes (Milch et al., 2018). However, the influence of past experiences regarding awareness and preparation decreases with time and is associated with how emotionally strong the past events were (Milch et al., 2018). In the case of Sint Maarten, when Irma hit the island, most people have forgotten what damage a strong hurricane can cause on the island as the last severe hurricane, Luis, struck the island 22 years ago, and this contributed to the lack of preparedness.

Evacuation Behaviour

Based on the destruction caused by Hurricane Irma, it is expected that people will respond more proactively to future hurricane forecasts, warnings and evacuation orders. However, based on the survey results and past trends, it is expected that evacuations to public shelters will remain unpopular as the general feeling among residents is that the shelters are not strong enough. It is highly likely that residents with concrete houses will not evacuate as they feel they will be safe.

Regarding evacuation to hotels, even though hotels were reported as one of the preferred locations to evacuate during Hurricane Irma, it is expected that the number of people evacuating to hotels in future hurricanes will decrease significantly. Most hotels sustained extensive damage as they were made of inferior construction materials for a hurricane-prone area, such as plywood, and because of their proximity to the shoreline, exposing them to storm surge damage (see Figure 2.18).

Another factor that played a major role regarding evacuation behaviour during Hurricane Irma was the announcement that public shelters would be closed before Hurricane Irma's landfall. The main reason provided by the authorities was that there was no guarantee for the safety of those who chose to evacuate to a public shelter. A new order to open some of the public shelters was only communicated to the public during the afternoon of 5 September. However, according to the findings of the fieldwork, that message did not reach the broader population of the island. The mixed evacuation orders will play a vital role in the lack of trust in institutions, warnings and evacuation orders (or advisory) for future hurricane seasons.

The information collected about the inhabitants' perception of existing public shelters is of great use for the Sint Maarten Disaster and Emergency Management group in the context of future evacuation behaviour if, as a result of the experience with Hurricane Irma, they decide to implement massive evacuations in future hurricane seasons.

Figure 2.18. Damage to hotel infrastructure. (a) Sonesta Maho (Simpson Bay) (b) Great Bay Hotel (Philipsburg)

Risk Perception

The perception of this research team is that residents of Sint Marten perceived they were not vulnerable to hurricanes prior to Hurricane Irma. This belief was based on the construction of "strong" concrete houses after the devastation caused by Hurricane Luis. Not experiencing hurricane disasters for two decades fed this misconception of safety and lack of proper warning information by authorities. Reaction and preparation by the inhabitants are expected to be high in the coming hurricane seasons, but this is an effect that is expected to fade through time. Therefore, regular hurricane awareness campaigns must take place in the coming years. It is important that parents pass their experience/knowledge of Hurricane Irma to the next generation as it was observed that younger generations of Sint Maarten did not know what a "real" hurricane can do to the island.

The semi-structured component of the household survey allowed the team to collect sensitive data to understand the formation of risk directly associated with the performance perception of the government during the emergency and disaster associated with Hurricane Irma. From this analysis, it was found that in general, the perception of the island's inhabitants is that the authorities performed poorly in all the phases of the disaster risk management. These include miscommunication and insufficient warnings about the severity of the hurricane.

The decision of the government not to open the public shelters before Hurricane Irma's landfall for those in need was highly criticized and viewed by the residents as one of the major failings during the emergency. The role of the government in the immediate aftermath of the hurricane was also criticized. Interviewees mentioned that the destruction to the island's economy due to the looting of supermarkets, shops, jewellery stores and other businesses in the immediate aftermath of Hurricane Irma was significant and considered by many as more critical than the direct destruction of the hurricane. This negative perception has had an undesirable and direct effect on people's trust in institutions for the future hurricane seasons. These findings need to be taken into consideration when communicating future hurricanes potentially affecting the island.

Hurricane Irma will create a positive effect on awareness and risk perception to hurricanes and as such the government of Sint Maarten should act with caution to not act based on fear and create false alarms in future hurricane seasons. Special caution should be taken in the lower-elevation areas which are prone to flooding and storm surges during a hurricane. Since Hurricane Irma was a relatively "dry" hurricane, some residents living in flood-prone areas may rebuild or reinforce their houses to withstand winds and pressures but neglect the effects of flooding in a possible scenario of a "wet" hurricane or storm.

2.5 Lessons learned during the survey

Surveying in the aftermath of Hurricane Irma on the island of Sint Maarten was as challenging task as initially expected. Firstly, some logistics issues needed to be addressed such as mobilizing the team to the affected area, which was only possible five months after Hurricane Irma due to airport damage. Accommodation on the island was also very limited as hotels and other buildings sustained significant damage. Overall, five months from the event and the start date for data collection could be considered optimal in terms of availability of resources on the island. Being in the reconstruction phase also served the survey by obtaining a reasonably good response rate as participants were in a better mood and willing to participate.

The adapted mix-mode survey, based on face-to-face interviews and the complementary web survey, proved to be beneficial in the context of a post-disaster assessment of vulnerability. By using the web survey, the team could reach areas that could not be covered within the timeframe of the fieldwork activities, and helped increasing the response rate of the study. Results are encouraging and suggest the mix-method as an adequate alternative as a post-disaster data collection method. It is important to note that the team sent reminders on social media about the survey every week and batches of responses were obtained in the following hours.

The preparation of maps using different sources to the ones used in this research was found to be beneficial during the fieldwork. It saved time by allowing the team to identify the targeted houses before the data collection and to prepare the daily work routes more efficiently. OpenStreetMaps has the potential to be used as a primary source of geographic information in areas with poor or no data in post-disaster assessment.

Proper identification of the survey team during the fieldwork was important, as a team with no government association played an essential role in the success of the field mission in terms of the response rate. This was especially true in those areas where the reconstruction activities were still underway or not yet started. Association with local authorities may have led to low response rates in some regions of Sint Maarten.

In a multicultural environment with different languages, the role of bilingual interviewers was very important for the success of the survey. In this particular case study, some areas are predominantly Spanish and French speakers and sometimes have limited or no English knowledge. Having a native Spanish speaker gave the team the possibility to perform some of the surveys in Spanish, which increased the response rate in specific sectors of Sint Maarten. In contrast, as the team members did not speak French, there was a lower coverage rate or limited interaction in the semi-structured part of the interview in the French-speaking regions. Regarding the web survey, having the survey and the invitation to participate in it only in English could have acted as a bias towards which regions of Sint Maarten could have access and willingness to participate in the survey.

The household survey performed on the island of Sint Maarten in the context of a post-disaster area after Hurricane Irma proved to be a good way for assessing vulnerability and risk at the local level, allowing the research team to include individual changes into a broader analysis of vulnerability based on statistical data. In addition, the household survey performed in the aftermath of Hurricane Irma explored more in depth how risk perception changed and how Hurricane Irma may have potentially modified evacuation behaviour for possible future hurricanes on the island. This information is relevant in the development of a disaster risk management plan on the island, and additionally it can be utilized for future raising awareness campaigns and the design of an early warning system.

3

SOCIOECONOMIC VULNERABILITY ASSESSMENT IN SIDS

This chapter presents a methodology to assess and map socioeconomic vulnerability in SIDS at a neighbourhood scale using an index-based approach and principal component analysis (PCA). The index-based vulnerability assessment approach has a modular and hierarchical structure with three components: susceptibility, lack of coping capacities and lack of adaptation, which are further composed of factors and variables. Applying the combined analysis of index-based approach with PCA allows us to identify the critical neighbourhoods on the island and to identify the main variables or drivers of vulnerability. Results show that the lack of coping capacities is the most influential component of vulnerability in Sint Maarten. From this component, the "immediate action" and the "economic coverage" are the most critical factors. Such analysis also enables decision-makers to focus their (often limited) resources more efficiently and have a more significant impact concerning disaster risk reduction.

This chapter is based on:

Medina, N., Abebe, Y. A., Sanchez, A., & Vojinović, Z. (2020). Assessing Socioeconomic Vulnerability after a Hurricane: A Combined Use of an Index-Based approach and Principal Components Analysis. Sustainability, 12(4), 1452. Doi:10.3390/su12041452

Sorg, L., **Medina, N.**, Feldmeyer, D., Sanchez, A., Vojinovic, Z., Birkmann, J., & Marchese, A. (2018). Capturing the multifaceted phenomena of socioeconomic vulnerability. Natural Hazards, 92(1), 257-282. Doi:10.1007/s11069-018-3207-1

3.1 Introduction

There is no unique definition of vulnerability in the scientific community (Paul, 2013). The definition of vulnerability for scientific assessment depends on the purpose of the study and can only be considered meaningful regarding a specific at-risk situation (Ciurean et al., 2013; Cutter et al., 2003; Nguyen et al., 2016). Regarding socioeconomic vulnerability to natural hazards, as the scope of this thesis, one of the most widely used definitions is the one given by the Intergovernmental Panel on Climate Change (IPCC). The IPCC defines vulnerability as "the propensity or predisposition to be adversely affected, and it encompasses a variety of three dimensions susceptibility to harm, lack of capacity to cope and lack of capacity to adapt" (IPCC, 2014). In summary, the IPCC concept of vulnerability is the degree to which a system is susceptible to, and is unable to cope and recover from the adverse effects (McCarthy et al., 2001).

Based on the IPCC definition of vulnerability, this research uses a methodology that allows capturing the three dimensions of vulnerability Susceptibility, Coping, and Adaptation. As it was discuss in Chapter 1, index based approaches are preferable in DRM to allow standardise concepts and to allow risk comparison among different implementations. Also, index based approaches are more flexible and allow to include local and relevant driver of vulnerability that are context specific.

An index based vulnerability assessment approach has been extensively used and reported for flood and weather-related events (Balica et al., 2009; Balica et al., 2012; Connor and Hiroki, 2005; Cutter et al., 2003; Kleinosky et al., 2006; Percival and Teeuw, 2019; Sorg et al., 2018; Vojinović et al., 2016). The results and conclusions of those studies suggest and support the feasibility of using an index-based approach for the assessment of socioeconomic vulnerability in the context of SIDS.

In accordance, we adopt as the starting point for the assessment of vulnerability the work we initially presented in Sorg et al. (2018), in which we developed an index-based framework to assess vulnerability. The assessment method called PeVI is based on multiple indicators and is composed of three major components on which we defined vulnerability. PeVI was conceived to be flexible and easy to adapt to properly reflect the information available and the needs of a case study.

In this chapter, we present the results of an expanded PeVI, in such a way that the index is able to capture key components of vulnerability in an island prone to frequent hurricane and floods (Vojinović and Van Teeffelen, 2007), and to capture how these components can alter the island vulnerability after a major disaster by incorporating a household survey in the aftermath of Hurricane Irma. Our approach, this is taking into account the changing in dynamics after a disaster, offers a different view on vulnerability assessment to extreme weather events and is aligned with recommendations in literature (Birkmann, 2008).

We have expanded PeVI by adding elements that can change after a disaster such as elements of risk awareness and perception and access to information. We also include elements related to the possible immediate actions to face the potential hazard, which are essential in the context of a small island due to the impossibility to move completely away from the possible threat. Finally, we use information collected in the aftermath of a hurricane that can be associated with the direct impact (building and infrastructure damage) and how the society was adapting to the disaster (speed of recovery and construction methods and materials).

The methodology presented in this study is applied in the case study of Sint Maarten, one of the Leeward Islands on the northeast Caribbean Sea. Despite the general agreement among stakeholders and academia on the importance of having a vulnerability assessment for small islands in order to have a proper strategy to reduce risk to climate associated events, the island of Sint Maarten lack such a study for the whole island to date. The need for a vulnerability assessment in Sint Maarten was evident after the disaster caused by Hurricane Irma in September 2017. Vulnerability and risk assessments are an essential input for disaster risk reduction and adaptation planning to climate-related hazards and to support the island's reconstruction efforts.

In addition to expanding the vulnerability index, we have extended the analysis and interpretability of results by combining the index-based result of PeVI, with the use of Principal Components Analysis (PCA) into the methodology. Aggregate indices of vulnerability, such as the one computed in this research are useful in identifying where the hotspots of vulnerability occur. Moreover, it gives decision-makers a powerful tool to focus their efforts for disaster risk reduction. However, the generation of a single composite vulnerability index can be problematic, because information regarding the relations between the original variables is averaged in the resulting aggregated index (i.e. from many variables to a single number). Two different locations may have a similar vulnerability index value, but the driving variables may differ (Abson et al., 2012). To overcome this issue, we use the PCA technique that allows returning to the original variables to understand and interpret the aggregate vulnerability index.

3.2 Methodology and Data

3.2.1 Data sources used for the vulnerability assessment

To perform the vulnerability assessment on Sint Maarten two different sources of information were used. First, we used the data collected during the fieldwork campaign which was extensively explained in Chapter 2. Second, census data was used to complement the survey data. The information available regarding census data in Sint Maarten was only available at the whole island scale. This level of information is not considered sufficient when performing vulnerability analysis. Vulnerability index computation at smaller scales is advisable as they help in the identification of the most critical areas and as such be beneficial for the local government to guide the reduction of vulnerabilities to disasters triggered by natural hazards more efficiently and to have targeted mitigation plans/measures (Balica et al., 2009; Rufat et al., 2015; Sorg et al., 2018).

According to the department of statistics of Sint Maarten, the island is divided into eight zones, 24 districts, and 54 neighbourhoods (Figure 3.1). For Sint Maarten, we managed to have partial access to information corresponding to the last population and census conducted in 2011 at the neighbourhood scale. Limitation to access the full extent of the census data poses a restriction to the number of variables and the scale we used in this study.

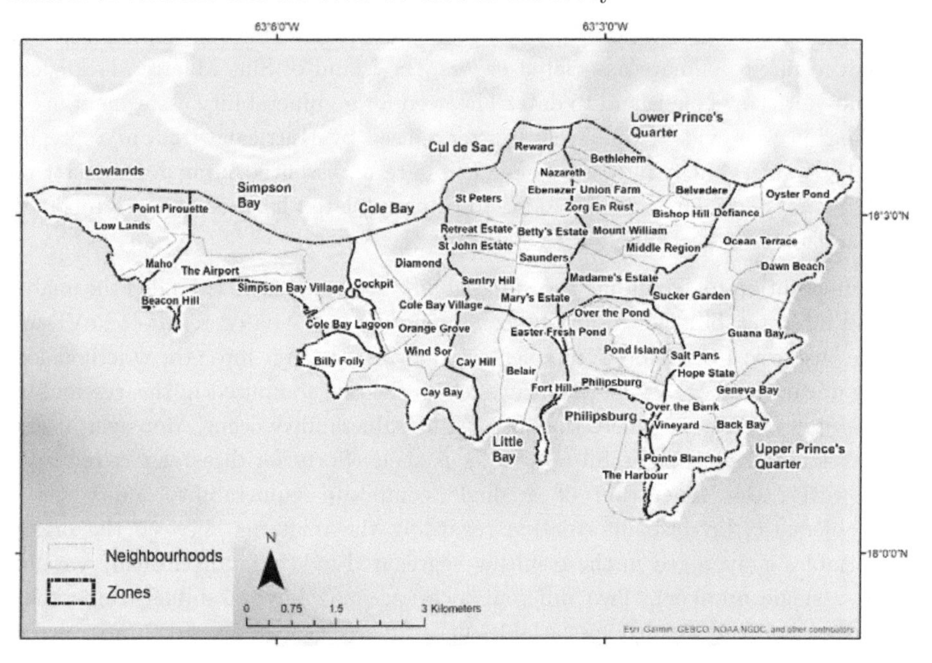

Figure 3.1. Administrative divisions of Sint Maarten at Neighbourhood and Zones scale. Texts in red are the zone's names, and texts in black are the names of the neighbourhoods.

3.2.2 Vulnerability Index

The framework selected for the computation of the vulnerability index in Sint Maarten is an extension on the work presented in Sorg et al. (2018). The PEARL vulnerability index (PeVI) aims to incorporate as many variables as possible to gain full insight into the vulnerability of a city or a region under analysis. PeVI has a modular and hierarchical structure with three main components: susceptibility, lack of coping capacities and lack of adaptation capacities. All three components consist of several factors which in turn are computed using a number of variables in a three-to-four-level hierarchy structure. The modular approach allows using any relevant and available information that captures the main components or drivers of vulnerability for the local conditions of Sint Maarten and to take into account not only the intrinsic and extrinsic factors of vulnerability but also takes into account the recent disaster caused by Hurricane Irma.

In the following subsections, the definition and computation of the three components of vulnerability, as well as the final vulnerability assessment, is presented. We recognise that some of the variables of a factor can be placed in another factor or component of the index; for example, the education variable may be used as an indicator for either the lack of adaptation capacities or in the awareness factor in the lack of coping capacities component. For this reason, we present in the following sections the explanations we used to support the rightness of use in each component based on literature review and expert knowledge. In addition, the supplementary material that accompanies the paper on which this chapter is based on contains a detail explanation on the computation of each variable and shows all the formulas, tables, questions and values we used to compute the index (See Medina et al., 2020, Supplementary material 1).

Susceptibility

Susceptibility in this research is defined "as '*the current' status of a society and its likelihood to be harmed*' (Sorg et al., 2018). In the second level of the hierarchy, this component has four factors: Demography, Poverty and Income, Housing and Infrastructure, as shown in Figure 3.2.

The **Demography** factor uses data from the 2011 census, and only one variable is used to compute it, the *Vulnerable Age Groups*. This variable has been extensively used in previous vulnerability assessments for natural hazards (Cutter et al., 2003; Fekete, 2009; Ogie and Pradhan, 2019; Sorg et al., 2018). In this group, it is suggested to include the segment of the population that is highly dependent (children younger than five years old) and the elderly population (older than 65 years old). These groups are more likely to require assistance, protection, transportation, financial support, and medications before and during disasters.

The factor **Poverty and Income** is a function of two variables *Dependency Ratio* and *Unemployment Ratio*, which are based on census data. *Dependency Ratio* is an economic parameter that captures the ratio between the population in a non-working age (i.e. younger than 15 years old and retirees) and the population in working age (i.e. 15 to 65 years old) (Nandalal and Ratnayake, 2011; Sorg et al., 2018). Higher values of this variable indicate higher pressure on the working group to be able to support the dependent one. *Unemployment Ratio* is the relation concerning the number of people register as unemployed and the number of potential workers (Fekete, 2009; Scheuer et al., 2010; Sorg et al., 2018). A higher rate of unemployment ratio reflects lower economic means to prepare appropriately for a disaster or to recover from its effects. This segment of the population may require more external aid from the government or other humanitarian organisations during the pre-disaster and during the recovery phases.

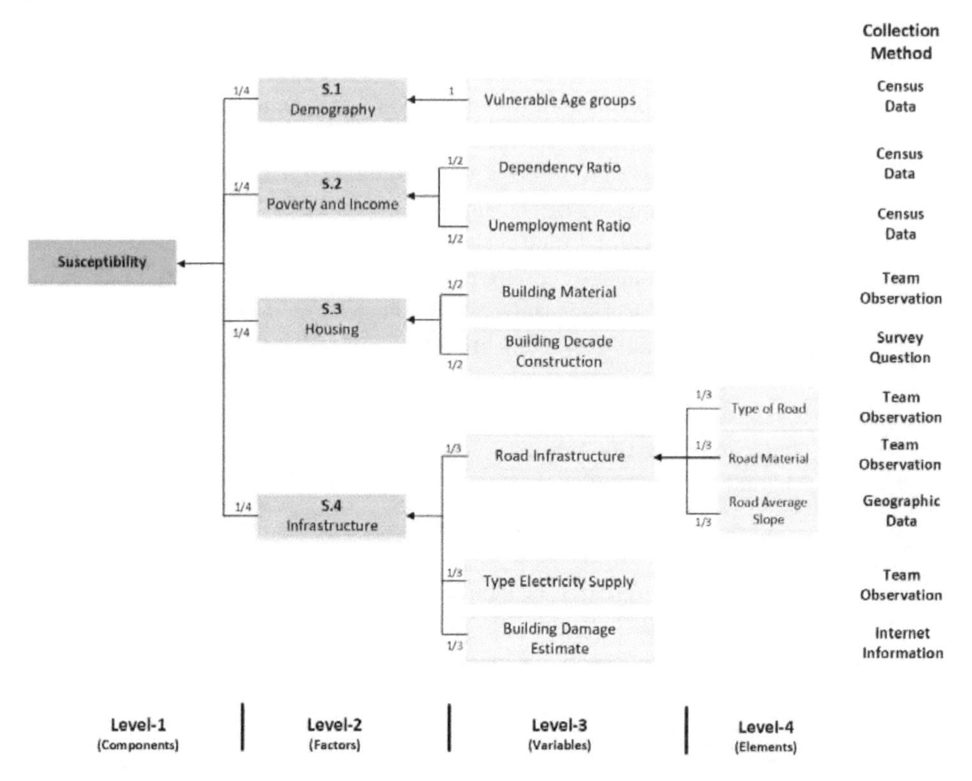

Figure 3.2. Structure of the Susceptibility component for the vulnerability index (PeVI), applied in the case study of Sint Maarten. The figure shows the four levels of hierarchy and the source of information used to compute each variable. The numbers next to the arrows indicate the weighting factor to compute the next level.

The **Housing** factor is directly related to the physical characteristics of buildings that increases or reduces vulnerability. In this study, the variables that define the housing factor are *building material* and the *decade of construction* of houses. B*uilding material* is computed using observations we made during the fieldwork. The data collected for this variable in the surveyed houses was the walls and roof primary material. This variable is directly related to the structural strength of the building to resist adverse extreme weather conditions. As such, concrete houses are expected to have better resistance (lower susceptibility) than wooden houses (higher susceptibility) (Ciurean et al., 2013; Kappes et al., 2012).

The *decade of construction* variable is of relative importance in Sint Maarten as it is a variable that has a direct relation with the construction method and material. We assumed that the older the house, the more vulnerable it is to natural hazards. As presented in Medina et al.(Medina et al., 2019), in Sint Maarten, it has been observed a significant change for better construction materials and better construction techniques after major disaster events such as those caused by hurricanes Dona (1960), Luis (1995) and Hugo (1998), and again after Irma (2017). Also, we assume that the older the building, the more susceptible a building is to withstand a natural hazard. The assumption was based on the natural process of material degradation, and also from field observation and data collection, where residents do not perform regular maintenance to their houses.

The susceptibility to **Infrastructure** factor includes three variables: *Road Infrastructure*, the *type of electricity supply* and the *damage estimate* to buildings caused by Hurricane Irma. *Road Infrastructure* is of vital importance during all phases of an extreme weather-related event, as they may get disrupted or highly damaged. *Road infrastructure* is vital for facilitating evacuation, emergency services, relief supplies, the flow of goods and clean-up activities (Berkoune et al., 2012; Markolf et al., 2019). To account for *roads susceptibility* three elements were used: *Type of road* (primary, secondary or tertiary), *road material* (Asphalt, concrete and unpaved) and *terrain slope* that is computed from the DEM as the average slope in percentage.

The type of road is extracted directly from *OpenStreetMap* attributes. Primary roads were considered more vulnerable since the few that exist are already working on full capacity and the limited redundancy on the transportation network make them almost mandatory to drive under any possible evacuation plan. This situation makes the primary roads more susceptible to collapse under an extreme weather event (Koks et al., 2019; Singh et al., 2018). It is essential to include the road material in the index because more susceptible materials such as roads built-in natural terrain or asphalt can be easily erodible during rainfalls. The slope of the roads is important because of the roads susceptibility increases in high steep areas due to the poor or non-existing drainage (Koks et al., 2019; Scawthorn et al., 2006), and the average slope of the road also influences the feasibility to access it (de Ruiter et al., 2017; Keller and Atzl, 2014).

The second variable used to compute susceptibility to **infrastructure** is the *type of electricity supply*. Electricity is a critical component in the recovery phase as societies depend significantly on the use of it, from household use to its vital use in other critical facilities such as hospitals and airports (Barben, 2010; Mohagheghi and Javanbakht, 2015; Ouyang and Dueñas-Osorio, 2014). The importance of this variable in Sint Maarten lies on the high destruction potential of hurricanes and floods to electric power system components, causing widespread outages over a long period of restoration and recovery. Also, blackouts are costly and entail considerable disruption to a society (Jufri et al., 2017; Panteli and Mancarella, 2017; U.S. Congress, 1990). In Sint Maarten, the type of electricity supply was collected during the fieldwork at the street level and later the length was measured in the office using a map of the island. The categories of electricity supply on the island are aerial and underground. Aerial distribution lines were considered to have high susceptibility value to weather-related events. Hence, areas with underground electricity supply have low susceptibility compared to areas with aerial supply. Areas with no electricity supply did not account in the computation of the variable. We acknowledged that underground electricity distribution lines could also be affected by floods. However, for the Sint Maarten vulnerability assessment, this is simplified to include only the effects of wind on the electric system based on the observed effects of Hurricane Irma.

Finally, the third variable on the susceptibility of **infrastructure** is the *building damage estimate.* The importance of using this variable is that it can be a reasonable estimation of the proper use (or not) of building codes and administrative capacity (and willingness) to enforce regulations and to some extent to be used as predictors of damage for future hurricanes (Taramelli et al., 2015). In addition, households that experience damages in the past may change their risk management behaviour to a most proactive reaction towards extreme events (Birkmann et al., 2016).

This variable is computed using the damage assessment for buildings done by Emergency Management Service, Copernicus (EU-JRC, 2017). The information obtained from Copernicus was a shapefile format of the buildings of Sint Maarten with the damage estimated in five categories for each building "Completely destroyed", "Highly damaged", "Moderately damaged", "Negligible to slight damage" and "not affected" by Hurricane Irma. Due to the rapid assessment performed by EU-JRC, (2017), the use of this information may have limitations of scale, resolution and data interpretation. Despite this disclaimer, the information was considered useful for rapid evaluation of the physical impacts of Hurricane Irma and how susceptible or not the building infrastructure was to the effects of a Category 5 hurricane.

Lack of Coping Capacities

The lack of coping capacities refers to "the strengths and resources for direct actions which potentially can lead to a reduction in the consequences of a hazardous event" (Sorg et al., 2018). In the PeVI, it is composed of six factors: Social Network, Immediate Actions, Government, Economic Coverage, Information and Awareness (see Figure 3.3).

The **Social network** factor is computed using two variables, *Household size indicator* and *Immigration.* From the census data, the average number of inhabitants per household in each neighbourhood is extracted to compute the variable *Household size.*

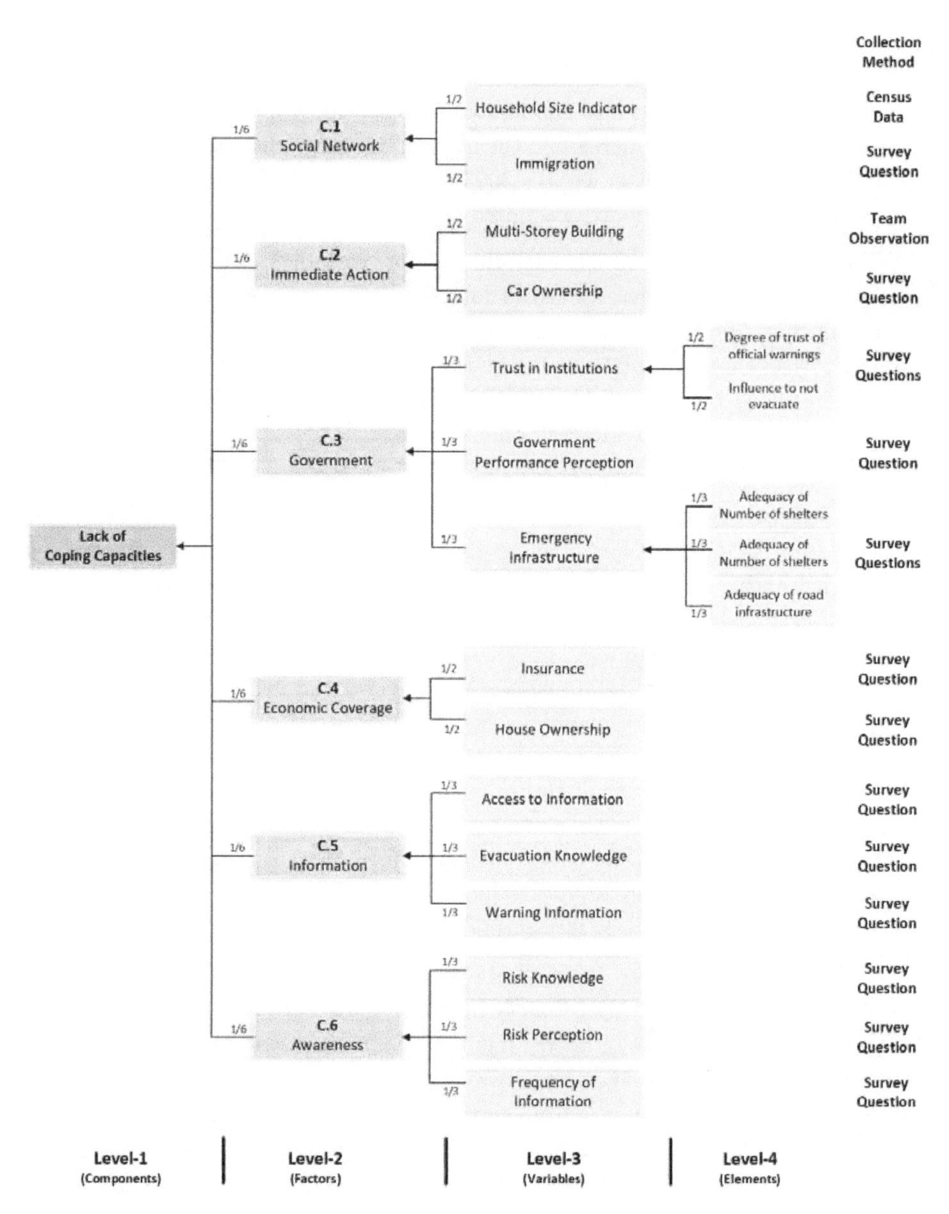

Figure 3.3. Structure of the Lack of Coping Capacities component for the vulnerability index (PeVI), applied in the case study of Sint Maarten. The figure shows the four levels of hierarchy and the source of information used to compute each variable. The numbers next to the arrows indicate the weighting factor to compute the next level.

Taking into account the formation of safety nets in the form of social networks, Welle et al. (2014) state that an increase in household size decreases vulnerability due to mutual help. The work of Lianxiao and Morimoto (2019), also suggests that the more people in the family, the higher the ability to respond. For this study, a household with only one individual is considered to have a higher lack of coping capacities. In households with four or more inhabitants, this variable is considered not to influence the variable negatively. We acknowledge that expanding the household size can also affect the vulnerability by increasing the scarcity of resources, an increase in the number of care of dependants and a higher population density (Cutter et al., 2003; Hashim et al., 2018). However, these associated adverse effects are accounted for in other variables of the PeVI.

To measure the lack of capacity due to the variable *Immigration,* we use a question from the survey. We asked for the number of years a respondent was living in Sint Maarten. It was decided to use the number of years lived on the island rather than the place of birth. Here we assumed that the more years a person has been living in a place could lead to a reduction of the vulnerability as they learn to cope and increase the knowledge of flood protection measures (Aboagye, 2012; Rufat et al., 2015).

The number of years in a place has been previously identified to increase the general knowledge of the city, such as the best places where to evacuate and also to navigate through the bureaucracy to request and receive help from the authorities (Dash and Gladwin, 2007; National Research Council [NRC], 2006; Rufat et al., 2015). The number of years living in a place can also facilitate tighter social networks (Depietri et al., 2013). A stronger social network can increase the coping capacity through economic, social and emotional support (Nakagawa and Shaw, 2004) as well as increasing knowledge about past disasters and exchange information about the risk of future events (Rufat et al., 2015). On the other hand, recent migrants (less than five years living in a place), can potentially have cultural, economic and language barriers, which in turn can affect access to warning information and access to post-disaster aid (Cutter et al., 2003; de Hamer, 2019; Depietri et al., 2013; Maldonado et al., 2016).

One crucial element to increase the coping capacities is the ability to take **immediate action,** getting to safety in a fast and secure way during a weather-related event. In the case of floods, having a *multi-storey building* allows to move quickly to a higher zone and in this way avoiding direct contact with the hazard and also to protect belongings from getting damaged from the floodwaters (Rufat et al., 2015; Sorg et al., 2018). During the fieldwork, we collected the number of floors of the surveyed houses. To compute this variable, we use the ratio between the number of houses with only one floor and the total number of houses in the neighbourhood.

A second variable for **immediate action** is related to the number of cars available in the household. It is a measure of the ability to evacuate during an emergency. We compute the variable *car ownership* based on a question from the survey. It is the ratio of the number of cars to the total number of inhabitants in the household. A ratio of 0.2 or bigger (i.e. having at least one car for each five-person), corresponds to a household with higher coping capacities. The smaller the ratio, the more vulnerable the household. Non-car ownership decreases the ability to move out of the hazard zone when required and closely related with low income and poverty factor (Colten, 2006; Rufat et al., 2015; Tapsell et al., 2002) and, not owning a car is highly correlated with non-evacuation behaviour (Karaye et al., 2019).

The **government** factor is computed using the variables, *trust in institutions*, the *performance perception* of the government during Hurricane Irma and the perception of the inhabitants about the quality of the *emergency infrastructure* on the island. All the variables of this factor are calculated using questions directly asked during the field survey. Previous studies such as Balica et al. (2012) also used the lack of trust in institutions as a variable that lower vulnerability. Vári et al. (2013), concluded that low levels of trust in institutions were highly correlated with variables that increase vulnerability such as low level of education, lower incomes and unemployed status as well as a strong relation with those who suffered the most damages. To address the *trust in institutions* variable, we used two questions of the survey. In the first one, we asked the participant that if based on their previous hurricanes experiences, they trust in official sources of warning evacuations on the island. The second one is related to those respondents that directly expressed they did not evacuate during Hurricane Irma because they did not trust the official warning. The higher coping capacities were assigned to those respondents that answer they have trust in authorities "to a great extent" and the lowest coping capacities for those respondents that answered they "do not trust at all" authorities. All the answers in between were assigned a proportional degree of vulnerability.

The second variable used in the **government** factor is the *performance perception of the government* in response to Hurricane Irma. Failures and inaction from governments are identified as a significant driver of present and future risk and can intensify the disaster impact (Birkmann et al., 2016; Khunwishit and McEntire, 2012). Low-performance perception has a direct relation to households with a lower income and low level of education, houses that have shown low or non-changes in risk management at the household level (Birkmann et al., 2016). Thus, they can be categorised as not being (fully) prepared in the event of new hazards events. *Government performance perception* is computed using a survey question indicating the relation between losses during Hurricane Irma and the responsibility of the government of the island. We use larger coping capacity in this variable for respondents that do not blame the government for the losses in the island and lowest coping capacities to those respondents that "*strongly*" blame authorities for the losses in the island.

As a final variable in the **Government** factor, we asked in the survey the perception of the respondent regarding the availability, location and accessibility to the existing *emergency infrastructure*. The questions used to build this variable were the sufficiency of shelters and if their locations were adequate, and if the road infrastructure was appropriate and sufficient to evacuate. A proper emergency infrastructure is vital for vulnerability and risk reduction. Emergency infrastructure acts as a way to mitigate the consequences of a disaster by potentially reducing exposure, especially among the socially vulnerable population (Karaye et al., 2019). In the PeVI applied in Sint Maarten, the higher the number of shelters available, the lower the vulnerability. For the computation, a strong agreement in the number of shelters or its adequate location or a proper road infrastructure is value as higher coping capacity (low vulnerability), and strong disagreement is ranked with lower coping capacity (higher vulnerability).

The fourth factor is the **Economic coverage** and is calculated using two variables *Insurance* and *House ownership*. Both variables are assessed based on survey questions, one directly asking if the household has insurance for disasters triggered by natural hazards and another if the respondent own or rent the house, respectively. Home *Insurance* for disasters triggered by natural hazards can be seen as one of the most effective self-protective actions at the household level as a preventive measure in the coping strategies dimension of vulnerability (Maldonado et al., 2016; Rufat et al., 2015; Sorg et al., 2018). Homeowners with insurance are less affected by disasters triggered by natural hazards as they can absorb, rebuild and recover from losses more quickly once affected by the disaster (Cutter et al., 2003; Khunwishit and McEntire, 2012). For this study, having insurance is rated with high coping capacity, whereas not having one is assigned the low capacity to cope with the effects of a disaster. In the households where participants do not answer the question or expressed lack of knowledge whether or not the house is insured we assigned an intermediate level of vulnerability. Those above, under the assumption that these households may not be insured and the question was avoided because in Sint Maarten it is mandatory to have home insurance when taking out mortgages (Medina et al., 2019).

House ownership has a direct relation with vulnerability to disasters triggered by natural hazards. First, house ownership is an indicator of available financial resources for adaptation and risk management (Preston et al., 2008). Second, it has been linked to increasing preparedness to weather-related events due to the sense of appropriation (Lechowska, 2018). Homeowners have shown more willingness to prepare their houses to withstand the expected magnitude of a specific hazard and more constant maintenance of the infrastructure. Also, according to (Steinführer and Kuhlicke, 2006), this behaviour is associated with the local attachment effect (the emotional bonds of an individual to a specific place). As a consequence, in this study, we associated the houses with their owner living on it, with a higher coping capacity and less vulnerable to disasters triggered by natural hazards. For those houses with renters, a lower coping capacity is used in the computation of this variable.

The factor **Information** is included as part of the coping capacities component. Warning information flow is essential to reduce vulnerability. Access to warning information needs to be received with sufficient time to react to a possible threat. The information also needs to be accurate, usable and understandable. We use three variables for this factor *Access to information*, *Evacuation Knowledge* and *Warning Information*. This factor is constructed entirely from survey questions.

In disaster risk management, one of the key drivers that negatively influences socioeconomic vulnerability is the lack of access to information (Cutter et al., 2003). Therefore, it is vital to acquire and disseminate the most accurate information in order to better utilise and target limited resources (Percival and Teeuw, 2019). Population in potential risk that has access to information has at least the theoretical opportunity to reduce its vulnerability by acting accordingly to the information received (Steinführer and Kuhlicke, 2006). Information in disaster management refers not only to have the means to distribute the warning messages to the whole population at risk but that the information transmitted contains sufficient elements that allow the population to act accordingly to minimise the impacts of a disasters triggered by natural hazards (Nguyen et al., 2016).

To compute the *Access to information* variable, we asked in the survey if the respondent knew where to get up-to-date information on early warning and actual evacuation news or instructions. We made no distinction between official sources of information and other sources. If the respondent answer that they know "to a great extent" from where to get access to warning information, we assigned a higher coping capacity value, and "not knowing" where to access information is assigned a low capacity to cope. The *Evacuation Knowledge* variable is computed based on a question asked to those who decide not to evacuate during Hurricane Irma. Low capacity to cope with the threat is given to the respondents that expressed that not knowing where to evacuate was an extremely influential reason to stay at home. We compute *Warning Information* with the number of days in advance (lead time) people receive warning information regarding the potential arrival of Hurricane Irma. The earliest awareness regarding Hurricane Irma the highest the coping capacity.

The last factor of the lack of coping capacities component is **Awareness**. Knowledge and risk awareness of a specific hazard are good indicators of the household levels of disaster preparation (Cutter et al., 2003; Percival and Teeuw, 2019). We measure this factor using the *risk perception* and the *risk knowledge* of the respondent and the *frequency of getting information* when a storm approaches. *Risk knowledge* plays a central role in vulnerability assessment as knowledge is a necessary precursor of preparedness (Cutter et al., 2003; Rufat et al., 2015). Knowledge of the hazard has been previously used as a measure of the coping capacities of a community, and it is recognised as a prerequisite to be able to trigger evacuation and coping mechanisms (Birkmann and Fernando, 2008). For Sint Maarten, this variable is evaluated using the number of hurricanes respondents remember that have hit the island directly while they were living on the island. Higher coping capacity is assumed to respondents that experienced more hurricanes because of the increase in risk knowledge based on first-hand experience. Similarly, the lowest coping capacity in this variable is for the respondents with no hurricane experience.

The variable *Risk perception* is considered crucial in vulnerability and risk reduction. It is defined as "intuitive risk judgements of individuals (and social groups) in the context of limited and uncertain information" (Slovic, 1987). Risk perception has the potential to either mitigate or enhance the potential of a hazard (Lechowska, 2018; Paul, 2013). There is a strong correlation between perceiving being at risk and vulnerability reduction behaviour. In contrast, low perception of risk in high exposed zones has proved to have catastrophic consequences in loss of life and high losses due to lack of preparation or protective behaviour(Messner and Meyer, 2006; Rufat et al., 2015). *Risk perception* has also been reported as one of the main reasons when deciding whether or not to evacuate during an extreme weather event (Dash and Gladwin, 2007; Medina et al., 2019; National Research Council [NRC], 2006). For those that did not evacuate during Hurricane Irma, we asked in the survey whether or not the decision to not evacuate was based on their feeling that Hurricane Irma would not be a real threat. Given the magnitude of the disaster caused by this hurricane, the minimum coping capacity value is for those respondents that ranked this question as "an extremely influential" reason not to evacuate.

How often an individual or group of individuals check for the latest updates regarding warning and evacuation information is a sign of increased awareness and readiness to cope with the adverse effects of a potential hazard. A positive effect on risk perception due to being regularly exposed to media has been extensively verified as reported in Hong et al. (2019). Staying up-to-date to the type of hazard allows citizens to adjust their behaviour when the hazard is approaching (i.e. stay home or go to a safer place) (Dieker, 2018; FEMA, 2013). *Frequency of Information* is incorporated in the coping capacities component using a survey question. We asked how often the interviewee checks for weather information when a hurricane or tropical storm is announced. Due to the high uncertainty in the path and the frequency of hurricanes in Sint Maarten, the lowest coping capacity is for respondents that check weather information with a frequency of less than once a day, and the highest one to those checking the updates throughout the whole day.

Lack of Adaptation Capacities

The lack of adaptation capacities "*is closely related to change and the ability to deal or recover from the negative impacts of a future disaster*" (Sorg et al., 2018). The four factors of this component are education, gender equity, level of investments and the vulnerability assessment of the critical infrastructure in the island. Each factor within this component was computed using only one variable (Figure 3.4).

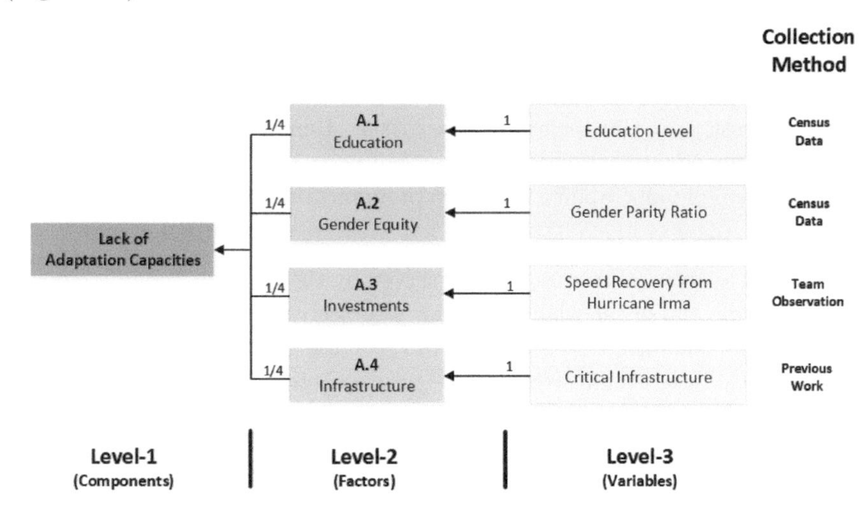

Figure 3.4. Structure of the Lack of Adaptation Capacities component for the vulnerability index (PeVI), applied in the case study of Sint Maarten. The figure shows the three levels of hierarchy and the source of information used to compute each variable. The numbers next to the arrows indicate the weighting factor to compute the next level

The *level of education* is the variable used for **Education** factor, and it is evaluated using census data by computing the ratio between the number of people reported holding at least high school degree and the population over 18 years old. We follow a similar approach as the one presented in Sorg et al. (2018) and Fekete (2009). Higher levels of education can be used as a measure of the economic capacities of a household as it may lead to better salaries. Wealthier households can prepare and mitigate better for disasters and are expected to recover faster, employing their economic status (Cutter et al., 2003; National Research Council [NRC], 2006; Rufat et al., 2015). Besides, people with higher formal education levels have shown more access to information (Steinführer and Kuhlicke, 2006). In contrast, people with lower *levels of education* has been observed to have less awareness or limited understanding of warning information towards the potentially catastrophic effects of an extreme event. Low education levels are also associated with less capability of adopting emergency measures and with limitations to access recovery information (Cutter et al., 2003; Welle et al., 2014).

We use the variable *gender parity ratio* in education as a measure of **gender equity**. Adopted from Sorg et al. (2018), this variable is calculated using the ratio of the number of females holding primary, secondary or tertiary education and the respective number of males with the same levels of education. A ratio of 1 on this indicator means equity in access to education and is the desired value; therefore, we assigned the highest adaptation capacity in the computation. Advantages for man in the parity ratio ranges from zero to one and larger than one represents an advantage for women. We assign low adaptation capacity to both of the extreme values of this variable. As summarised in Smith and Pilifosova (2001), it is frequently argued that adaptive capacity will have a more significant (positive) impact if the access to resources is distributed equally. Without equity, adaptive actions for vulnerability reduction may benefit only those sectors or individuals best placed in society (Adger, 2006). Hence, integrating elements of equity in the identification of vulnerability is key to achieve effective implementation of vulnerability reduction programs that include the marginalised sectors (Neil Adger et al., 2005).

The variable *speed of recovery* was observed by the research team five months after Irma impacted in the island. Though a subjective observation made by the field team, the compiled information is of great use to detect which areas were bouncing back faster (and stronger) in the reconstruction phase as a sign of adaptive capacities. The assessment of *recovery speed* was made for the entire Dutch part of the island, and averaged by neighbourhood and classified into five categories from very slow to very fast recovery, assigning from low to high adaptive capacities respectively. The capacity of a city to rebound from destruction has been used as a measure of resilience and adaptation capacities by several authors; a summary of those can be found in Gunderson (2010).

The variable *Critical infrastructure* is defined in the context of this research as physical assets that play an essential role in the functioning of the society and the economy. We include in this category facilities for electricity generation, access to water and food, public health, telecommunication, sheltering, education and transport. Damage to critical infrastructure can impede or limit access to disaster relief and are crucial in restoring essential services to normalise lives and mitigate the impacts of the disaster (Bach et al., 2013; Garschagen et al., 2016; Lovell and Mitchell, 2015). Hence, evaluating the vulnerability of the critical infrastructure of a city or region can be a good indicator of how fast the city will recover. For Sint Maarten, such evaluation already existed from a previous work of the research team in a total of 200 buildings (UNDP, 2012). Vulnerability to critical infrastructure took into account the physical condition of the buildings and the flooding potentiality. Each building then was assigned a vulnerability value in a five points scale low, medium, high, very high and extreme vulnerability

Vulnerability Computation

The implementation of the vulnerability index consisted in the computation in Microsoft Excel© of each one of the 27 variables described above. This process was performed in 49 out of the 54 neighbourhoods of Sint Maarten. Five neighbourhoods did not have enough information to compute the vulnerability index or its components; those were: Back Bay, Geneva Bay, Salt Pans, The Harbour and The Airport (see Figure 3.1 and Figure 3.7). Then, the variables are combined to produce every factor using the associated weight, and by combining factors, each one of the components is computed. Finally, by using equal weight, the three main components of the vulnerability index are added to produce the PeVI for each neighbourhood in Sint Maarten (Equation 3.1 and Figure 3.5).

$$Vulnerability = \frac{1}{3} Susceptibility + \frac{1}{3}\ Lack\ of\ Coping\ Capacitie + \frac{1}{3}\ Lack\ of\ Adaptation \quad (3.1)$$

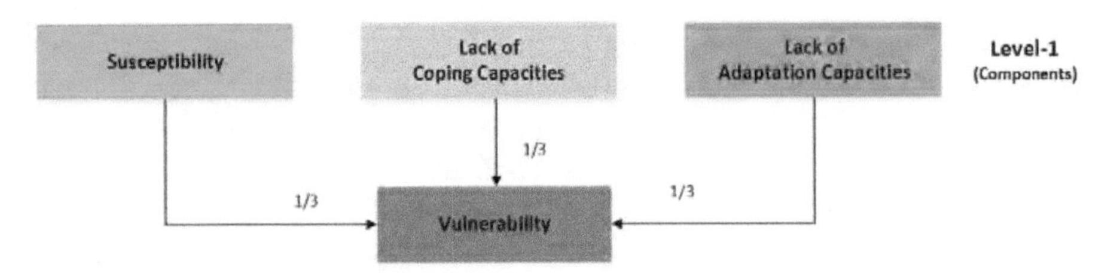

Figure 3.5. Composition of the vulnerability index (PeVI), for the case study of Sint Maarten with the three main components in the level-1 of hierarchy. The number next to the arrows correspond to the weight of each component in the computation of the PeVI index.

Variables used in the computation are of different nature and characteristics, ranging from quantitative to qualitative values and from different data sources (census, survey, observation and third parties). Such heterogeneity on the input data requires a standardisation of the data to ensure uniformity in scales and units (Percival and Teeuw, 2019). The data used in the computation of the PeVI can be categorised into four data types:

Type 1 – **Census data:** Variable of quantitative nature. Data (a number) in this category represents the total number of inhabitants in each neighbourhood for a specific variable; for example the number of people five years old or younger, and the number of residents per household.

Type 2 – **Categorical variables:** A variable that can take on one of a limited, and a usually fixed number of possible values; for example building wall materials (Concrete, brick, wood, others) and type of electricity supply (aerial, underground and no electricity).

Type-3: **Likert scale questions:** Measures how people feel about a question of the survey, based on a rating scale. There are three different ranges of this type of questions. One ranging from "Not at all influential" to "Extremely influential", other from "Strongly disagree" to "Strongly agree" and the third one from "To a great extent" to "Not at all". Examples in this category are the perception of the sufficiency in the number of shelters in the island and the influence of particular variables to not evacuate during Hurricane Irma.

Type-4: **Binary questions:** Questions in the survey in which the answer is of the type Yes or No. An example is house ownership.

All variables are standardised in numeric and dimensionless values, ranging from 0 to 100 (from lowest vulnerability to highest vulnerability). This standardisation allows to perform operations among the different units and magnitudes of variables, to weight them and to make comparisons. For **Type-1** variables, the standardisation is done by multiplying the ratios or numbers obtained in the computation by 100, except for gender parity ratio, which was obtained using a min-max normalisation method. For **Type-2** variables, a specific value of vulnerability from 0 to 100 is assigned to each possible answer the categorical variable can take. In the **Type-3** variables, for the answers "*strong agreement*", "*extremely influential*" and "*not at all*", a value of 100 is assigned. Similarly, a value of 0 is used to answers strongly disagree, not at all influential and to a great extent. With an exception in the variable emergency infrastructure where the Likert scale is inverted (strongly agree = 0 and strongly disagree = 100). The intermediate possibilities of the Likert scale are assumed to be evenly distributed over the minimum and maximum value. **Type-4** variables are standardised with a high value of vulnerability (80) to negative answers (absence of) and a low vulnerability value of 20 to affirmative answers (presence of). The complete set of formulas and values used for the 27 variables is presented in the supplementary material that accompanies the paper on which this chapter is based (See Medina et al., 2020).

3.2.3 Vulnerability Mapping

The PeVI for Sint Maarten and its primary three components susceptibility, lack of coping capacities and lack of adaptation, were cartographically displayed using GIS software (ArcMap 10.5). We mapped each one of the components in five classes of vulnerability (Very low, Low, Medium, High and Very high) using natural breaks classification method. The selection of the classification method is not a trivial choice. The resulting vulnerability maps and future decisions made, based on the maps, are very dependent on the classification method to be employed. The choice needs to be closely related to the aim of the vulnerability assessment (i.e. support for decision making, prioritising, and funding allocation). A poor choice in the method can be misleading (Papathoma-Kohle et al., 2019). We select natural breaks (Jenks) as a classification method for the spatial representation of vulnerability and its components.

With natural breaks classification, classes are based on natural groupings inherent in the data. Breaking points between classes are identified that best group similar values, and that maximise the differences between classes. The features are divided into classes whose boundaries are set where there are relatively significant gaps in the data values (Smith et al., 2018; Wei et al., 2020). The properties of the natural breaks classification method make it the most suitable for dividing neighbourhoods whose vulnerability is similar because it reduces the variance within each class (Vojinović et al., 2016). It is important to note that natural breaks is a data-specific classification method, and hence it is not useful for comparing multiple maps that use different underlying information.

For the reason mentioned above, it is not possible to make quantitative comparisons amongst the maps we produce. However, this does not necessarily mean that no comparison can be made, for example, all neighbours classified in the "very low" (dark green) group, viewed relatively, they all have a lower priority concerning the component or index it represents. Also having a "Very low" value in one of the maps does not mean that a specific neighbourhood is not vulnerable (or another component) it just has somewhat a lower priority than other neighbourhoods.

3.2.4 Principal Components Analysis (PCA)

PCA is a multivariate statistical technique that can be used to analyse several dependent variables (which usually are inter-correlated) in a dataset. PCA aims to draw conclusions from the linear relationship between variables by extracting the most relevant information in the dataset in the form of a (reduced) set of new orthogonal variables that are called principal components (Abdi and Williams, 2010).

PCA reduces the number of variables by identifying the variables that account for the majority of data variance, and by identifying the similarities between individuals for all variables, and by doing so, highlighting the main contributing factors to the phenomenon under investigation (Abdi and Williams, 2010; Husson et al., 2010). PCA works by performing an orthogonal linear transformation in an N dimension space to identify the vector that accounts for as much as possible of the total variability. The first vector is called the first Principal Component (PC-1). After the first principal components is extracted, the method continues building principal components that are also orthogonal and linearly uncorrelated to the previous component and each time accounting for as much of the maximum of the remaining variability as possible.

To run the PCA, we select the level 2 of the hierarchy structure, corresponding to the **factor** level. PCA was run using a total of 14 factors. PCA analysis in this research was carried out using a package in R called factomineR (Lê et al., 2008). The first step in the PCA analysis is the analysis of missing data, which was done using an R package called missMDA (Husson and Josse, 2016). This package uses an iterative method to impute data in the missing values by taking into account both similarities between individuals and relationships between variables. It works in a way that the PCA is constructed from observed data only (i.e. no contribution from the imputed data). (Husson and Josse, 2016; Husson et al., 2010). The dataset we used consisted of 49 individuals (neighbourhoods), 14 variables (factors), one quantitative illustrative variable (vulnerability) and one qualitative illustrative variable (administrative zones).

After this step, all variables were normalised using z-scores. It is advisable to perform such standardisation for comparisons of data across variables. Standardisation generates variables with a mean of 0 and a standard deviation of 1. Even when the units of measurement do not differ, this operation is generally preferable as it attaches the same importance to each variable (Husson et al., 2010; Oulahen et al., 2015). Then, we run the PCA on the standardised data using varimax rotation. This step is done to simplify the relationships among the variables and to clarify the interpretation of the factors (Fernandez et al., 2016).

3.3 Results and Discussion

3.3.1 Correlation Analysis and Selection of Number of Principal Components

Variables and Factors Screening

We screen all the variables of the PeVI for singularity and collinearity. This procedure is done in order to reduce the problems when analysing the data. The data screening on the entire set of original variables is done using the correlation and covariance matrices produced with the Principal Components Analysis (PCA). This step ensured that highly correlated variables are removed before the computation of the index. We found that a variable named economic status (of each neighbourhood) which was subjective based on field observations, had a strong correlation (R^2=0.86) with other variables of the index and therefore creating data redundancy. For that reason, we decided to remove it from the analysis.

We re-run the correlation test with the remaining variables to reassure their relevance. We found a strong positive correlation (R^2=0.76) between two factors of the coping capacities, immediate action (C.2) and economic coverage (C.4). This high correlation can be explained because the immediate action factor in this index has elements that could be directly related to the wealthy of households such as car ownership. Data from the survey reveals that in Sint Maarten the possession of cars on the island is not limited to wealthy households. Hence, we decided to keep both factors in the vulnerability assessment to adequately capture the effect of car ownership on the island's vulnerability.

For this study, 27 variables were used to compute the final PeVI of Sint Maarten. Eight for susceptibility, 15 in the lack of coping capacities and four in the lack of adaptation capacities. Based on this conceptual approach, the socioeconomic vulnerability to floods and hurricanes in Sint Maarten was computed and analysed from a spatial point of view, through a vulnerability index integrated into GIS, but within the limits of data availability from census, field data collection and data from third parties.

Selection of the number of Principal Components

The selection of how many components to include in PCA is an arbitrary decision but must follow some guidelines. For this research, we define the number of principal components in two steps. First, we decide to only keep factors with eigenvalues (contribution) greater than 1, using the Kaiser criterion as reported in Husson et al. (2010), and second, that the minimum number of principal components selected explain at least 60% of the variation of the original data. The first plane of analysis, composed by the first two dimensions of the PCA run in Sint Maarten dataset account for 40.7% of the total dataset inertia or individual's total variability. The percentage explained by the first two principal components, is an intermediate percentage, and the first plane represents only part of the data variability; thus, we consider the next dimensions (PC-3 and PC-4). The cumulative variability explained in the first four dimensions is 62.5%. Hence, for Sint Maarten, we select four principal components for the PCA (Figure 3.6).

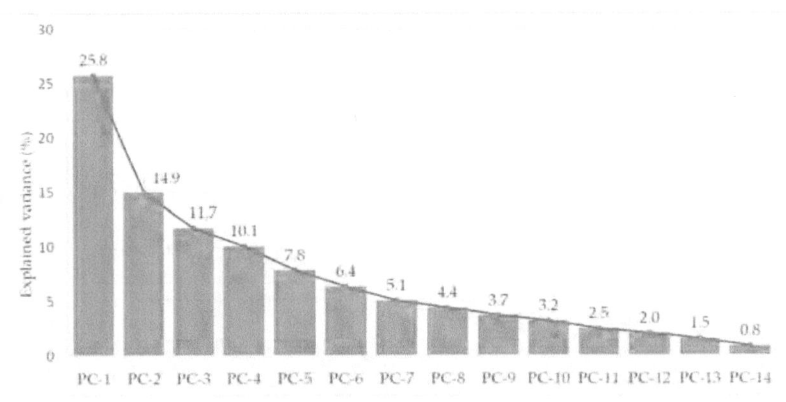

Figure 3.6. Scree plot with the decomposition of the total inertia on the component of the analysis. The first four principal components explain 62.5 % of the inertia in the data.

3.3.2 Vulnerability Index and components

The three main elements on which vulnerability was assessed are presented in Figure 3.7. Susceptibility (Figure 3.7-a), Lack of coping Capacities (Figure 3.7-b), and Lack of Adaption (Figure 3.7-c), and the resulting vulnerability index (PeVI) is presented in Figure 3.8. Furthermore, Table 3.1 we present the top 5 most critical neighbourhoods for the PeVI and for each component of the vulnerability index. The complete table with the computed values for all neighbourhoods is presented in Appendix B. Figure 3.9 presents a closer look at the top five driving factors for each one of the critical neighbourhoods.

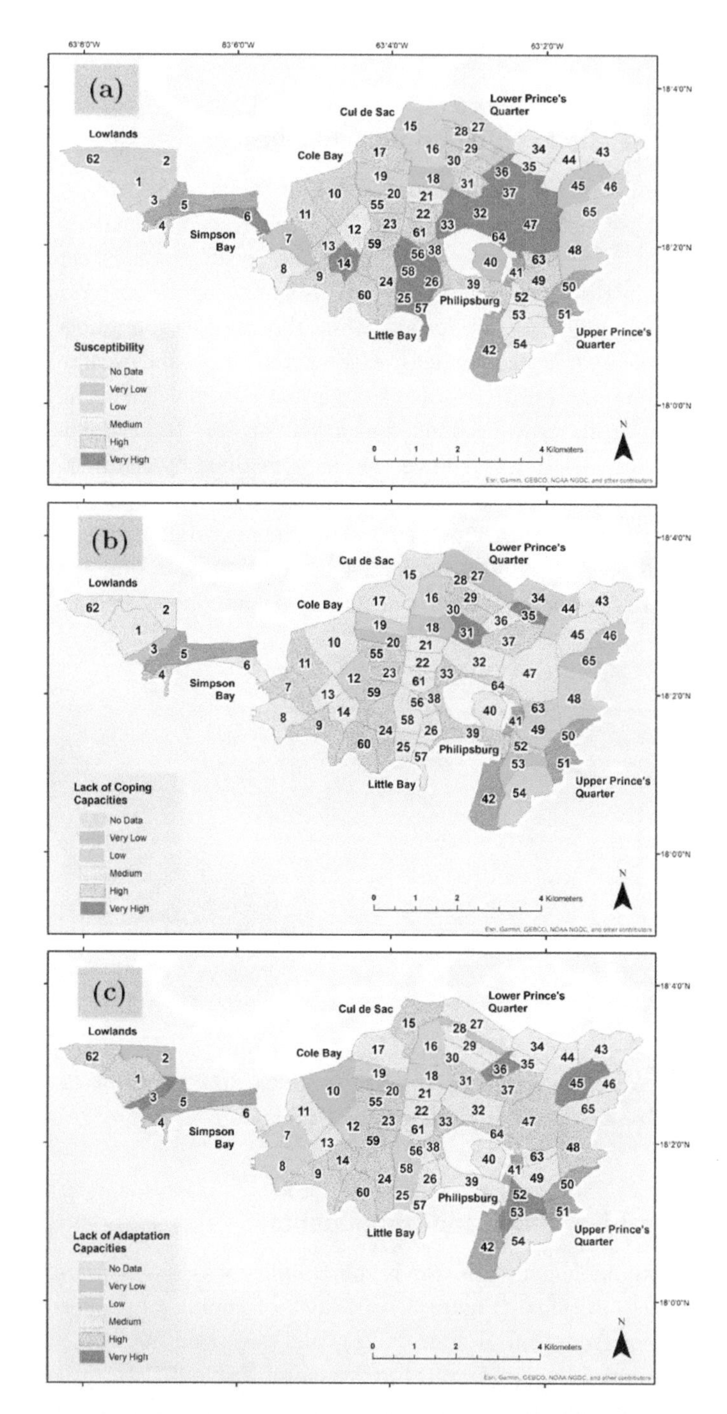

Figure 3.7. Susceptibility (a), Lack of coping capacities (b), Lack of adaptation capacities (c) for Sint Maarten at the neighbourhood scale. Numbers represent the identification (ID) of each neighbourhood, as presented in Appendix B and Table 3.1.

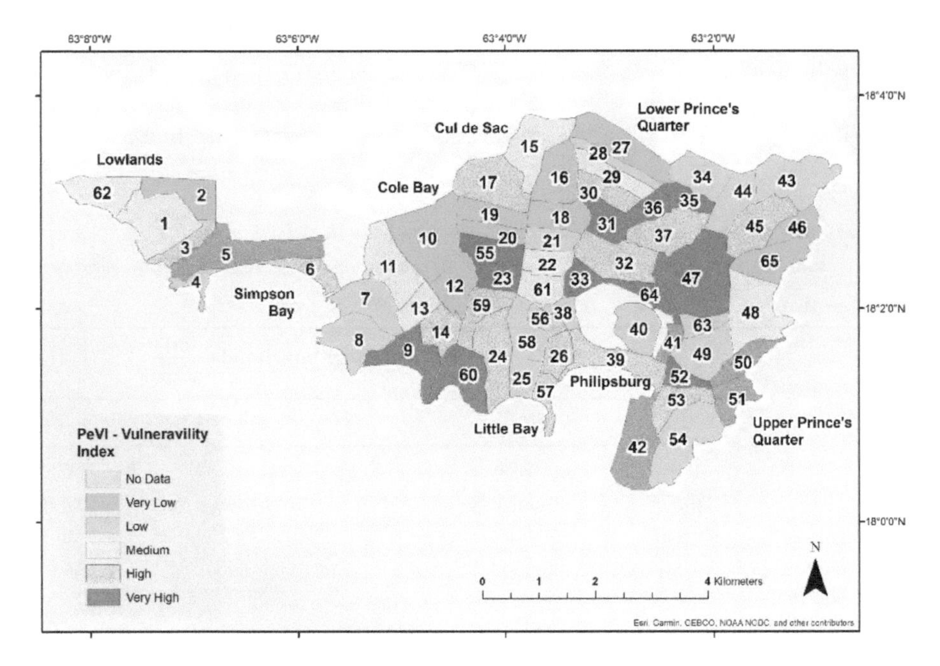

Figure 3.8. PeVI vulnerability index for Sint Maarten at the neighbourhood scale. Numbers represent the identification (ID) of each neighbourhood, as presented in Appendix B and Table 3.1.

In terms of **susceptibility** (Figure 3.7-a and Figure 3.9-a), for the top five more critical neighbourhoods (Table 3.1), *Housing* is the most important driving factor to increase susceptibility in these neighbourhoods, followed by the *infrastructure* factor. Inferior quality in the construction materials and the associated high level of destruction after Hurricane Irma in these neighbourhoods can explain why these areas of the island are the most susceptible ones. These neighbourhoods are also the place of residence of a considerable portion of the undocumented immigrants. This group has been struck especially hard by Hurricane Irma, as Irma damaged both their houses and their financial capacity. A successful plan to lower risk and vulnerability in Sint Maarten to extreme events will require a high level of compromise from the local authorities to improve and monitor the building codes in the island to accurately reflect the high potential hazard to hurricanes and floods.

Regarding *infrastructure* susceptibility, it is necessary to highlight that Sint Maarten is a very densely populated territory, the island is the most densely populated country in the Caribbean and the 12th in the world (United Nations, 2019). Population density pushes the limits of the island in terms of expansion or upgrade of the physical infrastructure such as roads and inadequately planned urban expansion in the riskiest areas such as hillsides (VROMI, 2015). The electricity supply in these critical neighbourhoods was very susceptible, as a considerable portion of the network remains aerial. The electricity company (or the government) cannot undertake the upgrade of the system in some of the neighbours because that requires intervention on private land and landowners may not allow it. It was observed during the fieldwork that some of the critical neighbourhoods were still in the restoration process of electricity lines.

Table 3.1. The five most critical neighbourhoods for PeVI vulnerability index and the three components: Susceptibility, Lack of coping capacities and Lack of adaptation capacities.

Component	Top 5 most critical neighbourhoods ID) / Name / (Value)				
	1	2	3	4	5
Susceptibility	36) Dutch Quarter (41.3)	6) Simpson Bay Village (34.5)	14) Wind Sor (34.4)	33) Over the Pond (34.36)	37) Middle Region (34.0)
Lack of Coping Capacities	35) Bishop Hill (63.5)	31) Mount William (55.2)	39) Philipsburg (46.7)	33) Over the Pond (44.9)	52) Over the Bank (44.1)
Lack of Adaptation Capacities	36) Dutch Quarter (57.3)	3) Maho (54.1)	53) Vineyard (52.56)	52) Over the Bank (51.3)	45) Ocean Terrace (50)
PeVI – Vulnerability Index	36) Dutch Quarter (44.9)	52) Over the Bank (41.1)	35) Bishop Hill (39.4)	33) Over the Pond (38.7)	31) Mount William (38.7)

In contrast, Ocean terrace, Betty's Estate and Dawn Beach are amongst the less susceptible neighbourhoods in Sint Maarten. These neighbourhoods have a large proportion of gated condominiums, some of which belongs to the time-sharing schemes that are abundant on the island. In Betty's state are located some of the wealthier houses we observed of permanent residents of the island; and according to findings during the fieldwork, the direction of the winds in this part of the hills favour the lesser destruction during Irma. Another factor contributing to low values of susceptibility is *demography*; the proportion between the number of people in working age and the most vulnerable age group indicates that there is a right balance in the number of the population to take care of the most vulnerable part of the population.

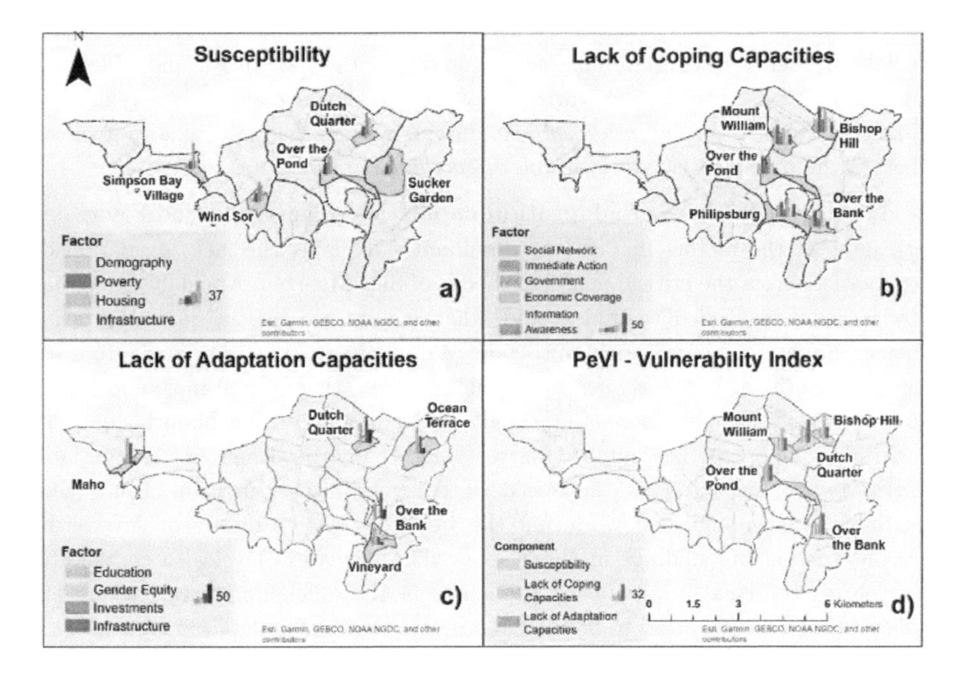

Figure 3.9. Top five critical neighbourhoods in Sint Maarten for each component of the vulnerability index. Susceptibility (a), Lack of coping capacities (b), Lack of adaptation capacities (c) and PeVI vulnerability index (d). The bars in the polygons represent the magnitude of each factor in the respective component or the component in the index. Labels in red represent the administrative zones and in black the neighbourhood names.

In terms of **lack of coping capacities** (Figure 3.7-b and Figure 3.9-b), *Immediate action* and *Economic coverage* are the critical factors associated with the low capacities to cope with the effects of hurricanes and floods in the most critical neighbourhoods. An explanatory variable that decreases the coping capacities is that in Sint Maarten house ownership is particularly low, with more than 53% of respondents living under rental agreements. Tenants in the island generally do not feel the responsibility to maintain and strengthen the houses they occupy and landlords are not also active in repairing the houses. The situation is even more complicated in those areas where the land is leased, in which usually it is forbidden (by the owner) to build a house with durable materials, which is against the construction code but a common practice in leased areas. Another explanatory variable is the low coverage of home insurance. We found that residents of Sint Maarten do not take insurance due to high rate of premiums, low trust of the insurance companies, not getting paid what the house is worth, slow claim processing, and poor client service.

On the other hand, Dawn Beach, Vineyard, St John Estate and Maho neighbourhoods have the highest coping capacities on the island. In these neighbourhoods, the driving variables of high coping capacities are related to higher *awareness* and *access to information* as well as the high capacity to react fast to the potential hazard (variable *immediate action*).

In terms of the **lack of adaptation capacities** (Figure 3.7-c and Figure 3.9-c), none of the factors in PeVI predominantly explains the lack of adaptation capacities across the critical neighbourhoods of Sint Maarten. An analysis location by location is needed to understand what factors are driving low adaptation capacities in the critical neighbourhoods. Maho, Vineyard and Ocean Terrace are especially critical in the *gender equity* factor. The ratio between male and female inhabitants with education is especially skewed in these neighbourhoods. The critical component in Dutch Quarter and Over the Bank is the variable *investments*, measured as the speed of recovery after Hurricane Irma. This variable was valued as low during the fieldwork as we observed slow or no reconstruction of buildings in these areas. The education level also contributes largely to the lack of adaptation capacities in these neighbourhoods where the literacy rate was reported to be the most precarious in the island. In contrast, the less vulnerable neighbourhoods in terms of having the best adaptation capacities were identified to be Point Pirouette, St John Estate, Diamond and Nazareth.

In terms of overall vulnerability index (Figure 3.8 and Figure 3.9-d), we identified that the most critical neighbourhoods are Dutch Quarter, Over the Bank, Bishop Hill, Over the Pond and Mount William. Generally speaking, all these neighbourhoods present relatively high values in the three components used to compute the vulnerability index. Four of the top five most vulnerable neighbourhoods are in the top five of the lack of coping capacities, being this component the most influential driver of vulnerability in Sint Maarten.

3.3.3 Principal Components Analysis (PCA)

The contribution values of each factor to the PCA analysis, as presented in Table 3.2 shows the degree of correlation between each component in the analysis and the dimensions or principal components. A coefficient closer to either 1 or -1 means that it has a stronger correlation to that component, positively and negatively affecting the component respectively (Oulahen et al., 2015). The more significant the contribution of a variable to a principal component, the more influential towards increasing vulnerability. This result is not only consistent with the PeVI but supports the conclusions that coping capacities and especially the variables immediate action and economic coverage are the key factors to have an effective vulnerability and disaster risk reduction plan on the Island.

Given the fact that in a small island state as Sint Maarten most of the population can be exposed during an extreme weather event, it makes sense that immediate action plays a vital role in the vulnerability assessment. It is essential that residents can move fast to a secure zone or even to have the possibility to fly out of the island and to have the means to protect their houses in case of an imminent disaster, but also the ability to bounce back faster from the effects of it. A wealthier economy can help to mitigate the impacts by increasing house ownership and more households that can acquire home insurance to protect their assets. Education can help to close the economic gap on the island as a mid or long-term solution.

Table 3.2. Contribution of each factor to each Principal Component for the socioeconomic vulnerability in Sint Maarten. In green, the main contributing factors to each principal component and in grey important contributing factors already accounted in another principal component.

Factor	PC-1	PC-2	PC-3	PC-4
C.2 Immediate Action	0.79	0.18	0.02	0.32
C.4 Economic Coverage	0.78	0.18	-0.08	0.40
A.1 Education	0.64	0.13	0.48	0.09
S.2 Poverty and Income	-0.20	0.63	0.56	0.34
S.1 Demography	-0.46	0.57	0.02	-0.16
C.5 Information	0.33	0.49	-0.06	-0.14
A.4 Infrastructure	0.31	-0.28	0.50	-0.37
S.4 Infrastructure	0.56	-0.25	0.45	-0.25
A.3 Investments	0.40	-0.54	0.27	-0.01
A.2 Gender Equity	-0.32	-0.53	0.13	0.55
C.3 Government	0.22	-0.40	-0.38	0.43
C.6 Awareness	0.54	0.37	-0.34	0.20
C.1 Social Network	0.57	-0.12	-0.54	-0.45
S.3 Housing	0.52	0.17	-0.03	-0.19
Eigenvalue	3.61	2.09	1.64	1.41
Variance (%)	25.77	14.93	11.72	10.09
Cumulative variance (%)	25.77	40.69	52.41	62.51

To understand better which components of the index are the most influential in the PCA and the overall vulnerability of Sint Maarten Figure 3.10 is presented. The first principal component (PC-1), is influenced significantly by the variables C.2 (Immediate action), C.4 (Economic Coverage) and A.1 (Education). The second principal component (PC-2), is receiving more influence from S.2 (Poverty and Income), S.1 (Demography) and C.5 (Information). In the third principal component (PC-3), the most contributing factors are A.4 (Infrastructure), S.4

(Infrastructure) and A.3 (Investments). For the PC-3 variables, A.1 (Education) and S.2 (Poverty and Income) also have considerable influence, but those are already accounted for in the most significant principal components. Finally, PC-4 is influenced mainly by A.2 (Gender Equity), C.3 (Government) and C.6 (Awareness) and also C.4 (Economic Coverage) that also influences PC-1.

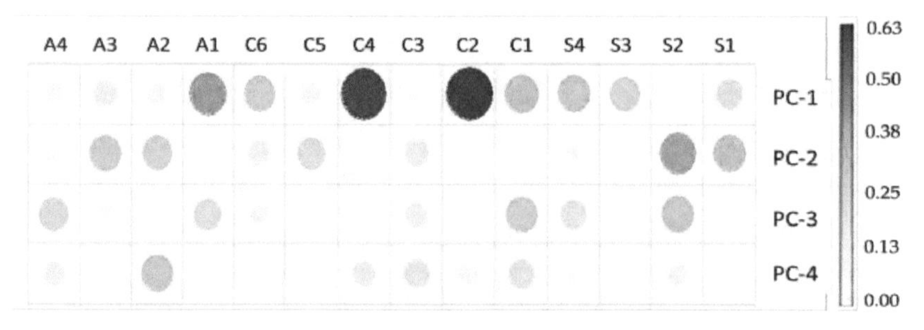

Figure 3.10. Contributing components to the first Principal Components in the PCA run for Sint Maarten vulnerability. The biggest and darkest the circle the most influential the component to the dimension and overall in the dataset. C2 (Immediate action) and C4 (Economic Coverage) are the most influential components for dimension one and the whole vulnerability index as well.

For the easiness of usability and interpretability of the principal components analysis, we used the dominant factors in each principal component to group them into four distinct categories that reflect the most important variables influencing vulnerability in Sint Maarten. The PC-1 can be labelled *Economic and education*, PC-2 *Demographic and information*, PC-3 *Infrastructure and investments* and PC-4 *Governance*. These are not precise categories but dominant factors in each principal component.

The spatial distribution of the Principal Components Analysis at the neighbourhood level is shown in Figure 3.11. The spatial distribution reveals that there is a geographic distribution of the different factor composing the vulnerability of Sint Maarten. PC-1, Economic and education category, is especially critical in the lower prince's quarter zone, in this category the neighbourhoods Bishop Hill, Mount William, Over the Bank, Dutch Quarter and Over the Pond are the most critical ones. The geographical distribution of PC-2, referring to the demographic and information category does not aggregate in a particular zone, instead is distributed along the island. Neighbourhoods St. John State, Union Farm, Nazaret, Defiance and Point Pirouette are the critical ones in terms of PC-2. The category Infrastructure and investments (PC-3), is more critical in two zones: Lower and upper prince's Quarter and distributed among Dutch Quarter, Over the Pond, Defiance and Over the Bank neighbourhoods. Finally, PC-4 distribution shows that governance category is also distributed on the island with not a clear critical zone. Four neighbourhoods are critical regarding governance, Ocean Terrace, Bishop Hill, Maho and Beacon Hill.

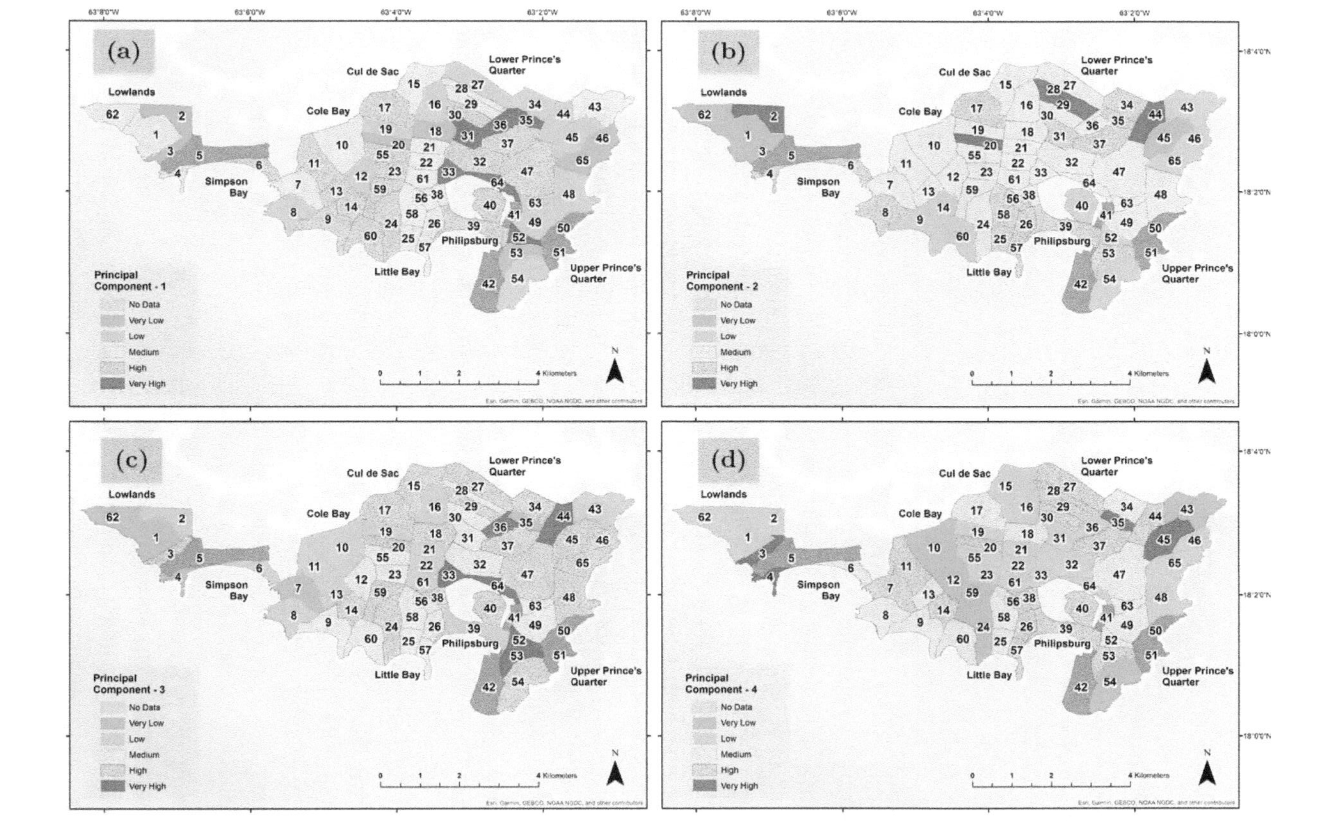

Figure 3.11. Principal Component 1 (a), Principal Component 2 (b), Principal Component 3 (c), Principal Component 4 (d) for Sint Maarten Vulnerability at the neighbourhood scale. Numbers represent the identification (ID) of each neighbourhood, as presented in Appendix B and Table 3.1.

3.3.4 Clustering Analysis

Furthermore, PCA allows performing a clustering analysis of neighbourhoods. The clustering can help decision-makers to see common drivers of vulnerability across the island. Clustering analyses can be used to evaluate where the potential impact of a specific measure to reduce vulnerability will have the most positive impacts. We run a classification method for clustering the neighbourhoods using the first two components. This process reveals five clusters of neighbourhoods in Sint Maarten (Figure 3.12).

Cluster 1 is made of two neighbourhoods Maho and Ocean Terrace. This group is characterised by high values for the factor Gender Equity (A.2) and low values for the factors Information (C.5), Social Network (C.1) and Housing (S.3). **Cluster 2** grouped Defiance, Nazareth, Point Pirouette, St John Estate, Union Farm and Vineyard. This group is characterised by high values for the factors Poverty and Income (S.2) and Demography (S.1) and low values for the factors Investments (A.3), Social Network (C.1), and Infrastructure (A.4). The **cluster 3** is made of neighbourhoods such as Betty's Estate, Dawn Beach, Low Lands and Oyster Pond. This group is characterised by low values for the factors Immediate Action (C.2), Economic Coverage (C.4), Housing (S.2), Education (A.1) and Awareness (C.6). Regarding **cluster 4**, this group have neighbourhoods such as Philipsburg and Simpsons Bay Village. This group is characterised by high values for the factor Housing (S.3) and low values for the factor Information (C.5). Finally, **cluster 5** is made of the most critical neighbourhoods in terms of socioeconomic vulnerability on the island such as Bishop Hill, Cay Bay, Dutch Quarter, Mount William, Over the Bank, Over the Pond and Sentry Hill. Neighbourhoods in this cluster are characterised by having high values for the factors Infrastructure (S.4), Education (A.1), immediate Action (C.2), Information (C.5), and Economic Coverage (C.4).

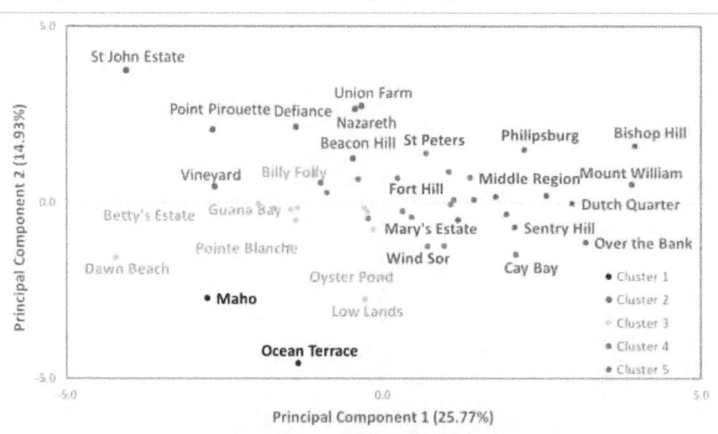

Figure 3.12. Clustering classification of neighbourhoods based on PC-1 and PC-2. The classification reveals five clusters in Sint Maarten.

A practical application of the clustering analysis is that it allows to evaluate in which neighbourhoods a specific measure adopted towards vulnerability reduction will have its greater impacts. For example, the government of Sint Maarten could evaluate how effective and where an invest in campaigns towards increasing awareness to natural hazards will have better positive impact. From the clustering analysis, such campaigns will be reflected more significantly in neighbourhoods within the **cluster 3**. In contrast, if the efforts to lower socioeconomic vulnerability in Sint Maarten are put on improving the economy of the residents of the island (improving factor C.4 Economic coverage), such effort will benefit the most, those neighbourhoods that belong to the critical **cluster 5**. One key step that can help improving people's economic coverage in Sint Maarten is the diversification of the economy, to not only depends on tourism, this will help to minimise the impacts of a disaster such as the one caused by Irma where the whole economy was severely hit, and its effect still can be felt in the island.

3.3.5 Recommendations for Vulnerability Reduction in Sint Maarten and policy implementation

The research revealed the specific needs of each neighbourhood which are necessary to lower its vulnerability (see sections 3.3.2-3.3.4). The government of Sint Maarten can use outputs from this research for disaster risk management activities on the island to address specific variables of vulnerability based on individual neighbourhood needs and to further develop existing policies and introduce new ones. The results also point to the most critical neighbourhoods which require additional focus, efforts and resources to lower risk and vulnerability, areas such as Dutch Quarter, Over the Bank, Bishop Hill, Over the Pond and Mount William. Within these areas, special attention needs to be placed on the economic and in the educational factors. In the above mentioned critical neighbourhoods, it is needed to improve the quality of housing, increase their ability to protect their assets, the ability to evacuate to safer zones or dedicated shelters and to increase the insurance coverage.

Our work identified that areas with higher number of undocumented immigrants are among the most critical ones in terms of socio-economic vulnerability. Undocumented immigrants in Sint Maarten have extra levels of vulnerability as they build their houses in the marginal lands of the hillsides using weak construction materials (wood walls and zinc roofs), they also have limited access to water and sanitation and less formal jobs or contracts. To address this issue, the government of the island should not only improve the outdated building codes and increase inspections but also assist in rebuilding both financially and technically across the island. The government should also review the land leasing model to implement more strict control over the quality of constructions in those areas identified as the most vulnerable during the household's survey (Medina et al., 2019).

In addition, the observed slow recovery pace after Hurricane Irma was directly related to economic issues of the island economy. As reported in (Medina et al., 2019), the government can address this situation among others with the implementation of a hurricane fund which can be implemented using a percentage of the taxes on the touristic sector in a yearly basis. Such fund will allow Sint Maarten to finance the reconstruction with less dependency on the Dutch government or other external financial organisations or donors and improving resident's wellbeing.

Finally, we see the use of PeVI in combination with the PCA analysis as a tool that can be easily used by the government to perform traceability and evolution of vulnerability in Sint Maarten once the authorities undertake policies and strategies to lower some of the drivers of the vulnerability identified in this paper. Alternatively, PeVI can also be used to evaluate what will be the possible impacts of a specific measure before its implementation.

3.4 Conclusions

We present in this research a methodology to assess and map the socioeconomic vulnerability of SIDS at a neighbourhood scale. We assess vulnerability using a vulnerability index with three major components, susceptibility, lack of coping capacities and lack of adaptation. The resulting index (PeVI) was then applied to the case study of Sint Maarten after the disaster caused by Hurricane Irma in 2017. To compute the index, we use census data in combination with data coming from a survey we performed in the aftermath of Hurricane Irma. Using the survey allowed us to expand the index to be able to capture elements that can particularly change vulnerability after a disaster, such as elements of risk awareness and perception and access to information in combination with information associated to the direct impact of the hurricane and the recovery in the island.

Vulnerability indexes, such as the PeVI and the associated maps, are a robust decision-making and communication tool. The index can be used to identify those areas more vulnerable to natural hazards (such as floods and hurricanes), and guide policymakers on where to focus the limited resources available to mitigate (or eliminate) the impact of a potential hazard. However, the representation of vulnerability and its components in a single number reduces the richness of the information that each variable used to produce such components and index can provide. Using PCA analysis as a complementary method can compensate for this trade-off between information richness (in the variables) and the robustness of communication of an aggregated index.

Vulnerability assessments based on the computation of indexes and vulnerability assessments based on Principal Components can be seen as complementary methods. The way we propose to use the methodology exposed in this paper is to compute first the vulnerability index to identify the most critical areas in terms of absolute vulnerability. Once hotspots have been highlighted, by using PCA, it is possible to determine which are the root causes or the most influential variables that contribute towards vulnerability in a specific area. PCA in this research allows us to increase the understanding of how multiple and often interdependent indicators of vulnerability vary in relation to each other and to understand the common drivers of vulnerability across different neighbourhoods.

Hurricane Irma was very catastrophic for Sint Maarten but offered an excellent opportunity to perform an in-depth analysis of some of the root causes of vulnerability and to incorporate new variables into the computation of vulnerability indexes that are only possible to observe and detect after the disaster has unfolded. To our knowledge, this is the most integrative study of this type and offer a framework to assess vulnerability in other similar areas with similar potential hazards and geographic characteristics.

The indexes and associated maps produced in this paper are the first of this kind for Sint Maarten despite the potential hazards they encounter each year during the hurricane season. Overall, we can state that we have offered a comprehensive and valuable static image of the vulnerability to hurricanes and floods in Sint Maarten. It is important to mention that we face limitations in data acquisition, access to the full extent of the census data was restricted, and we could not gain access to some areas on the island, especially gated condominiums.

4

MULTI-HAZARD MODELLING

In this chapter, we present the results of a multi-hazard assessment for the case study. We present the wind modelling results associated with Hurricane Irma and complemented with the pluvial flooding and storm surge using a synthetic (but plausible scenario). We coupled a 1D-2D simulation carried out for a rainfall of 100-year recurrence interval and a storm surge of 0.5 m. For the multi-hazards assessment, we use the gust winds modelled in combination with the water depth, extension and water velocities of the pluvial and storm surge modelling. The results of the models are then presented as average using the neighbourhoods as a representation scale.

4.1 Introduction

The hazard component was identified as the most studied element of the three components of disaster risk, hence not in this thesis's scope, and as such the purpose of this chapter is to present the modelling results for a multi-hazard assessment performed using Hurricane Irma as the hazard to be studied.

The hazard component includes models developed on the framework of this dissertation (wind model) and incorporates information and data from previous research or other researchers from the PEARL project in which this PhD took place (storm surge and inland flooding)(PEARL, 2018).

4.2 Hazard Modelling

Hurricanes can be considered as a multi-hazard threat. The main hazards associated with a hurricane are strong winds, inland flooding caused by heavy precipitation and storm surges. To assess the multi-hazard risk in Sint Maarten, we have selected Hurricane Irma. In previous chapters, we described that Hurricane Irma had been the most destructive hurricane that has impacted the island. The combination of Category 5 hurricane winds with the hurricane's path in which the hurricane's eye completely crossed the island contributed to the level of destruction observed in the aftermath of the disaster.

4.2.1 Wind Assessment

The most dangerous hazard during Hurricane Irma was the strong and dangerous winds. Irma crossed Sint Maarten in his peak intensity of Category 5 hurricane; unfortunately, the meteorological office lost the meteorological station before the peak intensity and not in-situ recorded winds are available. However, NHC aircraft observation near Barbuda reported wind gusts of 155 knots (79 m/s or 285 km/h) (Cangialosi et al., 2018). In addition, previous simulation available to this research estimated a wind speed above 70 m/s (Figure 4.1).

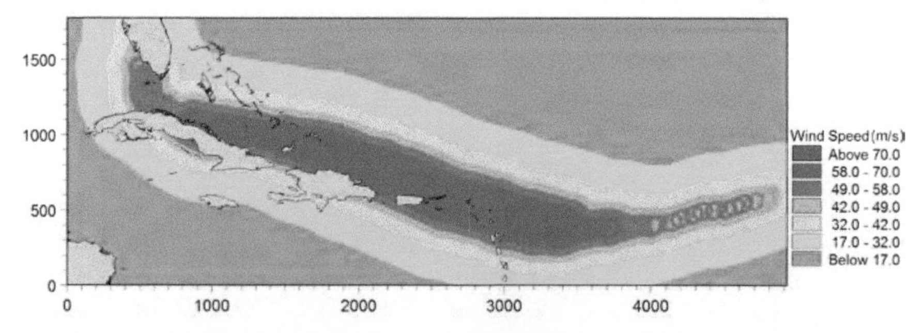

Figure 4.1 Maximum wind speed of hurricane Irma. Source: (PEARL, 2018).

However, as reasonable estimate these values can be to assess the hurricane's overall severity, it does not reflect the effects of attenuation or amplification that the hilly topography of Sint Maarten can play in the distribution of wind fields and gusts (Figure 4.2). Local variations of the wind help some areas not be as severely affected as other areas that were hit with full intensity.

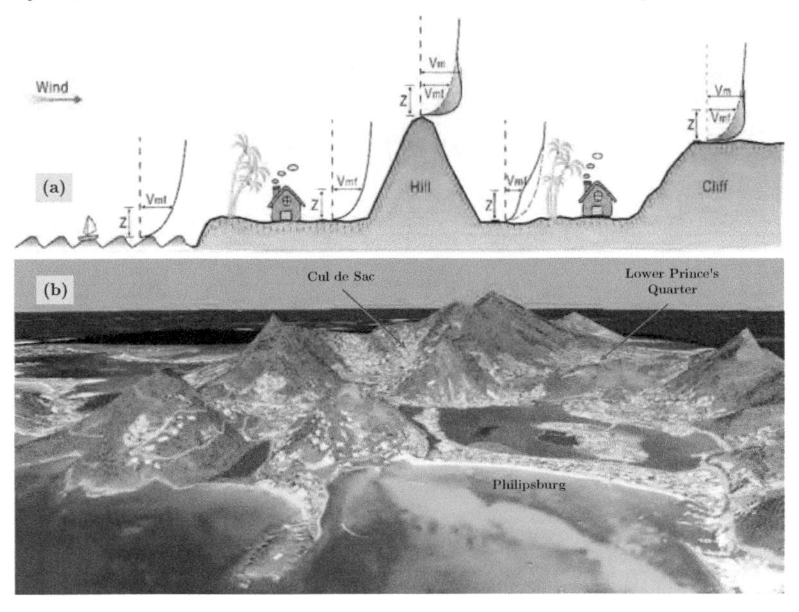

Figure 4.2 (a) Illustrative wind variation due to local topography (Source: Tan and Fang, 2018). (b) 3D representation of Sint Maarten to highlight the hilly topography.

To incorporate localised variation on wind speed, we used the software ERN-Hurricane (ERN Ingenieros Consultores, 2009), which allows simulating probabilistic gradient wind scenarios. The inputs to run the wind model are the tropical cyclone best track dataset of past hurricanes, wind speed and barometric pressures data, topography, bathymetry, urban areas and land use maps. In addition, some factors need to be assumed to run the model; these are topographic exposure factor, wind variation above profile, surface roughness associated with the land uses.

The model is used to obtain maximum wind gusts in the simulation space (Figure 4.3). According to the simulations, model outputs show the most impacted neighbourhood in Sint-Maarten during Hurricane Irma due to wind gusts. The east and most south neighbourhoods appear to have been the most affected ones, with peak gusty winds over 250 km/h. Some of them face winds of more than 320km/h, such as little cape bay, Cay Hill State, Belair and Pointe Blanche. It can be observed that some neighbourhoods received a mitigated wind due to the protection of the hills, such as those located in the Lower Prince's Quarter zone and to the north-east of Cul de Sac (see also Figure 4.2-(b)).

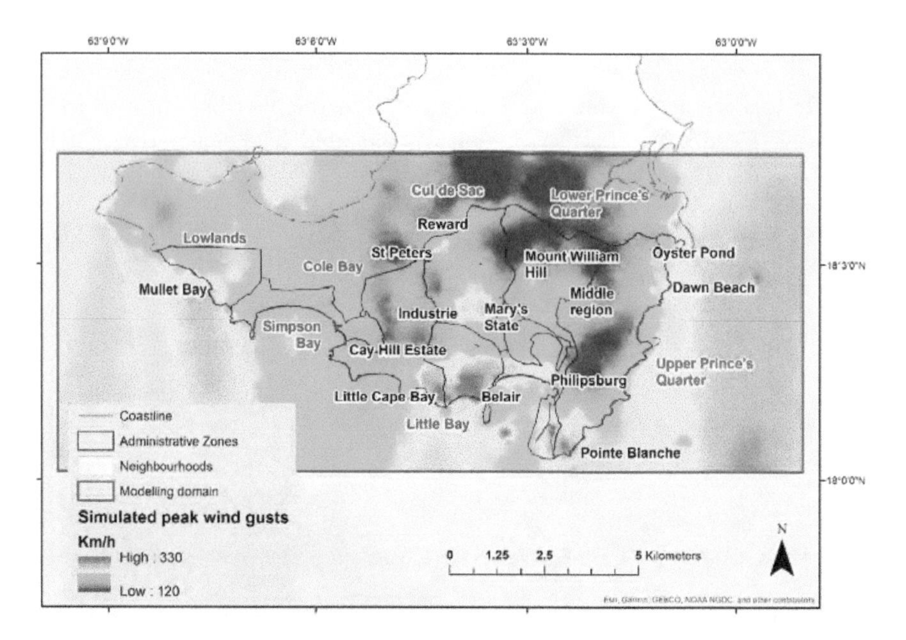

Figure 4.3 Simulated peak wind gusts during Hurricane Irma (km/h) using ERN-Hurricane model.

4.2.2 Storm Surge

The strong winds and the low pressure of Irma also caused waves and storm surges. The French part registered extremely damaging waves, which many reported that it killed people and damaged properties, especially in the French Quarter. The effect in Sint Maarten was milder, which damaged properties along the coast, especially hotels as reported in Chapter 2. Figure 4.4-(b) shows MIKE21 flow model results of the storm surge hazard, flood extent and depth, which affected low-lying areas such as Princes Juliana Airport and Simpson Bay in Sint Maarten and Sandy Ground, Grand Case, Orient Bay and Oyster Bay in Saint-Martin.

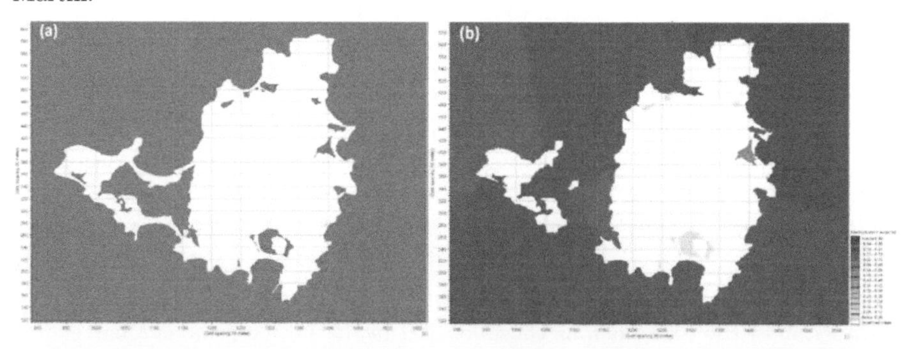

Figure 4.4 a) Initial water level (without storm surge) and (b) model result showing maximum storm surge during Irma.

4.2.3 Inland Flooding

As mention in Chapter 2, section 2.4, Hurricane Irma was a relatively "dry" hurricane. Irma did not bring extreme rainfall into the island. Also contributing to the low rain the fact that Hurricane Irma was a fast-moving Hurricane, it is estimated that Irma was moving at 22 km/h when it passes over Sint Maarten and it took around two hours for the hurricane to cross the island (PEARL, 2018). As a result, no inland flooding was modelled associated with Hurricane Irma.

Additionally, to incorporate the effects of inland flooding on the disaster risk assessment methodology, we included a synthetic (but plausible scenario). We coupled a 1D-2D simulation carried out for a rainfall of 100-year recurrence interval and a storm surge of 0.5 m. The simulation outputs are shown in Figure 4.5 and Figure 4.6 for the flood depth/extent and the flow velocity, respectively. The flooding in the Cul de Sac, Lower prince's Quarter, Little Bay and Madam's Estate is associated with the inland flooding whereas the flooding in Philipsburg, Cay Bay, Simpson Bay and Maho areas is mainly related to the storm surge flooding.

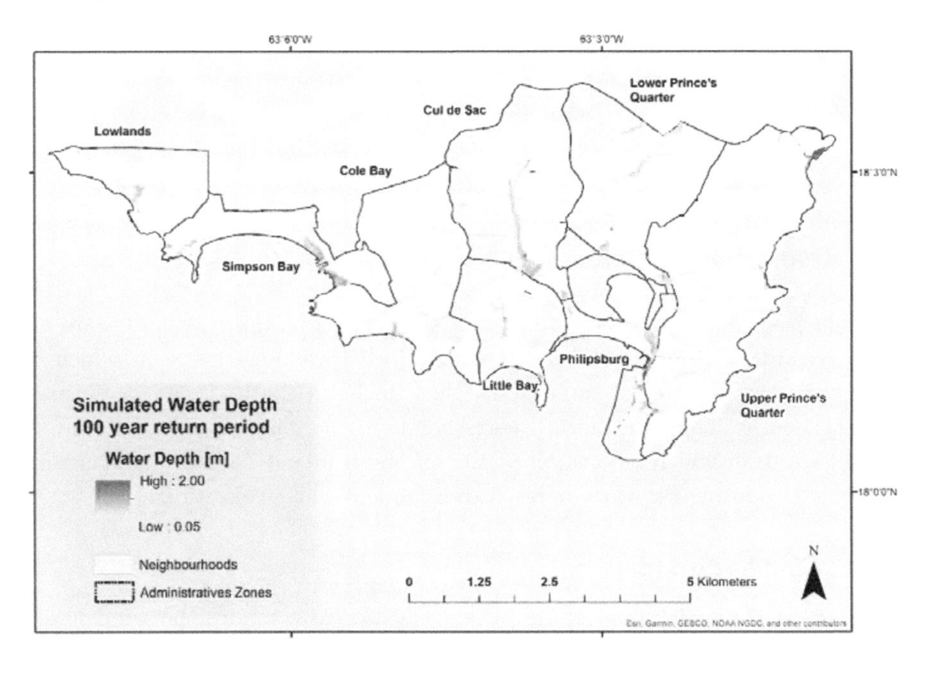

Figure 4.5 Maximum flood extent and depth for a combined rainfall event with a 100-year recurrence interval and a 0.5 m storm surge (PEARL, 2018).

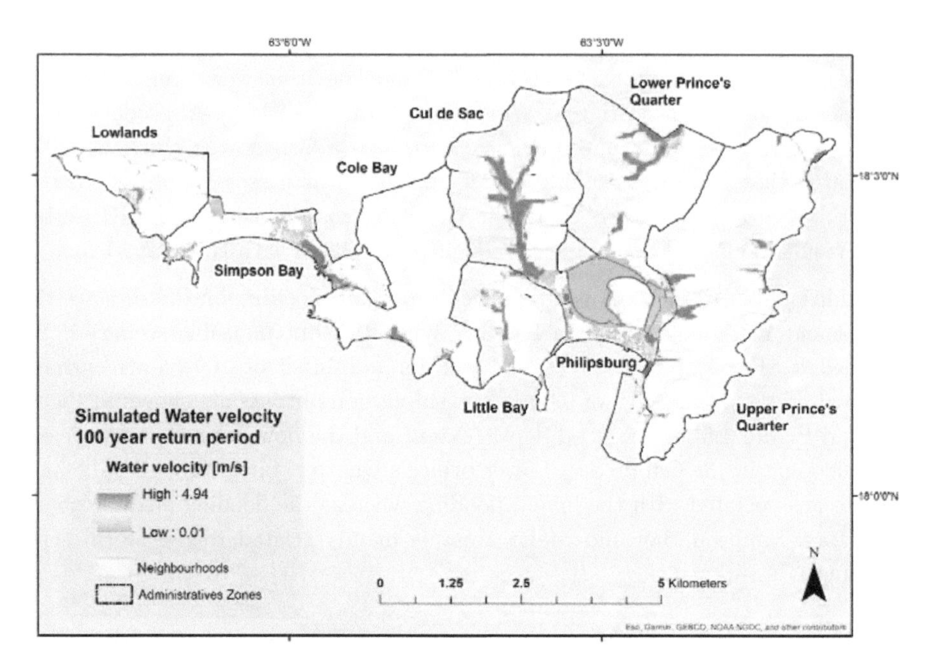

Figure 4.6 Maximum flood velocity for a combined rainfall event with a 100-year recurrence interval and a 0.5 m storm surge (PEARL, 2018).

4.3 Results - Hazard Mapping

The hazard analysis results are presented individually, this is, wind hazard and flood hazard, as well as the combined multi-hazard assessment in which we combined the effects of an event that includes both types of hazards simultaneously. The wind hazard corresponds to the gust wind model, and the flood hazard, in which we combined the effects of inland flood and storm surge flooding accounting for water depth/extension and water velocities.

4.3.1 Wind hazard map

We used the raster map of the wind velocities to compute the wind hazard (Figure 4.3). The methodology assesses the degree of hazard based on the intensity of the winds and its relation with the corresponding hurricane category. Table 4.1 shows the five classes in which wind hazard is represented and the expected impact associated with each category.

Table 4.1. Wind Hazard thresholds as a function of the hurricane category according to the Saffir-Simpson scale and the expected level of damage.

Value (Km/h)	Reclassified Value	Description[1]
0 – 153	Very Low = 1	Storm up to **Category 1** hurricane. Well-constructed frame homes could have damage to roof, shingles, vinyl siding and gutters. Large branches of trees will snap, and shallowly rooted trees may be toppled.
154 - 177	Low = 2	**Category 2** hurricane. Dangerous winds will cause extensive damage: Well-constructed frame homes could sustain significant roof and siding damage. Many shallowly rooted trees will be snapped or uprooted and block numerous roads
178 - 208	Medium = 3	**Category 3** hurricane. Devastating damage will occur: Well-built framed homes may incur significant damage or removal of roof decking, and gable ends. Many trees will be snapped or uprooted, blocking numerous roads
209 - 251	High = 4	**Category 4** hurricane. Catastrophic damage will occur: Well-built framed homes can sustain severe damage with loss of most of the roof structure or some exterior walls. Most trees will be snapped or uprooted and power poles downed. Fallen trees and power poles will isolate residential areas
> 252	Extreme = 5	**Category 5** hurricane. Catastrophic damage will occur: A high percentage of framed homes will be destroyed, with total roof failure and wall collapse. Fallen trees and power poles will isolate residential areas

[1] Based on Schott et al. (2019).

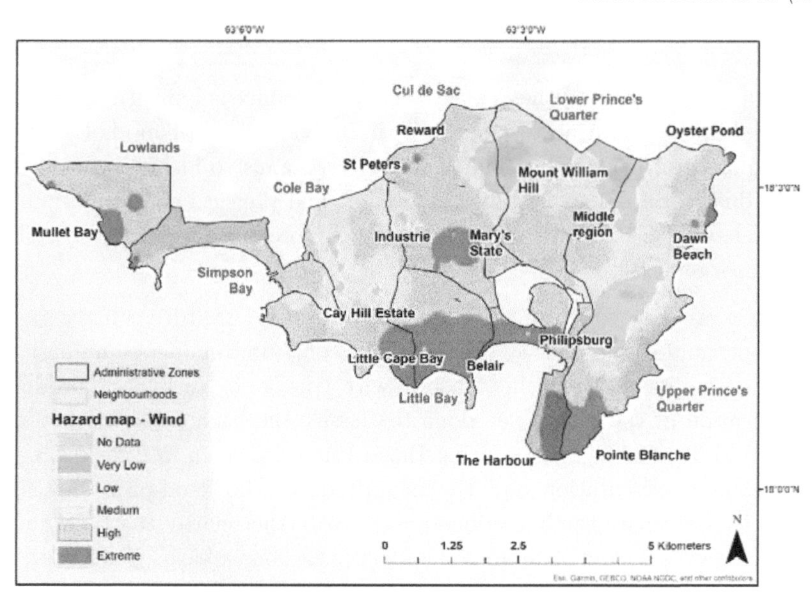

Figure 4.7 Wind Hazard map for simulated gradient gust winds.

4.3.2 Flood and Storm surge hazard map

To compute the flood hazard, the two raster maps corresponding to water depth and water velocity (Figure 4.5 and Figure 4.6) are combined using a modified version of the methodology presented in Priest et al. (2009). The methodology assesses hazard based on the probability of getting injured or killed as the product of depth and velocity. The associated levels of hazard based on water depth and the water velocity are presented in Table 4.2.

Table 4.2. Flood hazards thresholds as a function of depth and velocity. Adapted from (Kuntiyawichai et al., 2016; Priest et al., 2009)

Depth x Velocity (m^2/s)	Reclassified Value	Description
< 0.20	Very Low = 1	**Caution.** Hazard zone with shallow flood water or deep standing water
0.20 – 0.30	Low = 2	**Dangerous for vulnerable groups.** Elderly and children are more vulnerable due to deep or fast-flowing water
0.31 – 0.50	Medium = 3	**Dangerous for most people.** Deep or fast-flowing water Exposure to the hazard is likely to cause injuries or loss of life
0.51 – 0.60	High = 4	**Dangerous for all.** High danger in this zone. Exposure to the hazard is certain to cause injuries or loss of life. Poorly constructed houses may collapse.
> 0.60	Extreme = 5	**Dangerous for all.** High danger in this zone. Exposure to the hazard is sure to cause injuries or loss of life. All buildings in contact are prone to suffer damage or to collapse.

The range of values for the depth × velocity product is built using parameters recognised in the literature to play a role in individuals and households' stability when exposed to floodwaters. The range selected aims to incorporate the effects of individual variables such as height, weight and physical condition, and some hazard-related variables like water temperature, presence of debris, or buildings' structural stability.

GIS was used to compute the hazard. We multiply cell by cell the raster of water depth and water velocity (Eq 4.1). The computation offers a numeric value in each cell of the resulting raster with the two variables' product. A reclassification of the raster was done to classify the hazards in the five classes described in Table 4.2. The resulting Hazard map is shown in Figure 4.8. Given the scale of representation and the magnitude of the flood-prone areas, it is difficult to appreciate the hazardous areas. To further clarify the flood impacts, Figure 4.9 shows a closer look into the most hazardous areas.

$$Flood\ Hazard = Water\ Depth \times Water\ Velocity \quad (4.1)$$

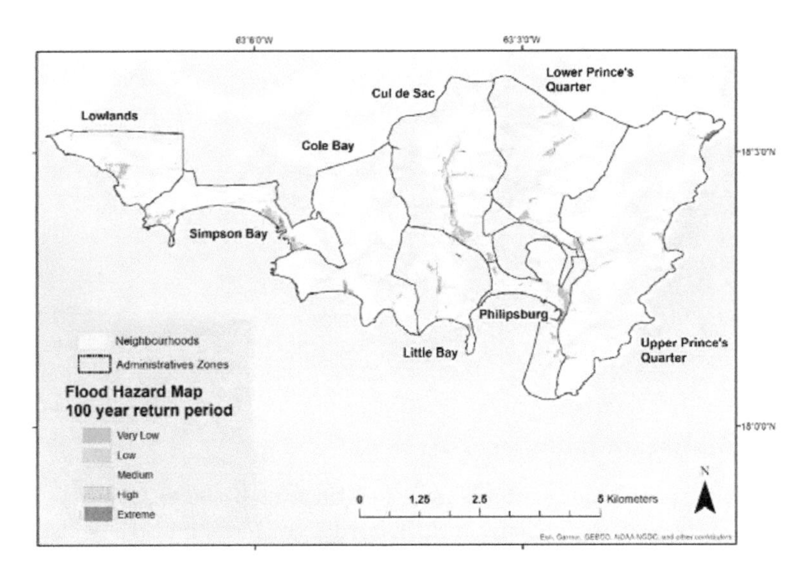

Figure 4.8 Flood Hazard map for a combined rainfall event with a 100-year recurrence interval and a 0.5 m storm surge.

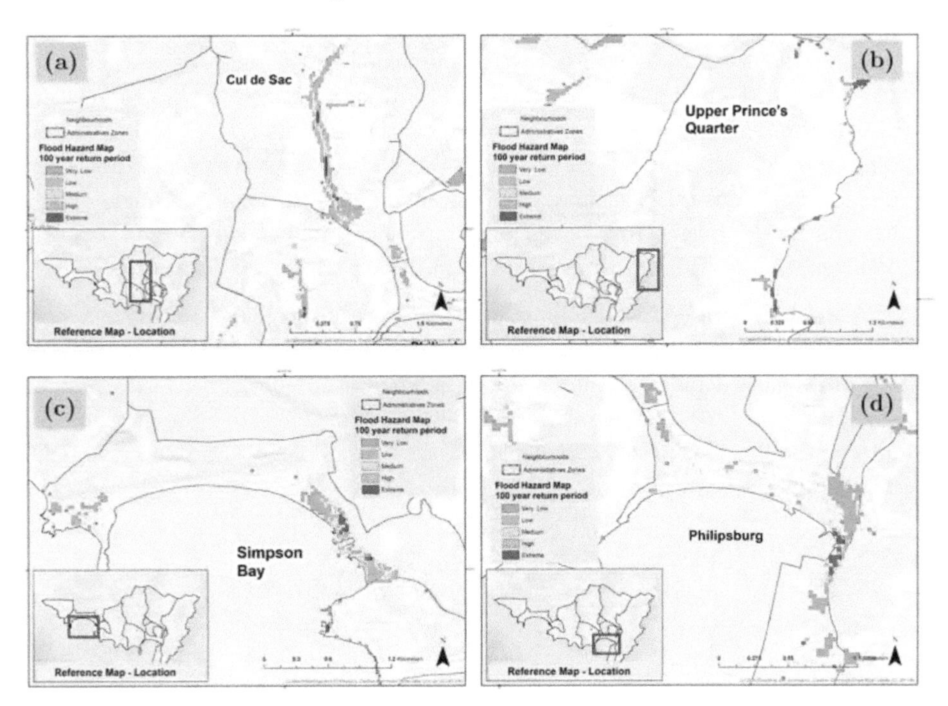

Figure 4.9 High and extreme hazards areas in (a) Cul de Sac district along the Zagersgut Canal, (b) Oyster Pond neighbourhood, (c) Simpson Bay along the Walfare and airport road, (d) Philipsburg area close to the Harbour area, (e) Dutch Quarter and (f) a region in Cay Bay in the proximity of GEBE power plant.

Figure 4.9 (Continuation)

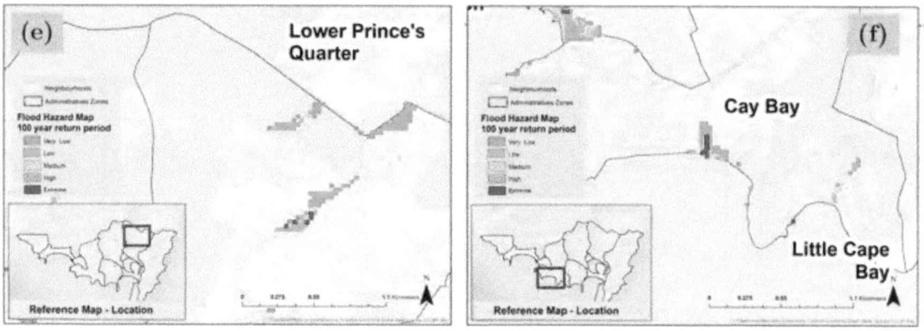

4.3.3 Multi-hazard Index

The multi-hazard index intends to reflect in a map the exposure to multiple natural hazards in specific areas. Multi-hazard assessment is of particular interest because the combined hazard can cause a more severe impact on infrastructure and cause significant loss of lives (Nachappa et al., 2020). The multi-hazard index for Sint Maarten provides an overview of the potential threats relevant to a specific neighbourhood and allows for analysis and comparison between neighbourhoods' hazard levels. Exiting methodologies to map hazards focus on single hazard mapping. Using the multi-hazard approach, we intend to present a measure that aggregates the individual potential hazards related to its location. Multi-hazard mapping was preferred in this thesis because it allows presenting the increased hazard that may result from the interaction of a single type of hazards for a specific area (e.g. hurricane winds and storm surge). For Sint Maarten, the multi-hazard integrates wind hazard with flood hazard.

A similar representation unit is needed before integrating the single type of hazard into a multi-hazard map. For Sint Maarten, the chosen unit to represent the multi-hazard index is the neighbourhood to keep consistency with previous computations (i.e. vulnerability index) and facilitate integration and interpretability. The flow chart depicting the methodology to produce the multi-hazard index is shown in Figure 4.10.

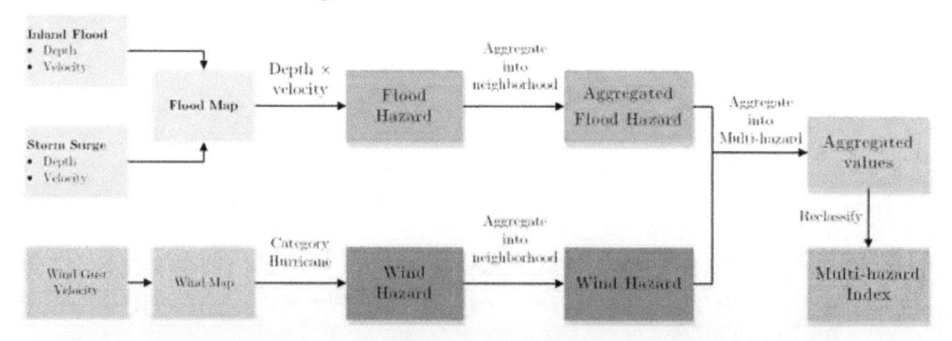

Figure 4.10 Flow chart of the steps required to produce the multi-hazard index.

As the aggregation method, the mean value was used. To aggregate the flood and wind hazard, we select the neighbourhood scale using the categorization of neighbourhoods used by VROMI, as presented in Appendix E. The aggregated flood hazard is presented in Figure 4.11-(a), most areas result on Very Low and Low hazard levels, and some areas are free of flooding hazards. In contrast, aggregation of the wind hazard (Figure 4.11-(b)), shows that the most frequent hazard level at the neighbourhood scale is high hazard, followed by extreme hazard. The contrast between the two categorisations of wind and flood hazard levels emphasises the multi-hazard index representation's feasibility and usability.

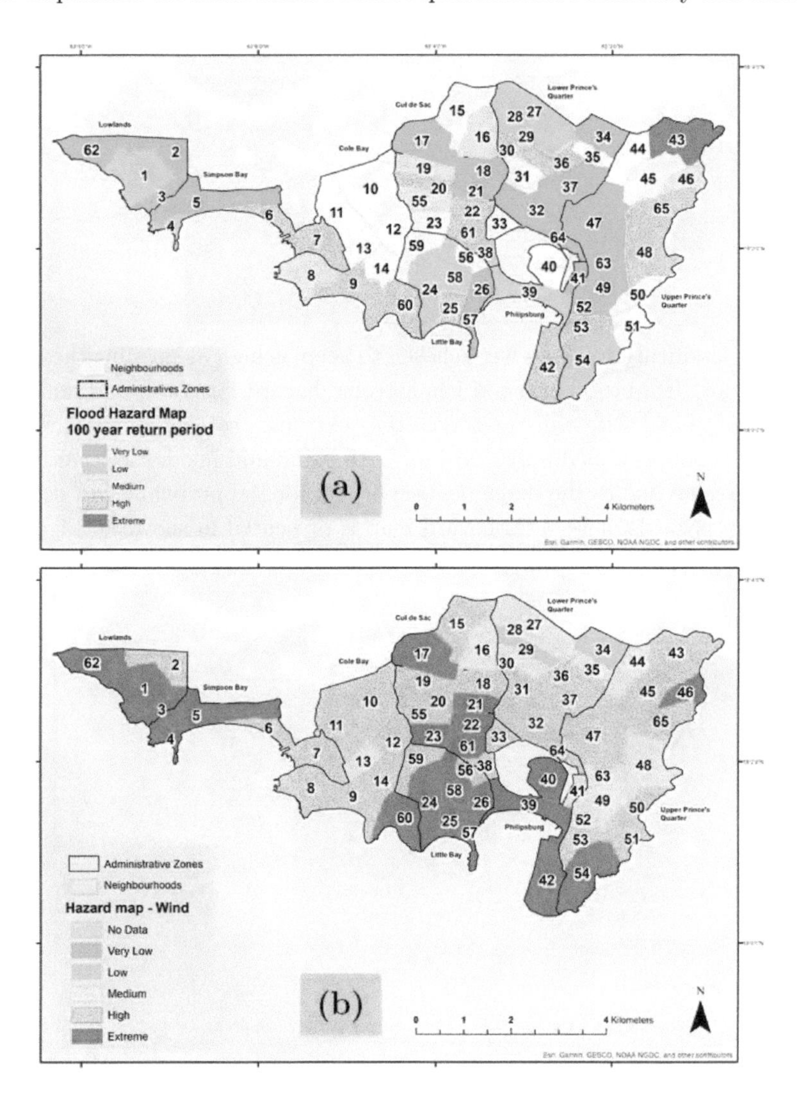

Figure 4.11 Hazard maps aggregated at the neighbourhood scale. (a) Flood hazard. (b) Wind hazard. Numbers represent the identification (ID) of each neighbourhood, as presented in Appendix E.

To aggregate the wind and flood hazard maps (at the neighbourhood scale), we add both rasters, wind hazard and flood hazard using GIS. Such operation values can vary from 0, representing no hazards in a particular neighbourhood, up to a maximum value of ten, representing a neighbourhood with both hazards being present at the maximum level (extreme). To reclassify the resulting aggregation, we used the values presented in Table 4.3.

Table 4.3. Values used to reclassify the aggregated values of wind and flood hazard to produce the Multi-hazard Map.

Aggregated value	Reclassified Value
1	Very Low = 1
2	Low = 2
3	Medium = 3
4	High = 4
5 - 10	Extreme = 5

The reclassification values were chosen to keep as high as possible the resulting hazard index. If a neighbourhood has extreme hazard for wind, and none or low hazard for flood, it will preserve the extreme category, similarly, if a neighbourhood has a medium hazard for both, wind and flood, the resulting value will be extreme due to the magnification of risk for the probability of occurrence of multiple hazards. The multi-hazard map is presented in shown in Figure 4.12.

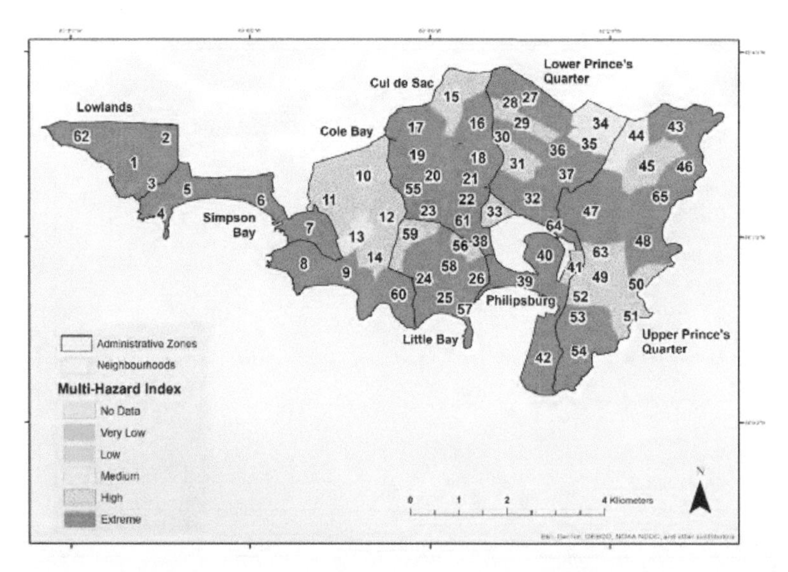

Figure 4.12 Multi-hazard index aggregated at the neighbourhood scale. Numbers represent the identification (ID) of each neighbourhood, as presented in Appendix E.

Analysing the results of the multi-hazard index (Figure 4.12), it can be observed the expected amplification of hazards in some neighbourhoods in Sint Maarten prone to the simultaneous occurrence of winds and floods or storm surges. The amplification is more evident in those neighbourhoods near the coastline, such as Oyster Pond (ID=43), Red Pond Estate (65) and Guana Bay (48) to the East of the Island (Figure 4.13); Simpson Bay (6), Cole Bay Lagoon (7), Billy Folly (8) and Cay Bay (9) located in the south-west part of Sint Maarten (Figure 4.14). Damages in the neighbourhoods mentioned above are associated with high storm surges and high wind gusts. These neighbourhoods shared high hazards in the wind component and were amplified due to at least low hazard for flooding.

Figure 4.13 Damages observed in Sint Maarten due to Hurricane Irma. (a) Oyster Pond. (b) Red Pond Estate.

Figure 4.14 Damages observed in Sint Maarten due to Hurricane Irma. (a) Simpson Bay. (b) Billy Folly.

Figure 4.15 Damages observed in Sint Maarten due to Hurricane Irma. (a) Middle Region (b) Zorg En Rust.

Another amplification worth analysing is in those neighbourhoods located down the Lower Prince's quarter zone, such as Dutch Quarter (ID=36), Middle Region (37). Madame's Estate (32) and Zorg En Rust (30). In this zone, some neighbourhoods are classified as medium or low to wind hazard due to its geographic location protected by the hills, but the risk of flooding makes the multi-hazard index increase in severity (Figure 4.15).

4.4 CONCLUSIONS

The proposed multi-hazard assessment probe to be a better representation of the actual level of risk in Sint Maarten. Identification of flood-prone areas allows us to quantify better the risk associated with the hazard component of the risk assessment. Considering the impact of only individual hazards may create a false sense of security in some regions, especially those prone to storm surges or pluvial flooding. Among the most flood-prone areas, special caution needs to be put in Neighbourhoods such as Philipsburg, that is in constant threat of pluvial and storm surge flooding; Welegen Road for its proximity to the only Hospital in the island and its potential disruption to the medical and emergency services during and in the aftermath of a disaster. Zagersgut road and Bush road will limit the connection of the east part of the island and Welfare road near Simpson bay as an important connection between the most western parts of the island.

This chapter's results identify as some of the most critical areas those located near the coastline associated with potential storm surges that were very damaging during Hurricane Irma. The effects of storm surges were most observed in hotels built near the shoreline to the south and west parts of Sint Maarten. However, it is important to note that most of the island is prone to high and extreme hazard levels due to Hurricane Irma's extremes winds. Future multi-hazard assessment should include other hazards such as landslides, tsunamis, and very important flash-floods.

Disaster risk managers on the island should pay particular attention in communicating the risk associated with floods, both, pluvial and storm surges for future forecasted storms. Residents must know whether they are located in such areas, which in turn can promote evacuation or any other type of protective behaviour before or on the arrival of a new forecasted hazard.

For Hurricane Irma, the pluvial flood was of less impact due to the low rainfall intensity ("dry" hurricane) and the hurricane's fast-moving pace when it was crossing the island. This was a "lucky" scenario for Sint Maarten. However, residents and the government should prepare the island for a future more "wet" hurricane when the effect of a multi-hazard such as the one presented in this chapter can be more devastating and potentially causing more loss of lives.

5

Assessing Exposure To Hurricanes using Evacuation Behaviour

One way to reduce disaster risk is by reducing the exposure of people or assets to the potential threat. Exposure reduction can be achieved in different ways, being evacuating the hazardous area one of those. However, despite the amount of literature that exists on evacuation behaviour, there is still a lack of agreement on which variables can be used as predictors for individuals (or households) to actually evacuate. This lack of agreement can be related to the many variables that can affect the evacuation decision, from demographics, geographic, to the hazard itself that may influence evacuation. Hence, it is essential to analyse and understand these variables based on the specifics of a case study. This chapter presents the most significant variables to be used as predictors of evacuation on the island of Sint Maarten, using data collected after the disaster caused by Hurricane Irma in September 2017. The results suggest that the variables gender, homeownership, percentage of property damage, quality of information, number of storeys of the house, and the vulnerability index are the most significant variables influencing evacuation decisions on the island.

This chapter is partially based on:

Medina, N., Sanchez, A., Vojinovic, Z., (2021). Emergency Evacuation Behaviour in a Small Island Setting; The case study of Sint Maarten during Hurricane Irma. Weather and Climate Extremes. Under Review

5.1 INTRODUCTION

In disaster risk management, one alternative to reduce risk from disasters triggered by natural hazards is by reducing the exposure of people and their assets to the potential hazard, which can be achieved by taking timely and effective protective actions. Amongst the possible actions, evacuation from the potential areas at risk is one of the most common and effective alternatives (Cutter et al., 2012). However, although evacuation has been hugely studied since the early 1960s, failure to evacuate is repeated every year across the entire planet, such as the disastrous consequences observed after Hurricane Katrina in the USA in 2005, Hurricane Irma across the Atlantic basin in 2017, and Hurricane Dorian in the Bahamas in 2019, among others. Understanding evacuation behaviour appears to be extremely important in order to mitigate the loss of life and have better and more realistic evacuation plans (Riad et al., 1999).

Evacuation behaviour during water-related disasters such as those caused by hurricanes, tsunamis and floods is affected by several factors, ranging from individual variables to group decision-making levels, such as demographics and socio-economic characteristics, and others related to the type and intensity of the hazard or to the geographical location (Baker, 1991). Understanding which variables are good predictors of evacuation is then a crucial element in the disaster risk management cycle (Thompson et al., 2017). On the one hand, it allows resources to be optimised through better design of instruments such as surveys to collect field data. On the other hand, it will help risk managers to understand which elements need special consideration to have better evacuation responses and thus reduce disaster risk (Dash and Gladwin, 2007; Huang et al., 2016).

Therefore, after the devastation observed on Sint Maarten after Hurricane Irma, it was important to understand the evacuation response of households on Sint Maarten in order to understand which variables could potentially predict the decision to evacuate (or not) taken by the population on the island. For this reason, we conducted a field survey in the aftermath of Hurricane Irma, aiming to capture information that could potentially explain the actual evacuation observed during Hurricane Irma. The understanding of predictors of evacuation could be used as input for a disaster risk management strategy that aims to reduce exposure to disasters triggered by natural hazards.

In this chapter, we will examine previous research on hurricane evacuation in order to identify and select which variables have the potential to be tested as predictors of evacuation. After the literature review, we present some findings of the fieldwork in order to understand better the evacuation behaviour in the context of Sint Maarten. Then we present a description of the methods and results used to assess the correlation of variables with the observed evacuation behaviours.

5.2 Materials and Methodology

5.2.1 Literature Review

The number of publications on evacuation due to disasters triggered by natural hazards is very large, with significant but often contradictory and inconclusive findings (Baker, 1991; Dash and Gladwin, 2007). It remains unclear which variables are good, bad or non-significant as predictors of evacuation, principally due to cultural and local differences as well as some hazard-specific variables. Researchers have tried to identify and categorise the critical variables that can be used as precursors of evacuation prior to a disasters triggered by natural hazards. To have a comprehensive view of these variables, we have selected four well positioned/cited review papers on evacuation behaviour, covering more than 50 years of research.

The main conclusions of these reviews are presented in Appendix C. The four review papers used were: (i) Baker (1991); this study presents a thorough analysis of predictors for hurricane evacuation based on information collected from 12 hurricanes from 1961 through 1989 in the USA. (ii) Thompson et al. (2017) summarise the main findings of 83 peer-reviewed articles published between 1961 and 2016 regarding evacuation from disasters triggered by natural hazards. (iii) Dash and Gladwin (2007) present a review of variables affecting evacuation decision-making using three main areas: warning, risk perception, and evacuation research. Finally, (iv) Huang et al. (2016) present a statistical meta-analysis that includes 49 studies on hurricane evacuation from 1991 to 2014.

From our review (of reviews), it can be concluded that there is no consensus on which elements are good or bad predictors of evacuation, which can be partially explained due to local, environmental and cultural differences, leading to evacuation rates that will vary from place to place under the same hazard, and will vary in time; this results in different hazards in the same location (Baker, 1991). It is important to mention that most of the cited studies have been performed in the continental and USA territories (Thompson et al., 2017), hence the lack of understanding of predictors in other areas is more significant. In addition, to the best of our knowledge, this is the first research focused on evacuation behaviour in SIDS.

Despite the lack of agreement on predictors, some conclusions can be extracted from the summary of the studies in Appendix C. Risk perception is the most accepted predictor of evacuation from a threat zone with robust and positive correlations reported continuously across studies. In addition, other widely accepted factors influencing protective behaviour towards evacuation are living in flood-prone zones, having evacuated under previous evacuation orders, having experienced losses in the past (injuries, loss of life and loss of infrastructure), and clear and direct communication of the evacuation order. Demographic variables have been extensively used to describe evacuation behaviour in the past, but there has not been consistency in their global usability as predictors has been found, yet only applied to some specific cases.

5.2.2 Disaster Risk Management structure in Sint Maarten

In terms of disaster risk management and evacuation, the island organisational structure is presented in Figure 5.1. As can be seen, at the top of the structure is the Prime Minister, who is responsible for the cohesion between the different action plans within the various Emergency Support Functions (ESF), and ultimately responsible for disaster management on Sint Maarten (de Hamer, 2019). He/she is supported by the Fire Chief, who acts as National Disaster Coordinator during a disaster, and also by the Section Head of Disaster Management, who in turn are both supported by ten Emergency Support Functions (ESF). In terms of evacuation, there are two ESF of interest. EFS-7 is responsible for evacuation, shelter, relief and mass care including humanitarian affairs, care for the elderly, and food and ration distribution for the general public. In addition, ESF-6 represents the Department of Public Health and is responsible for preventive and collective health during disasters; it is responsible for the evacuation of critical patients from hospitals, before and after a disaster.

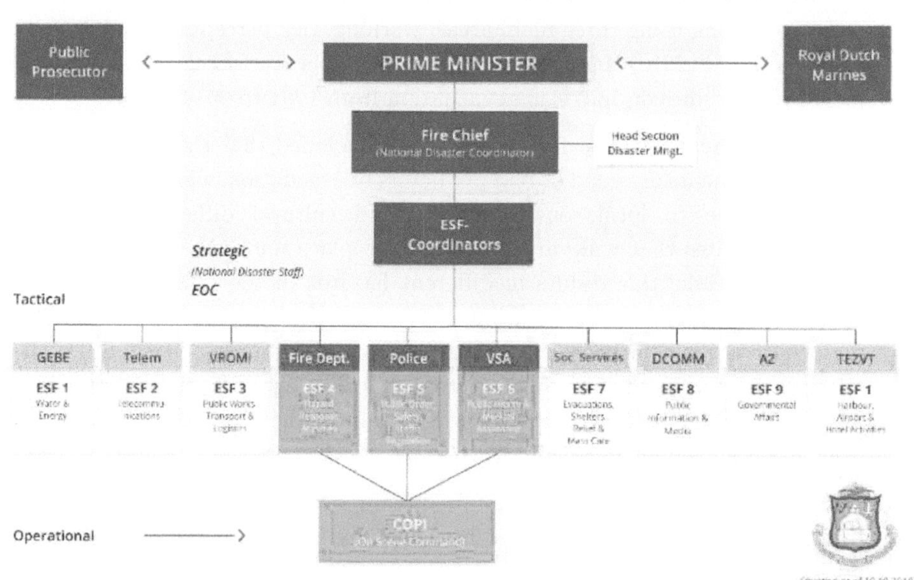

Figure 5.1. Sint Maarten disaster management organisational structure. Source: the official emergency website of the Government of Sint Maarten [3]

[3] Retrieved on September 10 2020 from: https:// https://sxmemergency.esimg.net/en/updates/relief/emergency-support-group/

On Sint Maarten, the scientific organisation responsible for issuing the early warnings in the event of any severe weather event is the Meteorological Department of Sint Maarten – most commonly referred to as the Met Office. During Hurricane Irma, the Met office used data coming from several weather services to produce the bulletins and special reports concerning the potential threat associated with the hurricane. The bulletins were posted on the Met Office's web page and its official Facebook page. In addition, the bulletins were also sent directly to airlines, airport management, the harbour/ marina, the government, radio stations, and hotels. As communication infrastructure was damaged on the island during the hurricane, only one radio station was almost entirely operational during and immediately after Irma passed. The Met Office had a continuous open-line communication with this station to update the residents and the general public of the island on the location and potential remaining threat of Irma.

5.2.3 Evacuation related Information Collected in the Field Mission

Some important information regarding disaster risk, evacuation and shelter management on Sint Maarten that we collected during the field campaign and that may have had a direct influence on evacuation behaviour on the island during Hurricane Irma is summarised here:

- Sint Maarten does not have, and is not considering having, mandatory evacuations.
- Sint Maarten has 11 buildings to use as shelters, at the beginning of each new hurricane season a list of shelters to be used in that particular year is released. In 2017, a total of nine buildings were selected.
- During Hurricane Irma, the government decided not to open the public shelters on the island; even though the reasons are not completely clear, there are indications that this was mainly due to the lack of resources and security.
- A late decision to open some of the official shelters was made on the morning of Tuesday 5 September. This mixed message created confusion among the island residents and may have influenced the evacuation rates reported during Hurricane Irma.
- There is a common perception among residents of the island that public shelters are weak structures. A reason given for this perception is that the government cannot invest money in shelter maintenance for private buildings, as is the case for those used as official shelters on the island. Also, poor or little maintenance is performed annually. This image was corroborated by two of the shelters losing their roof during Hurricane Irma.
- No discrimination on migration status was made in the shelters before or in the aftermath of Irma. On arrival at the shelters, people were asked only for nationality and name.

- Many tourists on the island decided not to evacuate prior to Irma's arrival based on personal motivation; some underestimated the power of a Category 5 hurricane, and others just mentioned they wanted to have the 'hurricane experience'.
- Shelters on Sint Maarten only offer roof protection; residents that evacuate to public shelters are requested to bring their own food, water, blankets and medicines.
- Sirens are installed on the island to communicate evacuation orders, but they were having technical issues and were not used during Irma. SMS and smartphone applications are being evaluated as an alternative to replace the outdated siren system.
- Schools and business on the island were only closed on Monday 4 September, as some of the operational members of the disaster management team only considered the real threat that day. Hence a sense of relaxation may have been transmitted to the community.
- Official warnings or evacuation instructions were transmitted only in English, but we detected large communities of inhabitants who only communicate in Spanish or French.

5.2.4 Predictors to be analysed

Based on the literature review, the fieldwork and survey, and the results of a socio-economic vulnerability analysis previously done by the research team (Medina et al., 2020), we have selected a set of parameters to be tested for significance and correlation with observed evacuation behaviour on Sint Maarten during Hurricane Irma. We classified the potential factors affecting evacuation decision-making behaviour into six groups: demographic, socio-economic, housing, information, place, and storm characteristics, and applied variables associated with the vulnerability index computed for the island. As shown in Table 5.1, the groups are composed of 20 variables and 76 categories to be tested as predictors. The hypothesis to be tested using the expected contribution towards promoting (+) or reducing (-) evacuation is also presented in the table. The last column shows the frequency of the respondents' answers.

Table 5.1. Variables and categories to be analysed as predictors of evacuation, the number of respondents and expected contribution towards evacuation.

[1]	[2]	[3]	[4]	[5]
Group	**Variable / predictor**	**Category**	**Expect to Contribute**	**Respondents (%).N= 255**
Evacuation behaviour	Actual evacuation during Hurricane Irma	Yes = 1	Dependent variable	80 (31.4%)
		No = 0		175 (68.6%)

Table 5.1 (Continued)

[1]	[2]	[3]	[4]	[5]
Group	**Variable / predictor**	**Category**	**Expect to Contribute**	**Respondents (%).N= 255**
Demographic characteristics	Gender	Female	(+)	140 (54.9%)
		Male	(-)	115 (45.1%)
	Age	18 – 30	(-)	26 (10.2%)
		31 – 40	(+)	40 (15.7%)
		41 – 55	(+)	84 (32.9%)
		56 – 65	(-)	32 (12.5%)
		> 65	(-)	31 (12.2%)
		No Answer		42 (16.5%)
	Car ownership	Yes	(+)	194 (76.1%)
		No	(-)	61 (23.9%)
	Household size	1 – 2	(-)	86 (33.7%)
		3 – 4	(+)	107 (42%)
		> = 5	(+)	62 (24.3%)
Socio-economic characteristics	Homeownership	Yes (Owner)	(-)	119 (46.7%)
		No (Tenant)	(+)	136 (53.3%)
	Job-status	Working. Fixed location	(+)	123 (49.4%)
		Working. Changing location	(-)	51 (20.5%)
		Retired	(-)	36 (14.5%)
		Unemployed	(-)	39 (15.7%)
		No Answer		6 (2.4%)
Housing characteristics	House construction material – walls	Bricks	(+)	22 (8.6%)
		Concrete	(-)	198 (77.6%)
		Wood	(+)	35 (13.7%)

Table 5.1 (Continued)

[1]	[2]	[3]	[4]	[5]
Group	**Variable / predictor**	**Category**	**Expect to Contribute**	**Respondents (%).N= 255**
Housing characteristics	House construction material – roof	Concrete	(-)	63 (24.7%)
		Metal sheets	(+)	176 (69.0%)
		Other	(+)	6 (6.3%)
	Insurance for disasters triggered by natural hazards	Yes	(+)	68 (26.8%)
		No	(-)	121 (47.6%)
		Does not know	(-)	65 (25.6%)
	Property damage due to Hurricane Irma	0-25%	(-)	154 (60.4%)
		26-50%	(-)	50 (19.6%)
		51-75%	(+)	20 (7.8%)
		76-100%	(+)	31 (12.2%)
Information Quality of message content	If a more direct and precise message is received, evacuation orders will be followed more	Strongly disagree	(-)	70 (27.5%)
		Disagree	(-)	36 (14.1%)
		Agree	(+)	78 (30.6%)
		Strongly agree	(+)	55 (21.6%)
		Other	(-)	16 (6.3%)
Place, geographical and storm characteristics	Number of storeys	1	(+)	143 (56.1%)
		Two or more	(-)	112 (43.9%)
	Length of residence in a place. Number of years living on Sint Maarten	0 – 10	(+)	39 (15.3%)
		11 – 20	(+)	59 (23.1%)
		21 – 30	(-)	55 (21.6%)
		31 – 40	(-)	53 (20.8%)
		More than 41	(-)	45 (17.6%)
		No answer		4 (1.6%)

Table 5.1 (Continued)

[1]	[2]	[3]	[4]	[5]
Group	**Variable / predictor**	**Category**	**Expect to Contribute**	**Respondents (%).N= 255**
Place, geographical and storm characteristics	Hazard awareness. Number of days aware of Hurricane Irma	0-3	(-)	58 (22.7%)
		4-7	(+)	138 (54.1%)
		8-14	(+)	49 (19.2%)
		More than 14	(-)	8 (3.1%)
		No answer		2 (0.8%)
	Perception of living in a flood-prone area	Yes	(+)	21 (8.2%)
		No	(-)	227 (89.0%)
		No answer		7 (2.7%)
	Previous hurricane experience	0	(-)	0 (0.0%)
		1-2	(-)	52 (20.4%)
		3-4	(+)	71 (27.8%)
		5-6	(+)	47 (18.4%)
		More than 6	(-)	85 (33.3%)
	Level of worry. Frequency of checking the storm information	Once or less a day	(-)	9 (3.6%)
		Several times a day	(+)	26 (10.4%)
		Every couple of hours	(+)	45 (18.1%)
		Throughout the whole day	(+)	169 (67.9%)
Vulnerability index components	Risk perception	Low	(-)	75 (29.4%)
		Medium	(-)	114 (44.7%)
		High	(+)	49 (19.2%)
		Very high	(+)	17 (6.7%)

Table 5.1 (Continued)

[1]	[2]	[3]	[4]	[5]
Group	**Variable / predictor**	**Category**	**Expect to Contribute**	**Respondents (%).N= 255**
Vulnerability index components	Government performance perception	Low	(-)	108 (42.4%)
		Medium	(-)	47 (18.4%)
		High	(+)	91 (35.7%)
		Very high	(+)	9 (3.5%)
	Vulnerability index	Very low	(-)	23 (9.0%)
		Low	(-)	44 (17.3%)
		Medium	(-)	44 (17.3%)
		High	(+)	85 (33.3%)
		Very high	(+)	59 (23.1%)

5.2.5 Model Analysis

In order to evaluate the relationship between the actual evacuation behaviour of Sint Maarten residents and the different factors or predictors, we have conducted a Multiple Correspondence Analysis (MCA). MCA is a well-known mathematical method mostly used to analyse data obtained through surveys; it is used to identify the associations and relationships between variable and categories. (Husson et al., 2017). The MCA analysis we conducted was based on principal components as the extraction method, a scree plot analysis in combination with eigenvalues allows us to determine the number of dimensions to consider in the analysis, which for this study will be limited to the first two dimensions.

The MCA results enables the six groups in which the predictors were grouped (Table 5.1 – column [1]), to be plotted in the first two dimensions in a biplot; the plot is used to evaluate if the groups are conceptually distinct constructs which can be analysed separately. It is expected that groups that are different will appear relatively separated in the biplot. The next step was to test for correlation between the variables (Table 5.1 – column [2]) to identify the degree of relationship between the variables and identify possible redundant variables, as well as to identify those more correlated to evacuation. Here, correlations above 0.7 are considered strong and should be removed before further analysis as possible explanatory variables; correlations around 0.5 are considered moderate relationships and around 0.3 are considered weak relationships between variables.

Once the groups and variables were proved to be independent, we used the results from the MCA analysis to evaluate whether or not each category contributes positively, negatively, or has no statistical significance according to our hypothesis presented in Table 5.1 -column [4].

The next step in the analysis was to identify which variables are more influential (or not) as predictors of the evacuation behaviour observed during Hurricane Irma. First, we conducted a χ^2 test to evaluate the relationship between the dependant variable *evacuation*, with all the other explanatory variables or predictors presented in Table 5.1 – column [2]. The results of this test are in the form of statistical significance for those variables that have a stronger relationship with the actual evacuation behaviour. Second, we studied the relationship between evacuation and the categories of the variables. We characterised each of the categories of the dependant variable evacuation (Yes and No) by using all the categories of the explanatory variables in Table 5.1 – column [3] to find which of them have statistical significance and can help to explain the observed household evacuation decision.

The outputs of both the MCA analysis and the correlation matrix allow us to determine the most significant predictors of evacuation for Sint Maarten. Using different combinations of the significant predictors, we developed three binomial logistic regression models to simulate the evacuation response.

5.2.6 Results and Analysis

The results of the MCA analysis of the six groups in which variables were categorised are presented in Figure 5.2. The distance observed between the different groups in the first two dimensions shows that all the groups can be considered conceptually distinct constructs. The validity of the use of the six groups is further supported by the correlation analysis, as presented in Table 5.2. The highest correlation among the scales was found to be moderate, with a correlation of 0.45 between the variables *home insurance against disasters* and *homeownership*, followed by *vulnerability index* and *risk perception* with a correlation of 0.41. No other set of variables has vulnerability greater than 0.40. Given the values of correlation obtained, the results suggest that all the variables in the analysis are distinct, hence we can use the complete set of variables and categories in the MCA analysis to evaluate whether or not they can be used as an indicator of the actual evacuation behaviour on Sint Maarten.

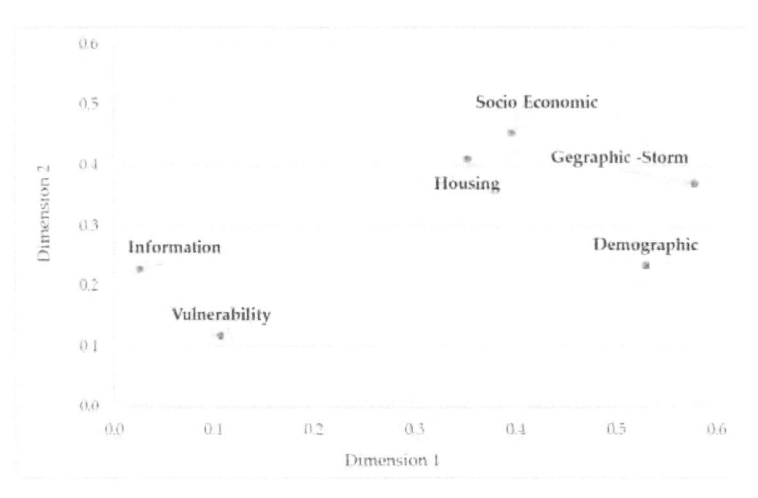

Figure 5.2. Group representation in the first two dimensions.

From the correlation matrix, there also seem to be some weak correlations involving some of the other variables. Gender and age appear somehow to be correlated to job status on Sint Maarten, with correlation R = 0.26 and 0.39, respectively. Age and length of residence (R=0.34), and the number of house storeys (R=0.31) influence to some degree the decision whether or not to take out home insurance. Property damage has a correlation with the material of the walls (R=0.35). The correlation matrix also provides signs of possible predictors of evacuation, the most correlated variables are property damage (R=0.31), the materials of the walls (0.28), the quality of information (R=0.21), the number of house storeys (R=0.21) and the insurance (R=0.21).

The statistical test we run for significance on the categories to evaluate our hypothesis of contribution towards positive or negative evacuation behaviour is presented in Table 5.3. From the initial 76 categories we tested, only 18 categories were found to be significant to a level of $p < 0.05$. All the six groups are represented in the selection of significant categories but only 11 variables from the initial 20.

From the categories' analysis, it can be inferred that households on Sint Maarten that have reported evacuating during Hurricane Irma share a high frequency for variables such as: high percentage of damage to their houses, the house is normally a one-storey building built with wooden walls, not having a car and not having home insurance, they generally rent the house (tenants), and most correspond to women. In contrast, those households that did not evacuate during Irma are normally men, who consider their houses are not located in a flood-prone area, their houses suffer low damage (0-25%), and are built with concrete walls and roof, and they are normally the owners of the house. They have insurance for disasters triggered by natural hazards, and there is a high frequency of having two or more storeys; in addition, they live in areas with a very low vulnerability index.

Regarding the hypothesised effects of the categories that were found to be statistically significant, almost all the categories had the hypothesised effect on evacuation behaviour, with the two exceptions of *car ownership* and *insurance for disasters triggered by natural hazards*. It was initially expected that households that have insurance and own a car would favour evacuation but instead the opposite was found; these two elements were found to have a negative effect on evacuation on Sint Maarten.

The quality of the content variable (in the information characteristics group) had one category with a negatively associated correlation effect on evacuation, meaning that those individuals that strongly disagree with the statement asked tended to be less likely to evacuate on Sint Maarten. Regarding those categories in the group of place and storm characteristics, we found, as expected, that one-storey building households tend to be more likely to evacuate in contrast to those living in a house with two or more storeys. Similarly, perception of living in an area not prone to floods also had a negative correlation with evacuation. From the vulnerability group, only those households with a very low vulnerability index were found to have a negative effect on evacuation, as initially hypothesised.

Table 5.2. Correlation matrix between variables

#	Variable	1	2	3	4	5	6	7	8	9	10	11	12	13	14	15	16	17	18	19	20	21
1	Gender	1.00																				
2	Age	0.15	1.00																			
3	Car ownership	0.08	0.28	1.00																		
4	Household size	0.15	0.23	0.20	1.00																	
5	Homeownership	0.01	0.24	0.16	0.19	1.00																
6	Job status	0.26	0.39	0.18	0.12	0.20	1.00															
7	House construction – walls	0.09	0.21	0.28	0.14	0.14	0.11	1.00														
8	House construction material - roof	0.03	0.11	0.13	0.14	0.18	0.14	0.21	1.00													
9	Property damage	0.12	0.11	0.21	0.07	0.13	0.13	0.35	0.21	1.00												
10	Home insurance	0.13	0.34	0.22	0.09	**0.45**	0.20	0.21	0.08	0.24	1.00											
11	Public information	0.10	0.14	0.14	0.09	0.19	0.15	0.14	0.16	0.14	0.20	1.00										
12	Length of residence	0.12	0.34	0.15	0.10	0.34	0.19	0.18	0.18	0.11	0.34	0.16	1.00									
13	Flood-prone area	0.07	0.12	0.01	0.07	0.11	0.08	0.01	0.03	0.15	0.15	0.15	0.09	1.00								
14	Number of house storeys	0.01	0.31	0.20	0.02	0.04	0.17	0.22	0.03	0.20	0.31	0.17	0.28	0.05	1.00							
15	Previous experience	0.03	0.24	0.11	0.04	0.26	0.16	0.11	0.14	0.12	0.14	0.09	0.28	0.05	0.14	1.00						
16	Concern level	0.08	0.12	0.16	0.15	0.18	0.13	0.11	0.15	0.11	0.11	0.15	0.11	0.22	0.15	0.10	1.00					
17	Hazard awareness	0.11	0.08	0.07	0.15	0.13	0.10	0.07	0.11	0.12	0.12	0.12	0.12	0.13	0.10	0.12	0.15	1.00				
18	Risk perception	0.13	0.15	0.05	0.12	0.13	0.14	0.12	0.10	0.12	0.13	0.10	0.16	0.06	0.15	0.10	0.10	0.15	1.00			
19	Government performance	0.08	0.17	0.06	0.12	0.12	0.09	0.11	0.10	0.13	0.10	0.07	0.16	0.19	0.08	0.07	0.09	0.10	0.24	1.00		
20	Vulnerability index	0.11	0.17	0.10	0.16	0.07	0.18	0.14	0.21	0.13	0.17	0.14	0.16	0.08	0.13	0.12	0.12	0.14	**0.41**	0.34	1.00	
21	**Evacuation decision**	0.14	0.07	0.16	0.10	0.16	0.08	0.28	0.14	0.31	0.21	0.21	0.04	0.06	0.21	0.14	0.06	0.05	0.10	0.08	0.18	1.00

Table 5.3 Variables and categories found to be statistically significant as predictors of evacuation. The expected and actual contribution towards evacuation is presented. The categories in bold had a different contribution to the one initially hypothesised.

Group	Variable / predictor	Category	Expect to contribute	Actual contribution (*)
Demographic characteristics	Gender	Female	(+)	(+) / (c)
		Male	(-)	(-) / (c)
	Car ownership	**Yes**	**(+)**	**(-) / (c)**
		No	**(-)**	**(+) / (c)**
Socio-economic characteristics	Homeownership	Yes (owner)	(-)	(-) / (c)
		No (tenant)	(+)	(+) / (c)
Housing characteristics	House construction material – walls	Concrete	(-)	(-) / (c)
		Wood	(+)	(+) / (a)
	House construction material – roof	Concrete	(-)	(-) / (c)
	Insurance for disasters triggered by natural hazards	**Yes**	**(+)**	**(-) / (b)**
		No	**(-)**	**(+) / (b)**
	Property damage due to Hurricane Irma	0-25%	(-)	(-) / (a)
		76-100%	(+)	(+) / (a)
Information characteristics	Quality of message content. If a more direct and precise message is received, evacuation orders will be followed more	Strongly disagree	(-)	(-) / (b)
Place, geographical and storm characteristics	Perception of living in flood-prone areas	No	(-)	(-) / (c)
	Number of house storeys	1	(+)	(+) / (a)
		Two or more	(-)	(-) / (a)
Vulnerability index components	Vulnerability index	Very low	(-)	(-) / (b)

(*)Significant at: (a) $p < 0.001$, (b) $p < 0.01$, (c) $p < 0.05$

As reported in Table 5.4, nine variables have statistical significance towards contributing to the decision by a household to evacuate during Hurricane Irma on Sint Maarten. Property damage is the most significant variable followed by the material of the walls, and third in importance is the perception of living in a flood-prone area. The number of house storeys was also found to be significant as well as the quality of the information received during an emergency. Also, gender in combination with having a car, having insurance and being the owner of the house seems to play a part in the decision whether to evacuate or to stay.

Table 5.4. Results of a chi-squared test to evaluate the link between the variables and Evacuation Behaviour

#	**Variable**	**p.value**	**df**
1	Property damage	0.000021	3
2	House construction material – walls	0.000050	2
3	Perception of living in a flood-prone area	0.000256	2
4	Number of house storeys	0.000965	1
5	Insurance for disasters triggered by natural hazards	0.004444	2
6	Homeownership	0.011574	1
7	Car ownership	0.012866	1
8	Information. Quality of message content	0.020459	4
9	Gender	0.028445	1

As the next step, we ran a binomial logistic regression model on the variables (Table 5.1– column [2]), to test them for statistical significance in the actual observed behaviour during Hurricane Irma on Sint Maarten. Regressing the evacuation decision against all of the other variables in the model (see Table 5.5, Model 1) showed that gender, homeownership and the number of house storeys had a negative effect on the observed evacuation behaviour. In contrast, property damage, quality of the information and the vulnerability index all showed a positive effect.

Similarly, Model 2 in Table 5.5 presents the re-computed regression results after removing the non-significant variables from Model 1. The changes in the regression coefficients of Model M2 were minimal and kept the associated positive or negative effect observed in M1. The errors in predictions related to Model M2 were minimal according to the residual deviance. M2 shows a superior balance between its ability to fit the data set and its ability to avoid over-fitting the model measured by the AIC score.

Table 5.5. Binomial logistic regression model. Prediction of the evacuation decision. Model 1 is built using all the variables. Model 2 is built using the statistically significant parameters of Model 1. The variables in bold were found to be statistically significant.

Variable	Model 1 (M1)			Model 2 (M2)		
	β	SE (β)	Odd ratio	β	SE (β)	Odd ratio
Gender	**-0.683**[c]	0.318	0.505	**-0.662**[c]	0.305	0.516
Age	0.176	0.183	1.192			
Car ownership	0.028	0.380	1.028			
Household size	-0.111	0.228	0.895			
Homeownership	**-0.658**[d]	0.369	0.518	**-0.676**[c]	0.308	0.508
Job status	-0.051	0.145	0.950			
House construction material – walls	0.036	0.368	1.037			
House construction material – roof	0.388	0.321	1.473			
Property damage	**0.449**[b]	0.158	1.566	**0.483**[a]	0.138	1.621
Insurance for disasters triggered by natural hazards	-0.170	0.242	0.844			
Information. Quality of message content	**0.259**[c]	0.130	1.296	**0.238**[c]	0.121	1.268
Length of residence	0.008	0.158	1.008			
Perception of living in flood-prone area	-0.002	0.560	0.998			
Number of house storeys	**-0.694**[c]	0.352	0.500	**-0.749**[c]	0.316	0.473
Previous hurricane experience	-0.197	0.157	0.822			
Concern level	-0.043	0.185	0.958			
Hazard awareness	0.033	0.217	1.034			
Risk perception	-0.121	0.193	0.886			
Government performance perception	0.120	0.171	1.127			
Vulnerability index	**0.266**[c]	0.135	1.304	**0.243**[d]	0.125	1.275
Intercept	-0.316	2.211	0.729	-0.213	1.041	0.808
Null deviance	317.25 on 254 df			317.25 on 254 df		
Residual deviance	264.76 on 234 df			270.87 on 248 df		
AIC	306.76			284.87		

Significant at: ‘a’ $p < 0.001$, ‘b’ $p < 0.01$, ‘c’ $p < 0.05$, ‘d’ $p < 0.1$

Furthermore, to gain an understanding of the predictors of actual evacuation, we computed the logistic regression models on the categories (Table 5.1 -column [2]). We developed three binomial logistic regression models. We built the models using a random sample of 80% of the dataset and leaving the 20% remaining of the data for validation purposes.

First, we ran a logistic model using the categories listed in Table 5.4; these are the categories with statistical significance in the chi-square test. The result of this model is presented in Table 5.6 -Logit-i. In this logit model, male gender, concrete walls, living in a multi-storey building and strongly disagreeing in the information content component all have a predicted negative effect on evacuation behaviour. On the other hand, being a tenant, suffering damage between 76-100% during a storm, and not having home insurance for disasters triggered by natural hazards all have a positive effect leading to evacuation.

The second model (Logit-ii in Table 5.6) was built using the categories in Table 5.3 with v.test $>$ 2.6 as those with more statistical significance. From Logit-ii, a value for the category property damage of between 76-100% was again found to be significant to engage in evacuation actions positively, and strongly disagreeing in the information component was again found to have a negative impact on evacuation.

The third logit model (Logit-iii in Table 5.6) was built with the variables in (M2) in Table 5.5. Logit-iii results show that strongly disagreeing with information has the most substantial negative effect, followed by multi-storey buildings and male gender. In contrast, property damage (75-100%) was found to have the greatest observed positive effect on evacuation, followed by the category tenant in the house ownership variable.

Concerning the values of the null deviance, it is relatively higher than the degrees of freedom (df), meaning that it makes sense to use more than a single parameter (intercept) for fitting all three models. In terms of the residual deviance for all three models, this is relatively low and close to the degrees of freedom, implying an appropriate and well-fitting model. Also, the values of the Akaike Information Criterion (AIC) allow us to compare the level of complexity between the three models. The model with the lower AIC score is expected to have a better balance between its ability to fit the data set and its ability to avoid over-fitting the model. Logit-I is the best of the three models in terms of the AIC coefficient, but the AIC values for all the models are relatively similar, which means there is not a clearly superior model.

Table 5.6. Prediction of Evacuation Decision. Binomial logistic regression models.

Variable. **Category**	**Logit-i**			**Logit-ii**			**Logit-iii**		
	β	**SE (β)**	**Odd Ratio**	β	**SE (β)**	**Odd Ratio**	β	**SE (β)**	**Odd Ratio**
Gender. **Male**	**-0.769**[c]	0.337	0.464				**-0.566**[d]	0.313	0.568
Car ownership. **Yes**	0.194	0.393	1.215						
Homeownership. **Tenant**	**0.976**[b]	0.370	2.654				**0.707**[c]	0.317	2.027
House construction material – walls. **Concrete**	**-1.025**[d]	0.540	0.359	-0.859	0.534	0.424			
House construction material – walls. **Wood**	0.225	0.692	0.798	0.032	0.687	0.969			
Property damage. **Damage 26 -50 %**	-0.292	0.411	1.340	-0.325	0.404	1.384	-0.297	0.393	1.346
Property damage. **Damage 51 -75 %**	0.003	0.593	1.003	0.115	0.595	1.122	0.490	0.542	1.633
Property damage. **Damage 76 -100 %**	**1.074**[c]	0.537	2.926	**1.051**[c]	0.534	2.861	**1.584**[a]	0.462	4.874
Insurance for disasters. **No**	**0.834**[d]	0.429	2.302	0.367	0.395	1.444			
Insurance for disasters. **Yes**	-0.214	0.512	1.239	-0.063	0.482	0.939			
Information. Quality of Content. **Disagree**	-0.565	0.527	0.568	-0.637	0.530	0.529	-0.481	0.511	0.618
Information. Quality of content. **Other**	-0.488	0.695	0.614	-0.449	0.681	0.638	-0.376	0.643	0.687
Information. Quality of content. Agree	0.026	0.429	0.974	0.019	0.428	1.019	0.049	0.404	0.952
Information. Quality of content. **Strongly disagree**	**-0.961**[c]	0.441	0.383	**-0.972**[c]	0.430	0.378	**-1.002**[c]	0.426	0.367
Perception of living in flood-prone area. **Yes**	0.121	0.565	1.128	0.224	0.542	1.252			

Table 5.6 (Continued)

Variable. **Category**	**Logit-i**			**Logit-ii**			**Logit-iii**		
	β	**SE (β)**	**Odd Ratio**	**β**	**SE (β)**	**Odd Ratio**	**β**	**SE (β)**	**Odd Ratio**
Number of house storeys. **Two or more**	**-0.603**[d]	0.36	0.547	-0.465	0.344	0.628	**-0.776**[c]	0.327	0.460
Vulnerability index. **Low**				-0.147	0.463	1.159	-0.113	0.462	0.893
Vulnerability index. **Medium**				-0.343	0.475	0.709	-0.239	0.459	0.787
Vulnerability index. **Very high**				0.378	0.415	1.459	0.406	0.399	1.501
Vulnerability index. **Very low**				-1.216	0.811	0.296	-1.187	0.804	0.305
Intercept	-0.553	0.800	0.575	-0.079	0.684	0.924	-0.596	0.420	0.551
Null deviance	317.25 on 254 df			317.25 on 254 df			317.25 on 254 df		
Residual deviance	249.05 on 237 df			255.98 on 236 df			264.94 on 240 df		
AIC	285.05			293.98			294.94		

Significant at: 'a' $p < 0.001$, 'b' $p < 0.01$, 'c' $p < 0.05$, 'd' $p < 0.1$

In addition, to evaluate the performance of the three logistic models, we estimate the prediction accuracy and the prediction errors of the models. The data was split randomly into training and validation data sets, using the rule of thumb 80%-20% of the data, respectively. The predictive power of the logistic models was then assessed by comparing the predicted outcome values against the known outcome values. Different metrics on model performance evaluation are presented in Table 5.7.

Table 5.7. Performance evaluation of the binomial regression logistic models.

Variable	Logit Model		
	Logit-i	**Logit-ii**	**Logit-iii**
Accuracy [%]	74.5	70.6	76.5
95% CI	60.4 – 85.7	56.2 – 82.5	62.5 – 87.2
Sensitivity [%]	52.6	47.4	52.6
Specificity [%]	87.5	84.4	90.6
ROC Curve [AUC]	0.822	0.773	0.825

The first value to evaluate the model performance is the overall classification accuracy; this value for the three models is relatively high with accuracies above 70%, with Logit-iii yielding the best results, correctly predicting the individual outcome in 76.5 % of the cases and with a confidence interval (95% CI) between 62.5% and 87.2%.

Model performance was also measured using the values of sensitivity and specificity. Sensitivity in the model assessment refers to the number of times the model was able to predict correctly the cases where a household performs an evacuation. In contrast, specificity refers to the number of times the model was able to predict correctly those households that did not evacuate. The importance of sensitivity and specificity parameters depends on the context; for evacuation processes, it is more important to have minimal wrong positive predictions (high specificity); this is forecasting that a household would evacuate, but in reality, it does not. Minimal wrong positive predictions translate into having a more precise picture of how many households decide to stay in their houses, potentially requiring assistance during or in the direct aftermath of a disaster. Specificity for all the models is above 80%, and the Logit-iii model rate is higher than the others with 90.6%. Regarding the sensitivity of the regression models it is around 50% for all three models.

Finally, the Area Under the Curve (AUC) from a ROC curve analysis summarises the overall performance of the prediction. The AUC metric varies between 0.50 (random prediction) and 1.00 (perfect prediction). Values above 0.80 are an indication of a good predictor. In our regression models, the models Logit-I and Logit-iii are considered to be good predictors.

The performance evaluation suggests that the model that better predicts the evacuation behaviour on Sint Maarten is the Logit-iii. The model is composed of six variables/predictors: gender, homeownership, percentage of property damage, quality of the information, number of house storeys, and the vulnerability index. The general equation from a logistic model is presented in Eq. 4.1

$$\log\left(\frac{p}{1-p}\right) = \beta_0 + \beta_1 x_{i1} + \beta_2 x_{i2} + \cdots + \beta_k x_{ik} \qquad \text{Eq. 5.1}$$

Where:

p = Probability of evacuation

β_0 = intercept

β_1 = Beta coefficient for parameter 1

x_{i1}= Value of parameter 1

β_k = Beta coefficient for parameter k

x_{ik}= Value of parameter k

The function depicted in Eq. 4.1 corresponds to a logarithmic function, hence the value of p will always be between 0 and 1. To assess whether the value of p indicates a household evacuating or not Eq. 4.2 and Eq. 4.3 are used.

$$\textit{From Eq } 5.1 \quad \textit{If } p \geq 0.5 \ \textit{ then Evacuation} = \textit{Yes} \qquad \textit{Eq. 5.2}$$

$$\textit{From Eq } 5.1. \quad \textit{If } p < 0.5 \ \textit{ then Evacuation} = \textit{No} \qquad \textit{Eq. 5.3}$$

Using the Logit-iii model, we computed the probability of evacuation for the completely Dutch part of the island. The result correspond to the prediction of evacuation for the households that were not surveyed.

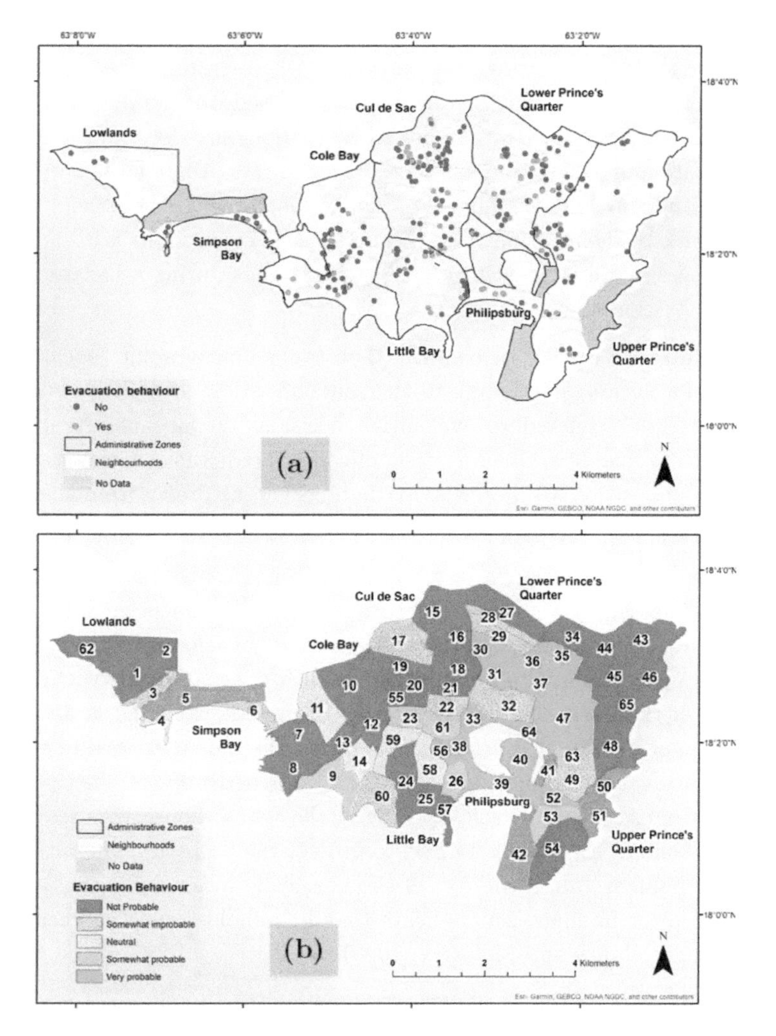

Figure 5.3. Evacuation behaviour in Sint Maarten. (a) Observed in the surveyed houses. (b) Predicted using Logit-iii model and results were aggregated at neighbourhood scale, green colours denote areas of high probability of evacuation, hence less exposure. Red colour denotes low probabilities of evacuation.

5.3 Discussion

Evacuation on Sint Maarten needs to be seen and understood in the context of a small island developing state (SIDS). Evacuation processes are challenging everywhere, but the context of a SIDS makes it particularly so in several ways. First, a SIDS is strongly associated with the low socio-economic status of its inhabitants (UNFCCC, 2005), which makes it almost impossible for a significant part of the population to flee the island no matter the severity of the hazard that is forecast. Second, given the typical size of the islands compared with the size of a major hurricane, this means that no matter where on the island an evacuee is, they cannot avoid to some extent being exposed to the hazard; this may considerably influence the decision not to evacuate. Third, for Sint Maarten, there is a significant number of undocumented immigrants, around 10,000 (Medina et al., 2020). Undocumented immigrants tend to not evacuate to official shelters to avoid the risk of deportation; also, limited social connections within the island limits the possibilities of evacuating to safer grounds during a hazard (i.e. flood, hurricane) (Wilson and Tiefenbacher, 2012).

The results of this study confirm that some demographic, socio-economic, housing, information, place and storm characteristics, as well as vulnerability factors, are closely related to evacuation decisions on Sint Maarten. Although there are not as many variables that significantly predict evacuation as those that were tested (11 out of 20), noteworthy findings can be drawn from the results.

Demographic characteristics have been reported as non-conclusive across multiple studies, as mentioned in the introduction. However, for Sint Maarten, it was found that female gender is a predictor that influences a household to evacuate and male gender is a predictor not to evacuate. Women have been reported in other studies to comply better with evacuation instructions and to have a better risk assessment (Bateman and Edwards, 2002; Dash and Gladwin, 2007; Huang et al., 2012). Furthermore, during the fieldwork, one of the reasons given for not evacuating included that they wanted to protect their homes from looting or from the storm itself. We found that some households leave at least one person behind, usually the father, to protect the property. This situation was reported to happen even if the rest of the household evacuates, behaviour that is contrary to prior findings that indicate that households tend to evacuate (or to stay) as a unit (Smith and McCarty, 2009).

We identify that looting is a big concern among the population of Sint Maarten, especially after Hurricane Irma, where shops and houses were heavily looted. The Disaster Risk Manager on the island should be concerned about this perception because it is affecting people's willingness to evacuate. Extra security in areas at high risk of looting should be guaranteed and communicated in time to promote timely evacuation in those areas.

We found that car ownership is a significant variable of a household's behaviour towards evacuation. The findings of this research are contrary to the hypothesised effect; it was expected that households with a vehicle would be more likely to evacuate (Sadri et al., 2014; Wu et al., 2012). However, we found that having a car on Sint Maarten is associated with those households that did not evacuate and that, in contrast, not having a car is correlated with households that did evacuate during Hurricane Irma. A similar counterintuitive result was found by Lazo et al. (2015), where people with lack of transportation was found to be more likely to evacuate. Not having a car promoting evacuation might be explained in two ways. First, households that do not have a vehicle may feel the need to evacuate early when a hurricane is forecast, and they do not want to be trapped in the middle of the hurricane in their houses. Second, public transport may be suspended during a forecast hurricane, making it difficult to evacuate when the hurricane is approaching.

On the other hand, having a car may create a feeling of non-urgency to evacuate, as they may (falsely) think they can evacuate whenever they want. Another explanation could be that the few roads available on the island have limited capacity, resulting in long traffic jams even in normal traffic conditions. Households may prefer to avoid driving during an evacuation for the discomfort associated with driving during an evacuation or to avoid being trapped in a traffic jam when the hurricane strikes (Lazo et al., 2015).

Our findings regarding homeownership indicate that tenants have a higher tendency to evacuate than homeowners (Hasan et al., 2011; Lazo et al., 2015). Owners may feel the "need" or "desire" to stay at home to protect the house during the storm or from looters (Baker, 1991; Lazo et al., 2010; Riad et al., 1999). Also, owners tend to do more regular maintenance on their house and hence feel more protected than in a public shelter or other destination. In contrast, tenants may not need the feel to properly maintain the household to a condition that withstands a hurricane force as they feel it is the owner's responsibility (Medina et al., 2019). This helps explaining why tenants on Sint Maarten evacuate more often. Furthermore, owners may not evacuate to avoid the discomfort and environment of public shelters (Baker, 1991), in contrast to some tenants who are low-income evacuees or undocumented immigrants for whom public shelters may be the main and sometimes the only place to evacuate to (Mesa-Arango et al., 2013).

Household construction material was found to be one of the strongest co-founders of evacuation behaviour on Sint Maarten. Households built with stronger materials (i.e. concrete walls and roof) tend to be less likely to evacuate than those living in houses built with weaker materials (i.e. wood). Perception of having a strong house was already reported in the literature to prevent people from evacuating. It has been reported that when households feel unsafe at their location and perceive their house as vulnerable to wind damage during a storm, this tends to increase their tendency to evacuate, and those who feel safe tend to stay (Baker, 1991; Lazo et al., 2015; Lindell et al., 2005).

Property damage for Sint Maarten was the predictor with the strongest statistical significance; it was found that those houses suffering the most (76-100% damage) were also the households that were more likely to evacuate. In contrast, households with lower levels of damage (0-25% damage) were found to be those less likely to evacuate. The expectation of damage or damage suffered in the past has been consistently reported as a good predictor of evacuation (Baker, 1991; Dash and Morrow, 2000; Gladwin et al., 2001; Whitehead et al., 2001). Prior research has found that when residents feel they or their relatives are at risk of death or injury, or that their house could face serious damage, they are more likely to evacuate.

Multi-storey residences on Sint Maarten tend to be less likely to evacuate than single-storey houses. It has been previously reported that multi-storey buildings have lower evacuation rates (Brown et al., 2016). Lower evacuation rates can be explained due to the possibility to look for higher grounds in the case of a flood event, or the possibility to protect valuables from potential floods. Besides, multi-storey buildings on the island are normally built with concrete which was explained above as a predictor of no-evacuation behaviour.

Alongside property damage and the number of house storeys as predictors of evacuation, it was also found that those households on Sint Maarten perceiving they are located in a flood or storm surge prone area are more likely to evacuate than those that reported they do not live in those areas. Houses having only a ground floor or located in flood-prone risk areas such as lowlands and the coastline are more vulnerable to flooding, property damage, and even suffering casualties in the past, which in turn is a predictor of future evacuation when a warning is received (Baker, 1991; Huang et al., 2012). It is also important to mention that risk terminology may be confusing for the general public, and sometimes those living in high-risk areas may not be fully aware that they are, creating confusion whether they need to evacuate or not under a possible threat (Huang et al., 2016). Furthermore, it will be necessary that emergency management officials on the island update the identification of areas prone to floods and storm surge, and perform awareness campaigns of the population at risk in these areas to prompt evacuation when needed in future evacuation scenarios.

Home insurance against disasters triggered by natural hazards was hypothesised as a precursor of evacuation, expecting that households with insurance will evacuate more as they feel they can leave the house and recover their losses through their insurance company. However, our results were contrary to this assumption, and insured households on the island tend not to evacuate. This finding can be a cofounder of homeownership, as presented in the correlation matrix (R=0.45). Homeowners are more likely to have home insurance (Huang et al., 2016), and on Sint Maarten homeownership was already explained as a strong predictor of non-evacuation.

We also found that the quality of the information that is sent in different phases of a disaster plays a role in household behaviour towards evacuation. Amongst the respondents of the survey, those that disagree with the statement that a more direct message will lead to complying more with evacuation orders are those that tend to be less likely to evacuate, and those that agree tend to be more likely to evacuate. Information content was also found to be an important predictor of evacuation behaviour in other studies (Baker, 1991; Huang et al., 2016; Huang et al., 2012). Prior studies have listed actions and information distributed by public officials as amongst the most important variables affecting the public response to evacuation. In addition, households are more likely to evacuate when they understand without question that an evacuation order applies to them, hence more custom-made ways of delivering the message may result in higher evacuation rates (Baker, 1991; Hasan et al., 2011).

The variable vulnerability was found to be statistically significant as a precursor of no evacuation for those households located in very low vulnerability index areas. Households on Sint Maarten associated with low vulnerability are normally those households with higher incomes, more education, stronger construction materials, more awareness of natural risk, and more possibilities to take immediate action to protect themselves against a natural disaster (Medina et al., 2020).

In addition, nine variables did not play any significant role in explaining evacuation intentions in the case of Sint Maarten. Variables associated with risk and vulnerability in our study were used given their strong positive effects found in prior research; these are the length of residence, hazard awareness, previous hurricane experience, level of worry, risk perception, and government performance. However, none of these variables offered a major influence on evacuation behaviour on Sint Maarten. One possible reason for this is that respondents may evaluate their risk and vulnerability with a more tangible measure such as house construction materials, or actual or expected damage from hurricanes.

5.4 Conclusions

This study explores the relationship between several variables with evacuation behaviour in the context of a small island developing state (SIDS) that was devastated by a hurricane. Hence, the findings and evidence provided in this research are valuable to understand what the variables are that can be used as predictors of evacuation in such a context.

Most of the findings of this study are consistent with the hypothesised effects. In this regard, gender, homeownership, a house's construction material, property damage, quality of the evacuation information, number of house storeys, perception of living in a flood-prone area, and vulnerability index were found to be influential factors. Car ownership and home insurance for disasters were also found to be statistically significant but with contradictory effect to that expected.

We found that on Sint Maarten people are most likely not to evacuate when a hurricane is forecast. This could be partially explained because the majority of residents do not fully rationalise the magnitude and potential consequences of a major hurricane. Several storms hit the island between the last major disaster in 1995 associated and Hurricane Irma in 2017, storms and hurricanes in which residents were relatively safe and no substantial losses were reported, creating a false sense of security and limiting households' willingness to evacuate. In addition, an evacuee under (almost) certainty of exposure to an upcoming threat, as in the case of Sint Maarten, may feel more comfortable or safe staying in their own house. This statement is not valid if the evacuee feels (or knows) their house is not strong enough or if they perceive that they live in a flood-prone or storm surge area according to our results.

People's perception of how strong their houses are in combination with past or expected damage assessment were found to be the strongest indicators of evacuation behaviour on Sint Maarten. Therefore, assessment of a household's perceptions of the structural vulnerability of their houses to disasters triggered by natural hazards will be beneficial to estimate future evacuations behaviours. This is important after Irma because we observed the ongoing reconstruction of houses across the island during the fieldwork, which may create a (false) sense of security for those houses that were rebuilt after the disaster.

If a major hurricane is forecast in the coming years, before the memory of Irma fades, the government of the island should be prepared to open public shelters with enough resources, as we forecast an increasing demand for facilities. The expected increase in number of evacuees is related with new evacuees due to the fear of a new disaster, also evacuation to hotels may decrease due to the extensive damage to the infrastructure in this sector, and also another segment of the population will always need a place to shelter: those with low income and undocumented immigrants that generally live in high-risk areas and in homes with poor construction materials. In contrast, there is a segment of the population that will never evacuate no matter how big the threat is, as long as they feel they are safe in their own houses or feel they need to stay to protect the house or to prevent looting. The government should be ready to assist those that may need immediate relief and assistance in the case of a major disaster.

Amongst the most important predictors of evacuation on Sint Maarten is the proper distribution of warning information according to our regression models of evacuation. This is an important finding for Sint Maarten disaster risk management because amongst the predictors we found statistically significant in this study, information and some components of the vulnerability index are the only ones that may be possible to influence directly without the need of investment in expensive infrastructure at the household or island level. The message content must reflect the need to evacuate; often official orders are misunderstood as advisory and may lead to non-evacuation.

6

AGENT-BASED MODELS FOR WATER-RELATED DISASTER RISK MANAGEMENT

Disasters triggered by natural hazards associated with climate change are increasing in severity and frequency. To address this challenge, Disaster Risk Management (DRM) has been evolving over the last decades towards a more holistic approach for water-related disasters such as floods, hurricanes and storm surges. Complex Adaptive Systems (CAS) and Agent-Based Modelling (ABM) have shown great potential to be used in DRM as it offers a holistic approach because those techniques can represent the interactions between the different actors involved and also the interaction with the environment. This chapter presents and discuss a comprehensive literature review on the use of ABM for water related DRM (WR-DRM). We have performed a systematic search to be able to compile the most relevant literature published in the field. An extended literature review on ABM was needed because this modelling tool will be used to incorporate the adaptive behaviour of humans into the exposure component of the proposed adaptive disaster risk assessment of this thesis (ADRA).

This chapter is based on:

Medina, N., Abebe, Y.A., Sanchez, A., Nikolic, I., Vojinovic, Z., 2021. Agent-Based Models for Water-Related Disaster Risk Management: A state-of-the-art review. WIREs Water. Under Review

6.1 INTRODUCTION

As stated in the introductory Chapter 1, disaster risk management needs to recognize and incorporate the interconnections between hazards, infrastructures, economic systems and the role of human factors in assessing and managing the risk. Incorporation of the above mentioned elements into DRM presents a large and evolving challenge that requires a more active engagement and innovative solutions by policy makers to better understand and be able to have a more effective management of existing and future risks.

To increase our understanding of the natural system and the ability to predict or observe the potential effects of DRM, complex adaptive system (CAS) theory and agent-based models (ABM) provide concepts and promising modelling tools (Mustapha et al., 2013). For example, ABMs have been used to model city evacuation under threat of a flood or hurricane, flood-related policy implementations, as a tool for awareness education to disasters triggered by natural hazards, to model resources distribution after a disaster, among other applications of interest. Despite the increasing popularity of ABM for DRM, a literature review in the intersection of both disciplines is still lacking.

This chapter aims to provide a systematic overview of applications of ABM for DRM, and to provide a starting point for further researches that works in both: ABM and DRM by reviewing current achievements and open challenges in ABMs applied in the context of DRM in water-related disasters. We specifically focus the review only on disasters associated with floods, tsunamis and hurricanes. The intention with our critical review is to report the main fields where ABM has being used for DRM, their main contributions to DRM as a field of study, the main methodological challenges, strengths and weaknesses of ABM for DRM. Also, we present some identified knowledge gaps, methodological issues and suggestions to enhance ABM applications as a novel tool in DRM, and we offer some recommendations and future direction.

6.2 THEORETICAL BACKGROUND

6.2.1 Complex Adaptive Systems

Complexity theory is a relatively new field that began in the mid-1980s at the Santa Fe Institute in New Mexico. The work at the Santa Fe Institute is usually presented as the study of Complex Adaptive Systems (CAS) (Miller and Page, 2007). The definition of CAS must start with the definition of its components or words. According to (Chan, 2001), something is "Complex" if results from the inter-relationship, inter-action and inter-connectivity of elements within a system and between a system and its environment. In the same way, systems in the context of CAS is defined as a set of interacting and interrelated elements that act as a whole, where some pattern or order is to be discerned (Van Dam et al., 2012). Last, "Adaptive" should be understood as a property of the individual and/or the system to adapt over time with a form of memory or learning (Crooks and Heppenstall, 2012). Given these definitions, it is possible to give a more comprehensive definition of CAS, as cited in Waldrop (1992), where John H. Holland defines CAS as:

"[...] a dynamic network of many agents (which may represent cells, species, individuals, firms, nations) acting in parallel, constantly acting and reacting to what the other agents are doing. The control of complex adaptive systems tends to be highly dispersed and decentralised. If there is to be any coherent behaviour in the system, it has to arise from competition and cooperation among the agents themselves. The overall behaviour of the system is the result of a huge number of decisions made every moment by many individual agents."

Given these definitions, a good model of CAS should be able to have the following characteristics: Multi-domain and multi-disciplinary knowledge, generative and bottom up capacity, and adaptivity (Van Dam et al., 2012). Following the general criteria for a good CAS model, Van Dam et al. (2012) presents a short review of the many modelling techniques available, amongst all the presented models, some of them most notable ones with potential application in DRM are: General equilibrium, system dynamics, discrete event simulation, and Agent-Based Modelling.

From the above, Agent Based Models (ABM) are considered to be the most suitable tool for modelling CAS, as presented in Van Dam et al. (2012), based on the work of Borshchev and Filipov (2004) (From System Dynamics and Discrete Event to Practical Agent Based Modelling). The main arguments for this assertion are that ABM enables the capture of more complex structures and dynamics, they provide for the construction of models in the absence of knowledge about the global interdependencies and Agent-Based modelling is most suitable when the system to be studied exhibits adaptive behaviour (Chan, 2001).

6.2.2 Agent-Based Models

There is no universal agreement on the precise definition of Agent-Based Models, although definitions tend to agree on more points than they disagree (Macal and North, 2009). ABM characteristics are difficult to extract from the literature in a consistent and concise manner, because they are applied differently within disciplines. Furthermore, the agent-based concept is rather an approach more than a technology, where a system is described from the perspective of its constituent parts (Castle and Crooks, 2006).

Agent-Based Models are founded on the notion of CAS that when modelling a system the result is not only more than, but very different from the sum of its parts (Miller and Page, 2007). To manage such systems, the systems or organizations must be understood as collections of interacting components. Each of these components has its own rules and responsibilities (Macal and North, 2009).

A good description of how ABM should work is presented in (Castle and Crooks, 2006), ABM is defined as models that contain multiple and interacting agents situated within a model or simulation environment. A relationship between agents must be specified, linking agents to other agents and/or other entities within the system. Relationships may be specified in a variety of ways, from simply reactive, to goal-directed among others. In addition, the behaviour of agents can be scheduled to take place at the same time or asynchronous. In addition, a good ABM also needs to define the environment, this is the space in which agents operate, serving to support their interaction with the environment and to other agents. Basically, an ABM consists of four elements: the set of agents; the set of agent's relationships, methods of interaction, and the agents' environment (Macal and North, 2010).

Regarding the benefits of using ABM or models where agents adapt their behaviour, is that they are very useful in the exploration of complex adaptive social systems, such as those present in DRM. From a modelling perspective, the use of adaptive agents provides the means to create models that can explore new domains of agent behaviour and complex outputs that exceed the usual constraints imposed by the modeller (Miller and Page, 2007) (Miller and Page, 2007). In addition Agent-Based models are useful as methods of getting deeper understanding of system characteristics, and exploring various institutional arrangements and potential paths of development to assist decision and policy makers (Billari et al., 2006).

6.3 Materials and Methodology

In this section, the methodology used for the systematic literature review is presented. It is based on a widely used review protocol proposed by Kitchenham and Charters (2007). The protocol includes objectives of the literature review, search and evaluation strategies, inclusion/exclusion criteria, data collection form and analysis.

6.3.1 Scope of the Review

Following the protocol, the first step is to set up the goals of the literature review. The objectives were defined as follow:

- to identify and classify the main application domains in which ABMs have been used to study water-related DRM
- to identify achievements and open challenges in ABMs applied for DRM, especially regarding model implementation and analysis

- to identify trends and commonalities among the different implementations of ABMs in DRM, and
- to present and discuss identified knowledge gaps and suggestions to advance the application of ABMs in DRM

Since the application domain of DRM can be vast, it is important to define the scope of the applications within this thesis. We limit the review only to papers that applied ABMs for disaster risk management concerning water-related hazards, specifically floods, tsunamis and hurricanes, and only those applied at a city or a regional scale. The type of hazards delimitation was based on the share of occurrence of different disasters triggered by natural hazards, as described

6.3.2 Search and Evaluation Strategy

The search strategy aimed to identify the primary and secondary studies related to the scope of the review. The search was conducted in two major citation databases – Scopus and Web of Science. Scopus database can be considered as the largest searchable citation and abstract source of literature (Bar-Ilan, 2007). Web of Science, on the other hand, is known as one of the world's leading scientific citation search engine and for searches in highly ranked journals, offering a broad search across multiple disciplines (Li et al., 2018). We decided not to include Google scholar as it is not possible to limit the results to peer-reviewed material; it is not clear the relevance of the results presented by the search, and it is subjective to the researcher to define how many pages to include as too many results pages may be offered.

We conducted the search using a Boolean expression that includes two sets of strings. The first string set defines the domain of interest, which for this review is agent-based modelling technique. However, given that different authors use slightly different wording, we have included in the search the most-commonly used alternatives, such as removing the hyphen (-) from the string. Alternative names such as multi-agent and individual-based modelling are also included in the first string set as they refer to the same technique as ABM. The second string set restricts the search within the scope of the article. Below, it is presented an example of the queries that were used for the search in the Scopus database, and in Table 6.1 is presented the complete set of search terms.

(TITLE-ABS-KEY ("Agent Based Models" OR "Agent Based Modelling" OR "Agent Based Modeling" OR "Agent Based Simulation" OR "Agent-Based") AND LANGUAGE (English) AND TITLE-ABS-KEY ("Flood" OR "Flooding")) AND DOCTYPE (ar OR re) AND PUBYEAR > 1997 AND PUBYEAR < 2020

Also, we imposed other restrictions on the search. Firstly, the string words were searched and limited to be present only on the title, on the keywords or in the abstract. Secondly, the document type was limited to include only research and review articles. Therefore, conference papers and books were excluded from the analysis as it is difficult to be sure whether a conference paper or a book is peer-reviewed. Thirdly, the search was limited to publications only in English, as this is the dominant language of international science (de Sherbinin et al., 2019). The final restriction was on the year of publication.

Table 6.1. Search terms used in the systematic literature review.

First string set (OR)	Boolean	Second-string set
"Agent Based Models" "Agent Based Modelling" "Agent Based Modeling" "Agent Based Simulation" "Agent-Based" "Multi Agent" "Multiagent" "Multi-Agent" "Individual Based" "Individual-Based"	AND	"Flood" OR "Flooding" "Hurricane" OR "Cyclone" "Risk analysis" OR "Risk Assessment." "Evacuation" "Disaster management" OR "Disaster-management" "Drainage" "Hydrology" "Early warning" OR "Early-warning" "Tsunami"

In addition, the search was limited to articles published in the period from 1 January 1998 to 15 November 2019 based on the results of a broad search of the term Agent-based model in the Web of Science. Figure 6.1 illustrates that ABM publications have very few records before the year 1998.

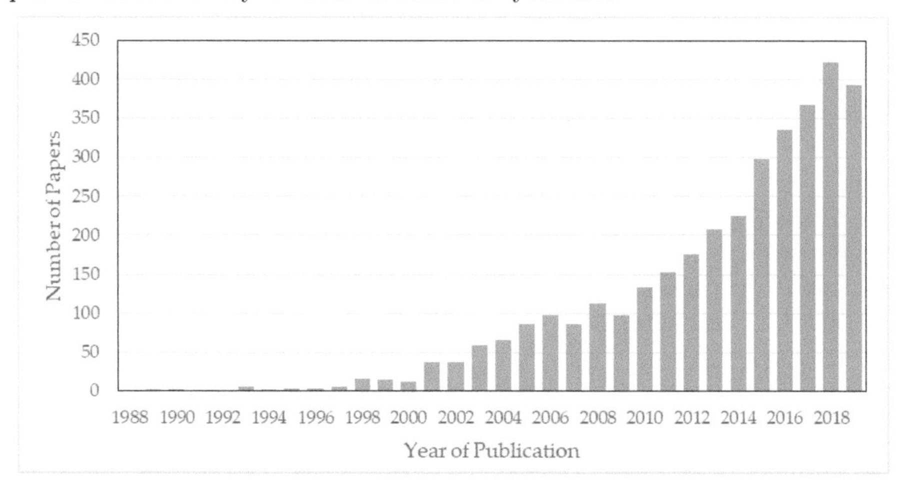

Figure 6.1. Number of publications with the words Agent-Based Model in the Title[4])

[4] Created using Web of Science Portal : http://www.webofknowledge.com

6.3.3 Selection and Exclusion Criteria

Initially, we found a total of 2125 papers using the queries presented in Table 6.1, 1351 papers were found in Scopus and 774 in Web of Science. After removing the duplicates across queries and the databases, the list was reduced to 1515 accounting for the 610 replicated papers. The remaining 1515 papers were screened for relevance based on the title and the keywords. We eliminated 1226 papers that were out of the scope of the current paper, including ecological modelling, disease spread, land-use change, transport optimization, supply chain, livestock evacuation of buildings in enclosed spaces, and other natural or man-made disasters such as earthquake, volcanic eruption, fire and terrorist attack. The high number of discarded papers in this step can be explained with the initial broad search we performed to make sure that we capture the most relevant studies to our objectives.

The remaining 289 papers were subject to a more detailed evaluation based on the contents of the abstract of each article. The relevance of the papers was assessed based on the subjectivity and experience of the research team. We further eliminated 174 papers mainly associated with the scale of application of the ABM or dealing with different disasters triggered by other type of natural hazards (i.e., different from floods, tsunamis and hurricanes). From the remaining 115 papers, two articles were not available; despite the queries restriction, one paper was not written in English; and four papers were either book chapters or conference proceedings. Thus, eliminating other seven papers, for a total of 108 relevant articles.

In addition, we extended the systematic review by including seven additional papers that were known by the research team with relevance to the objectives of this paper but were not found in the databases despite our broad and systematic search. Finally, a total of 115 articles were selected to be included in this extended, systematic literature review. While our search may have missed some potentially relevant literature, we believe that the systematic selection assures us to have the most representative papers concerning our objectives. The sample size is also sufficiently large and representative to be able to assess the objectives mentioned above. The list of the 115 final papers that were included is provided in Appendix D.

6.4 Results

We divide the analysis into two main components. First, we present some general characteristics of the reviewed papers in order to set the grounds for further analysis. Second, we analyse the selected papers for specific insights into the applications of ABMs in WR-DRM.

6.4.1 Characteristics of the reviewed papers

Publication Year

To explore how ABMs have been penetrating the world of water-related DRM (WR-DRM), Figure 6.2 presents the number of publications per year. In comparison with the general use of ABMs in the water sector (Figure 6.1) the use of ABMs in WR-DRM dates back to 2004 (in contrast with 1998) – only a couple of papers per year were found up to 2012. Starting in 2012, we can observe that there is a pronounced rising trend or interest in using ABMs to address the complex challenges of WR-DRM. The number of papers that applied ABMs for water-related DRM reached 25 in 2019, more than triple that of 2012.

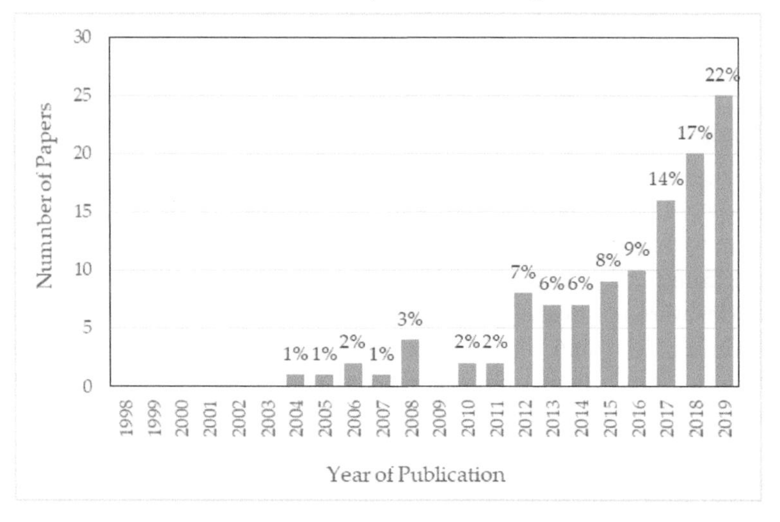

Figure 6.2. Number of papers selected for the systematic review per year of publication.

Geographic Coverage

The selected papers cover a wide range of geographic regions and countries (Figure 6.3). North and Central America (33.6%) and specifically the USA is the leading continent and country, respectively, in research and applied studies of ABM for WR-DRM. Europe follows accounted for 20.5% of the papers, where the UK is the country with the most papers on the continent. Eastern Asia (14.5%) is the third continent in the number of papers; where Japan and mainland China having the highest numbers. South America, with cases in Chile, is also well represented. Only Africa has non reported application in the studies founded.

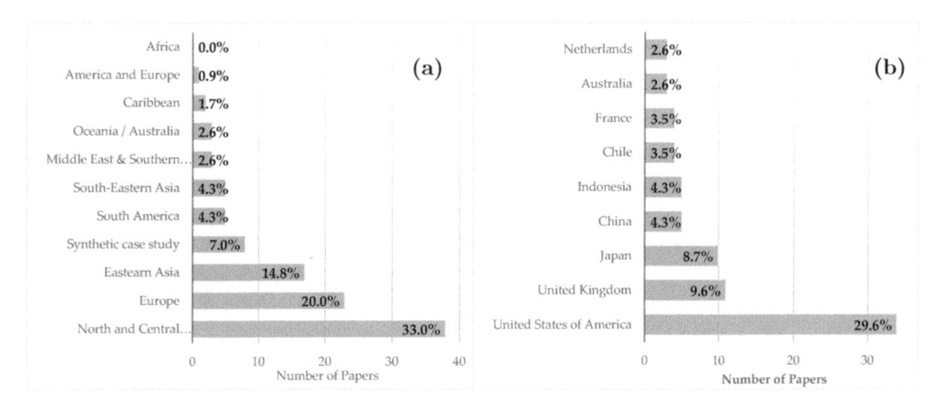

Figure 6.3. Geographic coverage of the application of ABMs in water-related DRM. (a) by sub-continent. (b) by country (top nine countries)

Software used

This section presents a summary of the main tools or software used to model Agent-based models within WR-DRM. An extensive review of the software used in the reviewed papers is out of the scope of this review paper. For a full review of the state of the art software to build ABMs, see Abar et al. (2017). From the selected papers in this review chapter, 23.5% implemented the ABMs in NetLogo (Figure 6.4). Whereas, 12.2% of the papers do not report which software they used. The second most used software in the reviewed papers is Repast Simphony (8.7%). We found seven papers that did not implement the ABMs (6.1%), these papers of theoretical value were used mainly by projects in the early stages, and only the conceptual framework is presented.

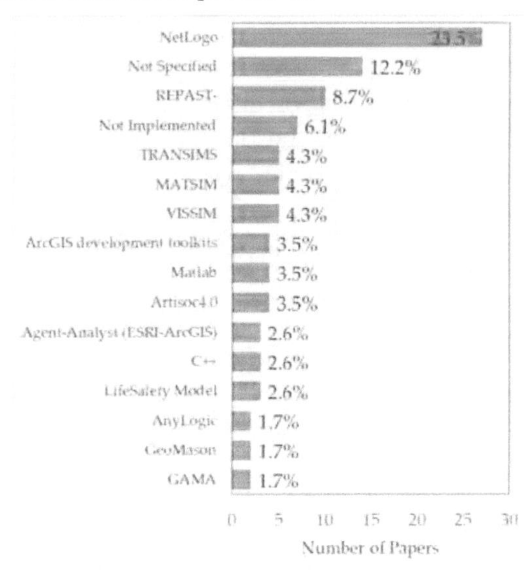

Figure 6.4. List of software used to implement ABMs in WR-DRM. The figure only shows the software that was used at least in two papers.

Regarding ABM software that was applied for traffic and evacuation scenarios, there are three leading software, each with five implementations (4.3% each). The first two, TRANSIMS and MATSIM, are open source and free software with proven capabilities to perform large city evacuations at a microscopic level by incorporating individual behaviours. They are also capable of assessing the impact on the traffic network from specific hazards such as hurricanes and floods. The third leading software in traffic modelling for WR-DRM is the commercial software VISSIM.

6.4.2 Critical review of Applications of ABM in WR-DRM

The scope of this section is to compare and document the current practices among the ABM modellers in the WR-DRM community. We focus on some of the main methodological challenges or concepts when using ABM as a modelling tool of social complex systems, such as problem formulation or purpose, the implementation goal, the design and parametrization of agents and which data sources are being used for this purpose. We also report on challenges related to verification and validation of ABMs in DRM, as well as the role and use of sensitivity analysis to make more robust the results and analysis of these type of models. We also reflect on the spatial scale and time scale representation of the models. In the following sections of this chapter, we will critically address how the reviewed papers handled the above-mentioned modelling challenges.

Model Purpose and Type of Implementation

A good practice for ABM modellers should be the definition of whether or not this is the appropriate modelling technique followed by the proper definition of the purpose of the model (Manson et al., 2020; Schulze et al., 2017). A standard categorization of the model purpose in ABMs is whether it is intended as a descriptive or as a predictive model. Predictive models have a general purpose of demonstrating and exploring ideas, and they are applied for testing hypotheses, whereas descriptive models are built to provide decision making, policy formulation and management support (Edmonds et al., 2019).

It is expected that ABMs with a predictive or explanatory purpose have a more realistic model set up of the environment, as they usually tend to solve case-specific issues, and the impacts on the real world from the simulated results can be observed and analysed directly. In contrast, the environment representation of a descriptive ABM model can be of less importance as their primary objective is to gain understanding on the system or extracting theories or general principles (Schulze et al., 2017), and as such less environment details are usually present, or even space is presented as a simplification in a stylized way. In our review, we recognized that a specific ABM might intend to be both, descriptive and predictive. In such a case, we assigned the category that we believe fits the purpose better. Also, we included an extra category for those papers that can be classified as review papers. Figure 6.5 illustrates the classification of the papers within the three categories.

We found that the majority of ABMs were predictive type (82.6%) and were designed as a model of a specific real system using observations or data. Predictive ABMs were applied to explore a wide variety of water-related DRM issues, including the effects of policy and management strategies implementation (e.g. Abebe et al., 2019a; Haer et al., 2019; Löwe et al., 2017), the role of spatial planning in risk management (e.g. Chandra-Putra et al., 2015; McNamara and Keeler, 2013; Mustafa et al., 2018), resources allocation (Manzoor et al., 2014), evacuation simulations focusing on traffic modelling (e.g. Dawson et al., 2011; Liang et al., 2015; Liu and Lim, 2018), evacuation modelling focusing on resources allocation (Chen et al., 2006), loss of life or infrastructure assessment (e.g. Coates et al., 2014; Naqvi and Rehm, 2014; Takabatake et al., 2018; Yang et al., 2018), risk insurance (e.g. Dubbelboer et al., 2017; Filatova, 2014), tsunami and city evacuation modelling (e,g. Cheff et al., 2019; Imamura et al., 2012; Mas et al., 2013) and risk communication during a disaster (e.g. Du et al., 2017a; Nagarajan et al., 2012; Rand et al., 2015; Watts et al., 2019).

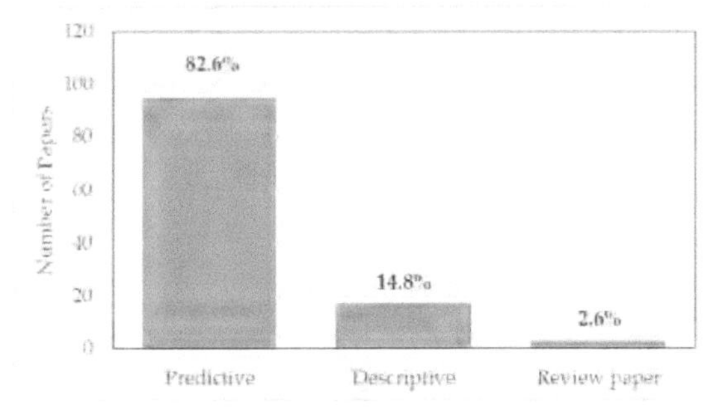

Figure 6.5. Number of papers based on model purpose.

On the other hand, descriptive ABMs accounted for only 14.8%. Similar to the predictive ABMs, the descriptive models were also utilized for different purposes such as implementations of decision support systems for disaster management (e.g. Eid and El-adaway, 2018; Ghavami et al., 2019), flood preparedness and mitigation (e.g. Coates et al., 2019; Li and Coates, 2016), policy analysis (e.g. Lipiec et al., 2018; Naqvi and Rehm, 2014) and for educational purposes (e.g. Becu et al., 2017; Sun and Yamori, 2018; Taillandier and Adam, 2018).

Model type distribution (Figure 6.6) shows an important characteristic of ABMs for WR-DRM- modelling case studies are predominant. A total of 90 papers (78.3%) are directly implemented to a case study with or without presenting a framework (last two bars on the graph). In addition, no generic models were found in our literature review. In terms of developing a framework, we found 26 ABM papers, of which 17 were applied directly in a case study and seven papers present only the theoretical framework. Regarding the use of synthetic or dummy case study, we found 11 papers (9.6%).

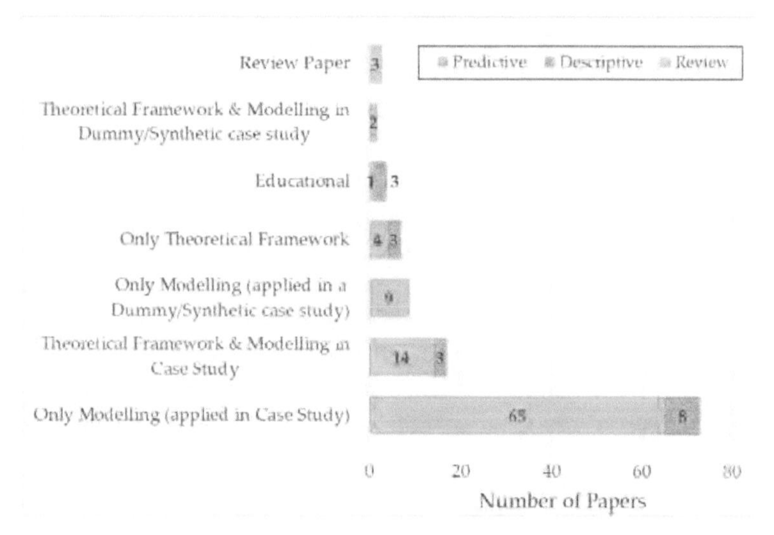

Figure 6.6. Number of papers based on the type of application of the ABM and categorized by model purpose.

Disaster Phase and Category

Traditional approaches to disaster management consider three major phases: *Before*, *during* and *after* the onset of the hazard. The definition of these phases is crucial for any research in DRM because people's cognitive and emotional states and behaviour will vary significantly across the phases. Hence, the design, parametrization, and purpose of an ABM could be different. The success or effectiveness of any behavioural model implemented in the ABMs relies on a good and solid understanding of the behaviour in each one of these phases. Equally important is to emphasise that failing in the characterization of behaviour can put the whole model at risk. Additionally, based on psychological sciences literature (Dos Santos França et al., 2012; Vorst, 2010), a new instance to the evolution of the disaster phases can be added – a *warning* phase. That is a moment between before and during disaster phases, when an individual becomes aware that there is a probability to encounter a hazardous situation.

The selected papers were categorized into distinct categories according to the different phases of the disaster cycle (prevention, preparedness, response and recovery) (Wehn et al., 2015). We include in a separate category those papers focus on policy for disaster management and other for the review papers. Table 6.2 shows the different categories and which disaster phase they may influence. Some categories affect one or more of the described phases. It is expected that a good parametrization of an ABM has a clear distinction of agents' behaviours as the disaster phase unfold. For example, in an ABM that focuses on evacuation processes, agents should have different behaviours in the three phases the model may cover, which are *before*, *warning* and *during* phases.

Table 6.2. Categorization of the selected papers in the different phases of a disaster.

Category	Disaster Phase			
	Before	**Warning**	**During**	**After**
Pre-disaster – Prevention and preparedness	X	---	---	---
Response – Evacuation	X	X	X	---
Response – Mitigation		X	X	---
Recovery	---	---	---	X
Policy / Disaster Management	X	---	---	X
Review paper	---	---	---	---

The result of the categorization is shown in Figure 6.7. Evacuation is by far the most representative type of studies presented in the selected papers, accounting for 50.4% of the total. Second in frequency is the use of ABMs for policy and disaster management, with 26.1%. The other categories are all underrepresented or less explored in our sample of papers having less than 10.0% of the publications. Concerning the review papers, we only found three papers. One of them focuses on advances of ABM in tsunami evacuations. The second paper discusses hydrological problems in flood-related disasters, while the third, also consider a perspective paper, reviewed the role of human behaviour in flood-related disasters.

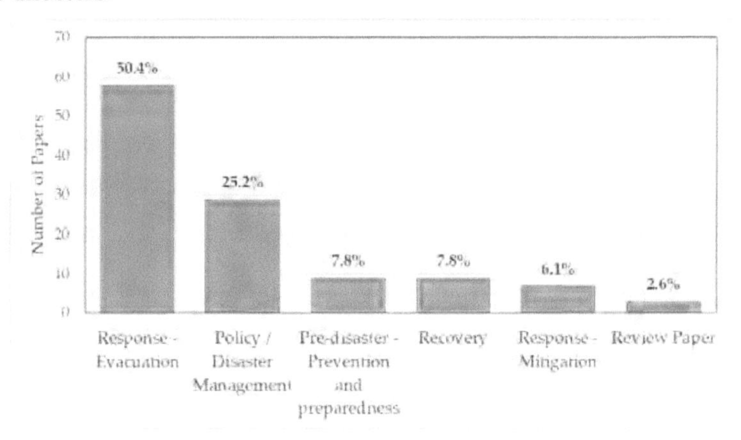

Figure 6.7. Number of papers based on the phase of the disaster cycle.

Model Documentation

Proper documentation of an ABM model is necessary in order to increase readability and reusability of the model. It is intended to provide supporting information on the model. The first way of documenting the model is by a proper model description. A good model description should include all the relevant information such as the purpose of the model, how it works, how it was designed and how decisions and behaviours (rules) are performed. It is advisable or good practice that the model description is prepared following a protocol as it helps modellers to communicate and readers to understand the models better. Protocols will also enhance the model development as other researches can see clearly what was done and build on top of an existing ABM.

Some of the ABM documentation protocols that have been reported in the literature include the TRACE framework (Schmolke et al., 2010), the Overview, Design concepts, and Details (ODD) protocol (Grimm et al., 2006; Grimm et al., 2010) and the upgrade extension ODD+D, which is used for describing human decision-making (Müller et al., 2013). Because human decision making is crucial and the central role in an ABM for DRM the ODD+D protocol seems to be the most appropriate standard to follow.

In our review, we found that documenting and sharing the model design and description is not a common practice (Figure 6.8-a). More than half of the publications (54.8%) did not use any standard protocol. Instead, models were just described using narratives. The ODD protocol was reported in 10 papers; this is 9.0% not including the four papers that were published before or in 2006 (year of publication of the protocol). ODD+D was described in only three papers (2.9% of the papers published after 2010).

It was also found that 15 papers that used industry or proprietary software (e.g. Anylogic, VISSIM, Life safety model), tend not to report either the model description, not the rationale behind human choices; instead, they report using the standard values in the model set up.

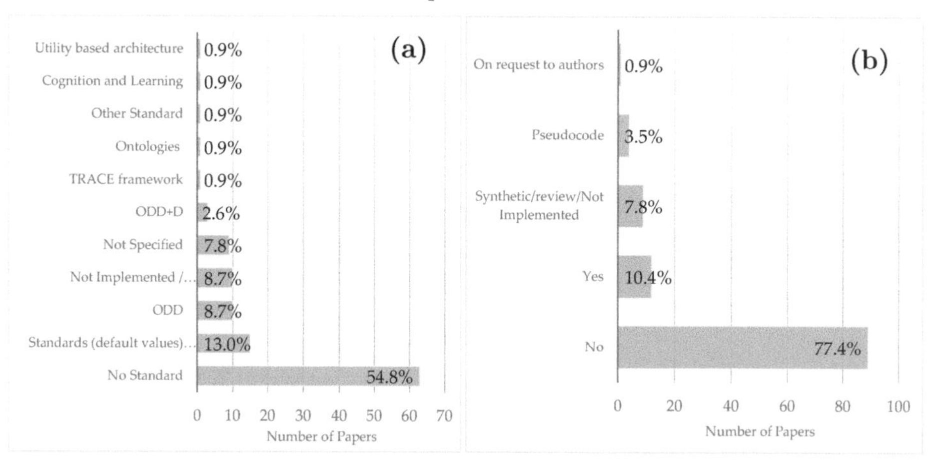

Figure 6.8. Number of papers based on (a) the type of model description and (b) based on source code availability.

Another way to improve model documentation is by sharing the source code. In terms of code sharing, it is evident that this is not yet a common practice among ABM modellers (Figure 6.8-b). Only 12 papers (10.4%) included the original code to be tested, adjusted or reused. Four papers (3.5%) shared the pseudocode using the ODD protocol, and from these, only one paper directly expressed the possibility to share the code on request by emailing the corresponding author. Those researchers that decided to share the code used mainly two repositories, CoMSES-OpenABM[5] and Github[6] with seven and three publications respectively.

Representation of Agents and data used to set up the model

- ***Agent representation***

The concept of an agent in an ABM model is just a simplification of the reality to represent components of a system. Agents act as autonomous entities that are capable of performing actions according to a set of rules and without the direct user intervention, that is able to interact with other agents and the environment and act (change state) based on these interactions (Abar et al., 2017; Nikolic and Dijkema, 2010). In complex systems where humans are at the core, such as is the case in social or ecological systems, the agent representation in the ABM model is not trivial and of major importance.

In our review of papers, agents in disaster risk management can take one of three forms, either as (i) individuals or (ii) the groups they form or (iii) institutions that can influence those individuals or communities. Agents were categorized either as individuals or as composite agents based on an adjusted definition presented in Ghorbani (2013). We consider agents as individuals if they represent single human, characterised by individual decision-making, such as pedestrians or tourists in evacuation models. In contrast, composite agents are those that represent a collection of individuals. We assumed two levels of these agents. Low-level composite agents represent a group of individuals that get together in the simulation and act as one unit, i.e. cars, households, housing developers, companies; and high-level composite agents represent institutional actors such as government entities, housing market, NGO's and insurance companies.

ABMs that consider individual agents and composite agents in the same model are expected to be more complex to conceptualize and program. The higher complexity can help explain why the majority of the papers we reviewed included only one type of agent, either individual agents (33.9%) or low-level composite agents (30.4%) as illustrated in Figure 6.9. Papers that model multiple types of agents are a minority in our review. For example, the papers that included all the

---

[5] CoMSES: Computational Modeling in Social and Ecological Sciences https://www.comses.net/about/

[6] https://github.com/

three types of agents and those that included both types of composite agents account for 6.1% and 5.2% of the total reviewed papers, respectively.

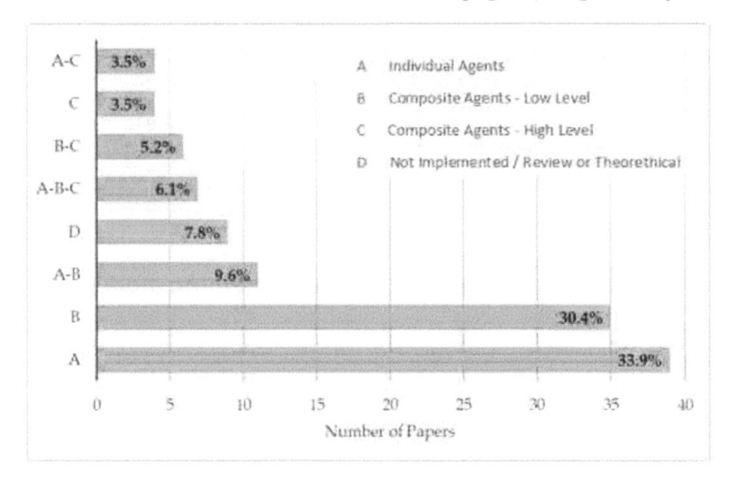

Figure 6.9. Number of papers based on types of agents in the ABM.

- ***Data for model set up and Agent parametrization***

How humans and their institutions are represented in an ABM will define the type and source of data needed to parametrize the model. The flexibility of an ABM is that it allows using both quantitative or qualitative data (Schulze et al., 2017). Examples of quantitative data are census data, geographic information systems (GIS) information and result from mathematical models. Qualitative data can be obtained from surveys, interviews, ethnographic fieldwork and desk studies. In Figure 6.10, we present the main sources of data reported in the papers to set up the ABM models.

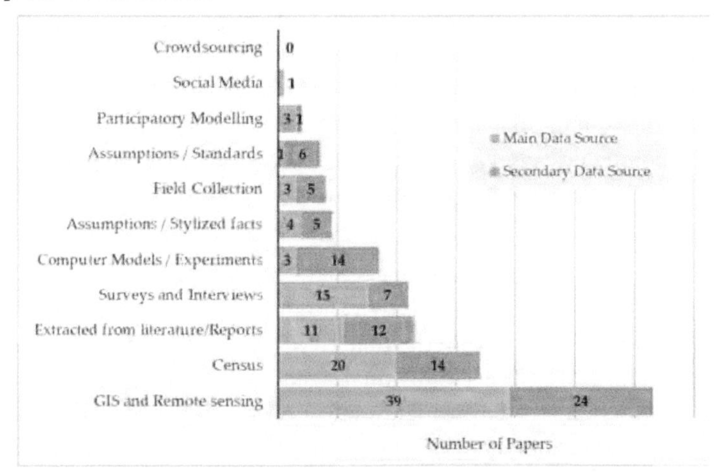

Figure 6.10. Number of papers based on the main and the secondary data source used to parametrize or run the ABM model.

From our review and the type of data used in the model, it seems that modellers of ABM for WR-DRM focus more on the description of how the environment is represented. GIS and computer models outputs representing the hazards (i.e. flood extension and duration) are the main concern of modellers while reporting data used to set up the model (environment representation). In 63 papers (31.0%), it was reported and extensively documented GIS or remote sensing data as the primary or secondary source of information to represent the physical space where the ABM takes place. In 17 papers (8.4%), the computer model to represent the hazard is extensively documented and described.

Regarding data used directly to parametrize agents' behaviour, census data is the predominant source with 34 of the selected papers (16.7%) using it as a primary or secondary data source to build the models. Census data has been widely used and accepted in ABMs due to its availability. Information extracted from literature is also one of the primary sources to parametrize ABM in WR-DRM, with 11.3%. Concerning using explicit behavioural data collected for parametrization of agents' attributes, surveys and interviews were used in 22 papers while social media (i.e. Twitter) was used in one paper (Rand et al., 2015).

Agent's decision making process

One key to the success of an ABM developed for WR-DRM studies is the way the agent's decision making is represented in the model (Yang et al., 2018). It is widely accepted and recognized that human decision making is very complex, but the way ABM models are built is instead a simplification (Balke and Gilbert, 2014). Modelling human decision making in ABMs should be able to encompass the properties and behaviour of the actors for specific goals of a particular model. With this in mind, one cannot categorize a human behavioural model as good or bad, just as if it is enough to represent the behavioural patterns of the agents modelled.

In ABMs for WR-DRM, humans at the centre of the decision-making process can be described using different theories and modelling architectures. For example, Balke and Gilbert (2014) defined the architectures in five main categories: (i) Production rule systems – a set of behavioural rules described using "if-then-else statements". Every rule will determine a particular agent behaviour based on the inputs from the model. (ii) Belief-Desires-Intention (BDI) and its derivatives – this architecture incorporate a "mental state" as the source of agents' rationality, which allows agents to be both reactive and able to "think" about intentions and modify them if required. (iii) Normative models – models that allow representing social norms into the decision-making process. These are agents' behaviours that are motivated by external rules, which can have a direct influence or limit agents' responses. (iv) Cognitive models – architectures that were created based on the cognitive structures, referring to a range of mental processes involving the acquisition, storage, manipulation and retrieval of information. (v) models based on psychology and neurology – they differ from cognitive models as these models take into account the presumed structural properties of the human brain.

In addition to Balke's categorization, other commonly used representations of human decision making include mathematical formulas or probabilistic rules. Also, a modified version of production rule systems has been broadly used in ABMs intended for evacuation purposes using an origin and a destination to each agent in the model.

In Figure 6.11, we present a categorization of how decision making is represented in the reviewed papers. From the 115 papers, 40 represented human behaviour using mathematical expressions. The other two commonly used decision-making methods are evacuation origin-destination model (35 papers) and production rule systems (15 papers). BDI and its derivatives were only used in four papers (3.5%). Participatory agent-based modelling in which real people directly tell the model what the agent would do under certain conditions was used in three papers (2.6%). Other architectures are under-represented or not used at all. In terms of evacuation modelling, we found that most of the papers (35 out of 58 papers) use a rather simplistic approach based on origin and destination modelling, by using the shortest path approach computed with the Dijkstra's algorithm (Dijkstra, 1959).

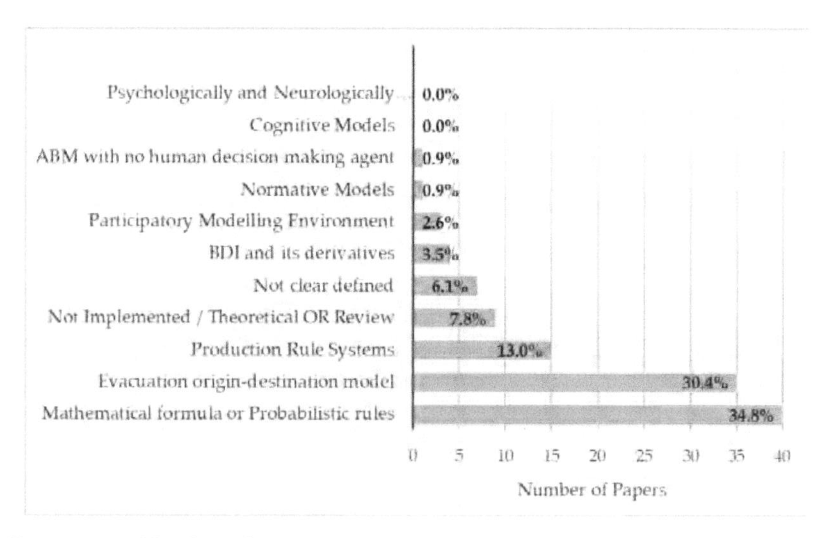

Figure 6.11. Number of papers based on the main architecture used to parametrize the ABM model.

Model Verification and Validation

Verification and validation are two terms widely used in computer simulations. Often misrepresented and used interchangeably, although they refer to two different moments on the modelling processes (David, 2006). Verification is a process by which a modeller investigates whether a computational software implementation correctly represents a conceptual model and that equations are solved correctly in the computer programme. In contrast, validation refers to the assessment of the degree to which simulated results mimic the real world that the model is intended to represent (Ormerod and Rosewell, 2006).

Verification processes in the selected papers (Figure 6.12) show a trend that researchers using ABMs for DRM do not give attention to performing verification (four papers), or they do not see the added value of reporting this step in the modelling process (93 papers).

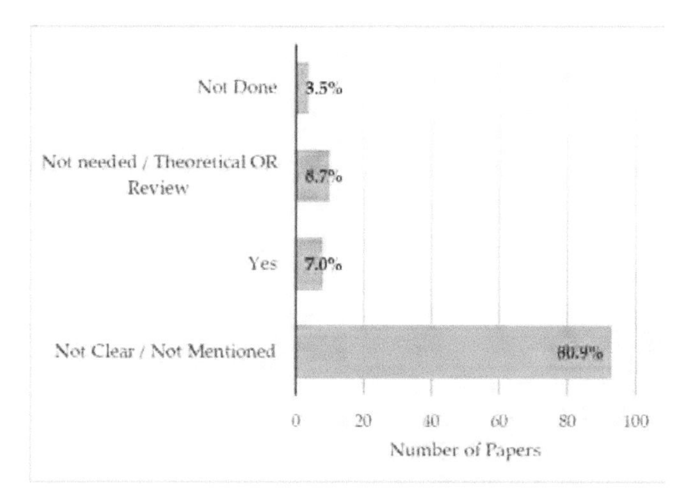

Figure 6.12. Number of papers based on whether or not model verification was conducted.

Validation of ABMs can be categorized in two main approaches: (i) a *constructivist/pattern approach*, in which the simulation results are compared with observations made in the real world, and (ii) an *expert validation approach*, where the quality of simulation outputs are evaluated based on the knowledge (or expectations) of the modeller or users of the ABM (Schulze et al., 2017). In our review, the expert validation technique accounts for more than half of the papers (53.6%) as illustrated in Figure 6.13. The use of pattern-oriented validation is low, but a representative number is found (28.7%). ABMs with no validation or no clear method reported are very low in our review (8.7%).

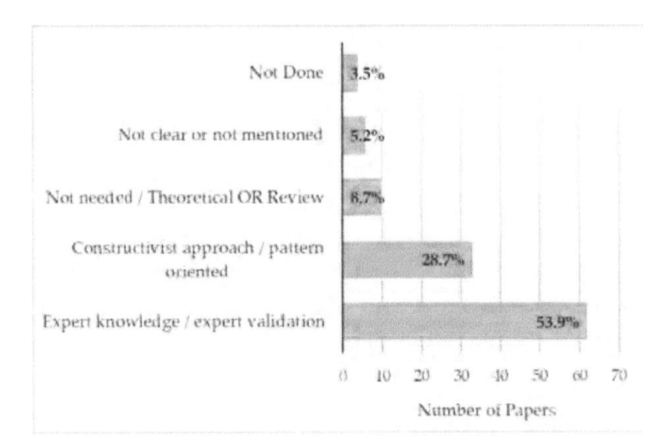

Figure 6.13. Number of papers based on the approached used for model validation.

Model Analysis

Model analysis in ABM can be achieved by using different methods and techniques. One that is regularly performed and becoming the standard to assess the quality of an ABM is sensitivity analysis (SA) (Augusiak et al., 2014). SA refers to the assessment of the sensitivity of the model outputs to changes in the most significant parameters used to set up the model or to examine the robustness of the model emergent properties (ten Broeke et al., 2016). A good model analysis will also probe that good matching results are not the result of "fine-tuning" parameters but that the theory behind the model set up is valid and holds for multiple set of input parameters (Augusiak et al., 2014).

In our review of papers, we found that model analysis was not a common practice among ABM modellers. Only 35 of the papers (30.4%) presented a model analysis based on sensitivity analysis (or another reported method), and 70 publications (60.9%) did not perform any type of model analysis (see Figure 6.14).

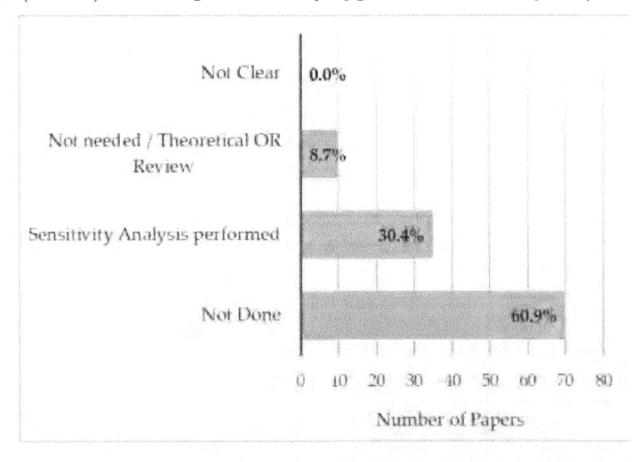

Figure 6.14. Number of papers based on whether or not sensitivity analysis was conducted.

Spatial, Scale and Time Representation

ABMs for WR-DRM have by its very nature or definition a spatial and time component either explicitly or implicitly given by the problem domain. Space and time are generally the two elements used to represent the environment of the ABM, where the agents are positioned and interact. Hence, it is relevant to analyse how these elements have been implemented in ABM for WR-DRM. In our selection, 101 papers (87.8%) explicitly uses both components, space and time. Furthermore, only five papers (4.3%) representing explicitly the time component, but space is used implicitly. An example of the latter category in the selected papers includes the work of Koning et al. (2019), where the space component is used implicitly as an attribute of the agents; in this case, the location of a house in different flood zones return periods. The remaining nine papers (7.8%) correspond to theoretical or review papers (Figure 6.15).

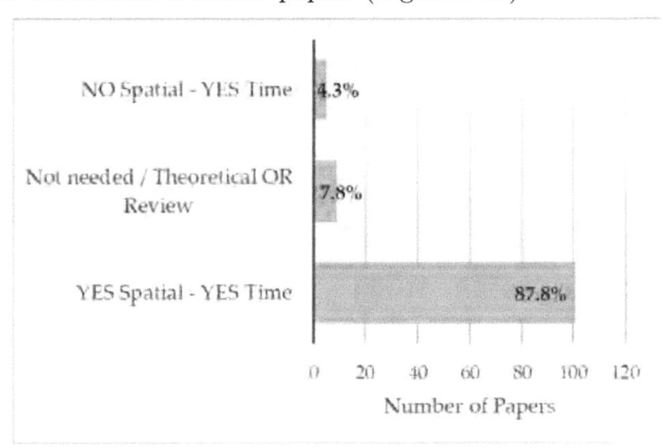

Figure 6.15. Number of papers based on reported use of explicit Spatial and Time dimensions.

The spatial scale or extent of the application is relevant in the design and implementation of an ABM for two main reasons. First, it will be the basis to properly understand the outputs as it defines the geographical limits of the application. Second, it can be directly related to the number of agents the space must contain in that specific area. We categorize the papers in the geographical extent from sector scale (An area of a city formed by one or more blocks but smaller than a neighbourhood) up to the continental application (Figure 6.16).

In the 115 selected papers, city-scale applications are the predominant ones, with 33.0% of the total number of publications, followed by regional-scale application with 20.9%. Smaller ABM application such as sector and neighbourhood scale accounts for 22.6% of the evaluated publications in total. Country or bigger scale to assess the impacts of disaster risk reduction policies are still very limited in our findings. The two applications on country scale were in Sint Maarten, a Caribbean island that is smaller than some of the regional application we found (Abebe et al., 2019a; Abebe et al., 2019b). The only study at continental scale was in Europe to assess the impacts of disaster risk reduction policies (Haer et al., 2019).

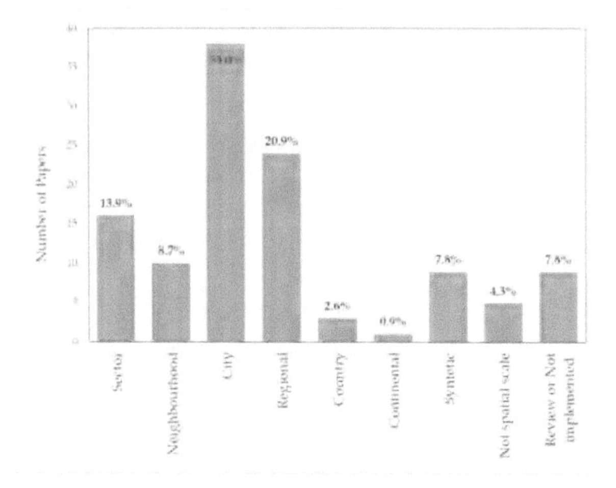

Figure 6.16. Geographical scale or extent of application of the ABMs.

6.5 Discussion

6.5.1 Characteristics of the studies

Publication Year

Given the observed rising trend on the number of publications that applied ABMs for WR-DRM studies, we can infer that researchers and to some extent practitioners, engineers and policymakers are now seeing the full potential and the benefits to incorporate an integrative modelling approach into disaster management. The barriers and concerns regarding the use of ABMs are being broken down, and general acceptance is increasing. An investigation of the possible drivers of the rapid increase in the use of ABMs for water-related DRM should be further considered, but that is beyond the scope of this review.

Geographic Coverage

The geographic coverage of the studies shows the widespread use of ABMs for WR-DRM, with almost all continents being represented. When looking at the countries where the majority of studies have been implemented a shared feature appears, they all have in common that they are repeatedly affected by physical hazards such as hurricanes in the USA, cyclones and tsunamis in Japan and Indonesia, or recurrent floods of the UK and China. In South America, Chile and Peru have the highest number of case studies in relation to recurrent tsunamis alerts. Despite an improvement in DRM practices, impacts of water-related hazards in those countries continue increasing in terms of economic damages and loss of lives. Researches working in those countries may be inclined to try new methodologies to address DRM from a more holistic approach, and this can help explaining the relative higher use of ABMs. Also, the constant threat from natural hazards makes these countries rich in data, which is necessary and sometimes a limitation to set up and validate ABMs.

Software used

The preference for using NetLogo can be explained because it incorporates easy to use interface and has a fast learning curve. Hence, NetLogo cab be a right choice for novice users or non-specialists. It has been extensively reported as an ideal solution for the implementation of model conceptualization and for testing early stages of model implementation. NetLogo also provides several libraries and examples that can be re-used and adjusted to the modeller needs.

Repast-Symphony is gaining wide acceptance among ABM modellers because it is a robust software that allows handling bigger and more complex ABM models, in terms of the number of agents, interactions and how the environment is represented. Despite having a steep learning curve, it can be considered as a rapid prototyping tool.

In the selected papers, commercial software was chosen as they provide an easy-to-use environment, and no programming experience is needed. Such software can be used with a set of default parameters, which make them an option to explore and answer specific questions such as the effects of different strategies in an evacuation.

It was surprising the significant amount of papers we found not sharing details about which software was used to implement their ABMs. Not sharing a detail as the software of implementation can be considered as a bad practice and one that can harm the general acceptance of ABMs for WR-DRM, it makes more difficult to believe or trust the outputs of the models.

6.5.2 Applications of ABM in Disaster Risk Management

Model Purpose and Type of Implementation

Despite ABM for WR-DRM being a relatively new field of study, the high number of direct applications we found shows the acceptance of ABM as a prominent and feasible tool to model complex systems such as those involved in WR-DRM. Hence, frameworks or methodological papers are not the focus and sometimes not needed or wanted in research articles with the primary objective on DRM, and can help explain the low percentage of papers belonging to the descriptive nature category in our review.

Case studies application are the most common use of ABM in WR-DRM. We acknowledge that the work that has been done in case studies have helped the reputation of ABMs as a suitable tool for DRM. Those papers have set the grounds to make ABMs more robust and widely accepted by researchers and to some extent to stakeholders and policymakers. However, ABM models in DRM should not be limited to case studies. Modellers should reflect more on how their work can contribute to improving theories or methodological issues such as model design, validation and verification, agents parametrization and model analysis (O'Sullivan et al., 2016). By focusing (or reporting) primarily on the case study, a broader lesson from the ABM may be missing.

We did not find generic models in our selected papers; this is an ABM which structure is generic enough and intended for use in a wide range of different projects with relatively minor restructuring. Generic models are possible at the expenses of having more degrees of freedom in the model set up and the parameters. Setting up such models is possible only if the modellers have a deeper understanding of the systems under analysis. Modelling case studies limit the degrees of freedom. However, the models are generally easier to set up but at the cost of not able to generalize or extrapolate the results to other case studies or propose a framework or theory.

In terms of papers presenting a framework or DSS, it was initially expected that ABMs proposing or testing a framework would be first applied in synthetic cases before scaling the theories to a bigger case study. One can infer from the categorization done in the our review that disaster risk practitioners using ABM models already consider their theories are robust and can directly be tested into case studies. It also can mean that the interest to create a framework or that the ABM will be used for the final user is not (yet) prevailing; and hence, participatory ABMs are currently not fully explored.

Going beyond academic and desk studies will be necessary to harness the full potential of the ABMs for WR-DRM. It will be necessary to go beyond testing the capabilities on case study applications and use the models to inform decision-making or to be used as a social learning tool. ABMs for DRM can have a more profound impact with the active involvement of end-users (e.g. policymakers, stakeholders) during model set up and validation phase. This can be achieved using participatory modelling, surveys and workshops, which, in turn, can give a degree of familiarity with the models and increase the chances to either use the outputs of the simulations or become the end-user of an ABM tool. Involving stakeholders, however, needs to take into consideration some limitations such as bias in the selected group, conflicts between them, and how accurate the obtained results of a control experiment are in comparison with those expected in "real-life".

Disaster Phase and Category

The majority of the papers we reviewed focus on evacuation modelling, showing an interesting fact about WR-DRM practices, where currently most of the effort to reduce the impacts of a hazard focus on response actions, rather than focusing on protective and preventive measures. In this case, by studying the effects of limiting the number of people potentially exposed to the hazard. In our review, evacuation models were used to answer a wide range of research questions; including optimization of mass city evacuation, estimation of the total time needed to evacuate a city, identification of critical infrastructure (i.e. bridges, roads, shelters), evaluation of potential loss of life under different mitigation measures, dissemination of information, city planning and testing different evacuation strategies.

The few studies found in the categories of pre-disaster planning, recovery and response suggest that more research is needed to evaluate the feasibility of using ABMs to answer particular questions that may arise in these phases, such as effective distribution of teams and resources in the response or recovery phases; or to evaluate the effects of individual adaptations to floods or other disasters.

In the selected papers, and to the best of our knowledge, there is not yet a research that incorporates all categories into one model. Also, based on the advanced search strategy, to date, no review paper focuses on the big picture of ABMs in WR-DRM, making the contributions of this paper relevant.

Model Documentation

Model documentation using standards or protocols was found to be a limiting factor in the development of ABMs for WR-DRM, with a small percentage of the papers we reviewed doing so. The papers that described their models using the ODD or ODD+D protocols have greater readability and re-usability. Those papers are easier to understand in comparison to the ones that did not follow a standardized format. One way to get familiar with the use of these protocols is by checking others works as templates. In our review, we believe that some good examples of the model description using the ODD protocol are presented in the work of Abebe et al. (2019a), Dubbelboer et al. (2017), Watts et al. (2019) and Haer et al. (2017). Similarly, ODD+D good practices can be seen in Walls et al. (2018), Magliocca and Walls (2018) and (Dressler et al., 2016).

Code sharing should be the standard in ABMs for WR-DRM not only as a good practice but for transparency, model verification, use and re-use of the models. Such practices will help to advance faster the use of ABM for water-related DRM as it will stimulate stakeholder perspectives towards this modelling paradigm.

Representation of Agents and data used to set up the model

The differentiation of the agent type is crucial when designing the ABM and the rules that govern agents' actions and interactions. It is compulsory to have a suitable representation of human decision-making for the model outputs to be useful. In ABM for WR-DRM, set of rules for individual agents and low-level composite agents may be expressed based on personal experiences, beliefs and desires or cognition architectures, whereas high-level composite agents commonly involve policy and norms (Balke and Gilbert, 2014). ABM models that are able to integrate individuals and institutions may be more suitable for WR-DRM. Models integrating both types of agents may offer a better representation of the real world because decisions that affect the environment and that may restrict or promote certain individual behaviours are taken at higher societal levels. For example, the decision to open or not shelter during a hurricane will change the individual behaviour or whether or not to evacuate and where they will evacuate (Medina et al., 2019) or a national policy on flood insurance can affect the individual risk at the household level (Dubbelboer et al., 2017).

The low number of papers using either surveys, interviews and participatory modelling to parametrizes the agents' rules suggest that modellers are mostly interested in rapid developments, testing and obtaining results. One reason for the low number is that collecting this type of data is resource-intensive and may require multiple field visits or workshops. Modellers find it challenging when there is a resource limitation, or stakeholders are not willing to participate (Manson et al., 2020), leaving no other option to the modeller than to use what they have at hand.

We believe qualitative data should be at the centre of the design of the rules of ABM for WR-DRM. By collecting cultural context-specific data and document specific human (or institutional) behaviours, it will be possible to add an extra level of understanding of the system by including the underlined motivations of the decision-making process. Qualitative data can help answer context-specific questions such as who evacuates when a tsunami warning is issued or under which circumstances a household would prefer to stay at home. At the same time, quantitative data, especially census demographic data, should remain a standard to support the model development.

ABM for WR-DRM can significantly benefit from non-traditional sources of information such as participatory modelling, social media or using crowdsourcing, whose use remains unexplored. Besides, we did not find papers that benefit from the use of big data to parametrize models, and only one paper used cell phone location data to validate an evacuation model (Yin et al., 2019). These non-traditional sources have huge potential in getting information about daily mobility, exposure and vulnerability to natural hazards.

Agent's decision making process

The predominant use of mathematical or probabilistic rules to represent agent's decision making in ABMs papers we reviewed can be explained on the ease of implementation of such equations into ABM code but also has been proven to represent some types of human behaviour properly.

In terms of evacuation modelling, the use of origin and destination as the rationality of the agent decision making assumes that an agent will choose to evacuate to the closest safe point to the origin and that act on bounded rationality where the agent knows the best (usually the closest) destination and the best path (roads) to get to safety. This modelling approach may be valid to study specific problems, such as testing roads, bridges or shelter capacity, total evacuation time and the effects of different evacuation strategies under a critical scenario.

For example, Imamura et al. (2012), estimate the total time needed by all the evacuees to leave the potential hazard zone; Mostafizi et al. (2017) identify the most important roads and intersections during mass evacuations; Chen (2012) investigates the effects on the traffic load of stage evacuation strategies on the only bridge available to exit the Galveston islands in Texas. However, if a more realistic evacuation is needed according to the research question, a different parametrization of human decision-making is needed – one that allows accounting for individual decision making and individual preferences such as if agents will actually evacuate, and if so, where is the preferred location.

An example of a more sophisticated behavioural model is presented in Watts et al. (2019), A BDI type of model is used to represent the agents' decision making whether or not to evacuate. It takes into account individual interpretations of warning and evacuation orders based on past experiences and trust on the institutions in charge of DRM, as well as individual interpretation of the environmental hazard. Similarly, a rule-based example is presented in Yang et al. (2019) using a series of conditional behavioural statements (if-then-rules). The researchers modelled the transmission of warning messages from authorities to households and later replicated among different households. Those rules define the probability of evacuation for every individual household.

In addition, two well-known protocols or frameworks to describe agent behaviours that allow to include a more "realistic" type of human behaviour or reasoning did not appear in the papers we reviewed. These are the BDI architecture (Norling, 2009), and the PECS model (Urban and Bernd, 2001). Architectures that appear to resemble BDI are reported in four papers. Although the authors did not report directly using the BDI framework, their implementations can fit into its definitions. None of the papers we reviewed use PECS (or any other cognitive or neurological model). None of the papers has implemented high level cognitive or neurological architectures such as SOAR (Laird, 2012). Reasons for the lack of use of more sophisticated representations of human behaviour are associated with high complexity on the conceptual framework and its implementation, and the associated high computation demand.

Comparisons of different behavioural models to evaluate the effects of such theories in the final model results are reported in only one of the papers. Haer et al. (2017), implemented three economic models, (1) Expected Utility Theory; (2) Prospect Theory with bounded rationality and (3) Prospect Theory with changing risk perceptions and social interactions, in an ABM to evaluate the effects of human behaviour in flood losses reduction through household investments. Making this type of modelling in other ABM papers may help to ground theories beyond an implementation. Testing the effects of different formulations on the decision-making modules will allow selecting the best theory for a particular phenomenon of interest.

Model Verification and Validation

The complex nature of the processes modelled with ABM in WR-DRM makes it very challenging to verify such implementations. Hence, errors in code implementation can be easily present, especially considering that ABM modellers may not be expert programmers. Despite the challenges, not reporting model verification process, as in the case of most of the reviewed papers, could diminish the credibility of the implementations. Ways to move forward may include testing the models in smaller scales or simplified versions and start scaling up the model (Galán et al., 2009) or proper documentation or sharing the code to allow for reimplementation and external verification (Grimm et al., 2017).

We cannot assume that all the papers that did not report model verification did not actually did not do it. It can be that modellers performed an expert judgment of the simulation results and were satisfied with them. In such a case, the problem is communicating the method used to verify that their models fit reality for the problem purpose (Augusiak et al., 2014). Modellers of ABM for WR-DRM should include a section where explicitly the model verification was performed either quantitatively or qualitatively, and this will help other modellers on different methods on how to perform (and) report their verification process.

Regarding the eight papers that described model verification, one performed the verification using built-in functionalities in the proprietary software Artisoc[7], which allowed the researchers among others to check whether or not the agents were moving at the desired speed (Takabatake et al., 2018). Others reported which parameters of the model they tested. For example, Gehlot et al. (2019) reported they verified that the average time of an evacuation trip would increase as road congestion also increases, Eid and El-adaway (2018) reported they applied regression tests to ensure that the agents perform their designed procedures. Abebe et al. (2019b) explained how they used an evaluative structure to assess if

[7] Artisoc web page: https://mas.kke.co.jp/en/artisoc4/

the values of selected variables match agents' actions (i.e. certain policy implementations leading to more elevated houses in the case study). One only study reported they have done model verification without explaining how the verification was carried out (Nagarajan et al., 2012). Others mentioned they performed model verification without providing additional information on how they did it. But, they share the source code and extra documentation that will allow verification if someone is interested in verifying or re-using it (Dubbelboer et al., 2017; Jenkins et al., 2017).

On the topic of validation of ABM for WR-DRM, the most applied validation method in the papers we reviewed was based on the expertise of the modellers. Validating model results using expert knowledge is not a bad practice by itself, especially in ABM modelling, where data to perform a traditional pattern-oriented type of validation is not always feasible to obtain due to resource constraints. A modeller cannot wait for a hurricane to strike a city in order to measure the real evacuation behaviour or the effects of implementing a new policy may not be reflected immediately but in a timespan of years or even decades. Relying on expert knowledge for model validation of ABMs for WR-DRM is, therefore, not only necessary but sometimes the only option available, with the associated level of subjectivity and degrees of freedom in model design and validation.

Here it is perhaps the most significant challenge faced by ABM to be accepted as a valid tool for WR-DRM, where engineers and stakeholders have been exposed to deterministic type of models such as hydraulic and hydrological models where validation is performed with actual recorded data, and model predictions are based on physical (and measurable) variables. Involving external experts to assess the model outputs, or involving local stakeholders, through surveys, interviews, or any kind of participatory modelling to help validate the results is a way to advance in the acceptability of such type of expert knowledge type of validated models (Heckbert et al., 2010 ; Schulze et al., 2017).

Even though non-ABM modellers may accept more widely pattern-oriented validation, some limitations are also present in this type of validation, for instance, can a model validated to a specific event may hold for future similar events?, or is it that human learning will change the outcomes (behavioural patterns)?. Modellers need to be aware of these limitations and ask themselves what type of questions can be answered with patterned oriented calibrated models and act with caution about what type of predictions can be done with such models.

In our review, pattern-oriented modelling has been used in those cases where data was sufficiently available to do so. We found a great potential of this validation approach in hurricane or flood evacuations. Such cases include the evacuation patterns in the 2011 riverine floods in Brisbane (Liu and Lim, 2018), the records of evacuation observed in the Florida Keys during Hurricane Georges in 1998 (Yang et al., 2019) and the evacuation patterns observed during evacuations of Hurricane Katrina in 2005 in the USA (Handford and Rogers, 2012; Liang et al., 2015).

Pattern oriented validation model approach has also been used in policy analysis, specifically using the effects of different policies in land use or real estate market. Examples of pattern-oriented validated models can be found in Chandra-Putra and Andrews (2019), the paper uses historical land-use data and insurance claims to understand how humans adapt to the impacts of climate change in a coastal community affected by Hurricane Sandy in 2012 in New Jersey, USA. Mustafa et al. (2018) used an ABM for urbanization and densifications of urban areas and used historical data to simulate observed land uses.

Model Analysis

Our review revealed that ABM modellers in the field of WR-DRM focus more on scenario analysis or even single output models from their models, rather than analysing the robustness of the model and the sensitivity of the parameters used. Not performing SA on ABM models could lead to adding more uncertainty to the results, possibly making the final users or policymakers to question how reliable are the results or how appropriate the ABM is to make new "predictions" on how the world they are modelling will look like based on the outputs of the simulation.

Scenario analysis is necessary to answer to the research questions that may arise in WR-DRM, but focusing only on a single type of answer may cause to miss some results of interest, we believe that in addition to this type of modelling, modeller's should always perform an appropriate type of model analysis, and SA can be used for that purpose. SA helps modellers understanding what are the key and more influential parameters on their models, resulting in a better understanding of the model and what can be inferred from it (Augusiak et al., 2014). A good starting point for ABM modellers into the theory of SA for ABM is the work of Lee et al. (2015) and to assess what type of SA may suits better their models the work of ten ten Broeke et al. (2016).

Most of the SA reported in our review used the OFAT (One-Factor-At-A-Time) method. We believe it is because the method is relatively easy to perform and yet give valuable insights. For example, Tonn and Guikema (2018 - Supplemental online material), assessed the sensitivity of seven input parameters using the OFAT approach. Similarly, Yang et al. (2018) present a SA on six strategic variables for flood loss assessment. Also, Baeza et al. (2019) use SA to evaluate the degree of sensitivity of governance scenarios in the model outputs for different regions in Mexico City. Only two papers from our review performed global sensitivity analysis (GSA): in Erdlenbruch and Bonté (2018) applied GSA to evaluate the dynamics of individual adaptation to floods, and Dressler et al. (2016) applied GSA to assess the robustness of sub-models over an extensive parameter range.

Spatial, Scale and Time Representation

One of the biggest advantages of using ABM tools for WR-DRM is the possibility to have the process being modelled on an explicit spatial and time domain. More realistic and explicit representation of the environment is possibly a more powerful way to communicate the message to stakeholders and the community in general, as it allows them to see the impacts of different scenarios over a particular region of interest (Manson et al., 2020). Suitable representations of time and the spatial environment depend on the modelling question(s), and the degree of accuracy (realism) sought.

Implementations of DRM where space and time are explicitly analysed and where these two elements play a major role in the impact on the model performance can include evacuation processes or densification of urban areas. On the other hand, policy implementation or educational ABM may or may not have (nor need) the explicit geographical or spatial component, for example, evaluating how urgent information flows in social media (Rand et al., 2015), or representing policy process among stakeholders (Valkering et al., 2016).

In the ABMs we reviewed, time is represented by a time step in minutes, hours or years using an equal-duration/time approach in which the actions performed by the agents are updated at the same time based on the rules of the ABM. In terms of time step selection, ABM modellers should think carefully an adequate time scale that adequately captures decisions, changes or actions based on the objectives of the implementation across the different WR-DRM applications but keeping in mind computational performance.

Equally important is the total simulation time (TST); this is the time range that the ABM runs. TST vary according to the purpose of the ABM and should be large enough that allows to observe the changes in the system that are intended to be captured with the ABM. For example, ABM for hurricane evacuation can be modelled in days or weeks as the needed time to evacuate a large-size city, or it can be decades as the time needed to observe the effects of policy implementation for disaster risk reduction.

In terms of space, most ABMs in WR-DRM represent it as a virtual simplification of the study area; this is the neighbourhood, city or region. Those ABMs added a certain type of projection and coordinate system that create a reference into which agents can be mapped and move through, and in most of the cases, using GIS data as input for representing the environment. Some different approaches for space representation were found in the work of Naqvi and Rehm (2014), they represented cities as interconnected nodes in a so-called stylized layout representation of the region under analysis. In an ABM set up for evaluation of variation in behaviours in response to flood warnings Du et al. (2017b), used a simplified representation of transportation networks using a direct graph approach, in which edges and nodes represent the routes and intersections of the road network respectively, and weight in the edge represents the cost of using the route it represents. The direct graph is then mathematically represented as a matrix. Also, the work of Erdlenbruch and Bonté (2018) presents a simulation of the dynamics of individual's adaptation to floods using a network representation of closest neighbours.

GIS representation in ABMs usually takes one of two forms raster or vector representations. Raster representations are very common in ABM models. We found that older applications of ABMs for WR-DRM rely mostly on raster data due to software and hardware limitations. Raster data models have the advantage of a faster visual representation of the model, as well as faster computational processing of the outputs. Vector or shapefile representations can be more powerful to communicate the outputs of an ABM for WR-DRM compared to raster representation because it has higher ability to represent the real-world elements closely, and is gaining popularity among ABM modellers in WR-DRM domains. Also using satellite images or 3D representations can sometimes enhance the transmissibility of the message but at expenses of more computational resources. A balance here is then required between computational demands and accuracy of the representation of the reality when selecting the type of data to represent space.

In evacuation models, vector data and specifically network data models are gaining in popularity. Network data models can be seen as a set of links (roads), and nodes (junctions) that form a road network where the agent move and interact with other agents and with the environment. They also contain some physical characteristics such as distance and location that are crucial in evacuation processes (Manson et al., 2020). The networks in the papers we analysed were either represented using a rather simplistic notion with one lane representing the space where agents move (Chen and Zhan, 2008; Mostafizi et al., 2017), or using more sophisticated properties such speed limit, the number of lanes, material or slope (Chen, 2015; Liang et al., 2015; Liu and Lim, 2018), or the inclusion (or not) of specific features such as bridges, stop signs or traffic lights (Chen, 2011; Naghawi and Wolshon, 2010). An interesting work is presented in Gehlot et al. (2019), where roads were modelled as agents because they wanted the roads to be able to change "behaviour" through the simulation such as lane closure, change in speed limit and blockages.

In terms of the scale of application, the fact that most of the studies in our review use city-scale reflect what has already been discussed in this chapter, that most of the application we found are descriptive models based on solving specific issues in a particular case study rather than elaborating the theory of WR-DRM using ABM tools. Regional-scale in the selected papers typically corresponds to papers that are expanding previous models to evaluate impacts in a bigger geographical scale, such as the impact of neighbouring towns based on decisions taken by its neighbours or the impacts on the transportation and shelters in regional evacuation schemes.

Smaller scales (i.e. sector, neighbourhood), were typically used in WR-DRM to probe concepts and as a preliminary assessment of a particular theory. For large scale applications (i.e. country and continent), were very rare in the 115 papers we analysed. This scale of application may be limited because theories of DRM are difficult to generalize and to scale-up, as well as some limitations with the computational tools and higher computational demands.

6.6 Recommendations and Future Directions

6.6.1 Theoretical Recommendations

The use of standards and protocols to describe and document the design of the ABM should become common practice on ABM for WR-DRM. We believe that ODD+D should be used. It has been extensively used in other fields (Lee et al., 2015; Schulze et al., 2017), and also has proven its feasibility to document ABM models that need to include the description of human behaviour component. (Lee et al., 2015; Schulze et al., 2017).

One area that would benefit significantly from a framework or generic model is the use of ABMs for city evacuation. A generic model, this is a model that allows reuse for a wide range of different projects with relatively minor changes and that can be tailored for specific needs will allow not only to speed the model implementation but also will to perform a comparative analysis between different implementations. Generic models would require a strong understanding of the system under analysis in order to be able to extract the commonalities that are not case-specific.

However, some question will remain open to discussion. How standardisable are the results of a stakeholders approach? Can a modeller in Asia use the results of a model built in America or Europe? Most likely, the answer to these questions is No, but if ABMs are appropriately documented they can be used as guidelines to extrapolate results or as guides to be used in other contexts. To overcome this challenge, it will be imperative that modellers document properly how they transform their qualitative data into quantitative data that is usable in coding the ABM.

6.6.2 Modelling Recommendations

Software

Based on the findings of this systematic literature review, we offer here a reference that can guide professionals, researchers and academics in selecting an appropriate ABM software for designing and developing their system models and prototypes, which account for both their expertise and the specific requirements of WR-DRM. The selection of the software should not be trivial. It can affect the generality, usability/reproducibility, robustness and modifiability/replicability and scalability of the results. Using open software and sharing the code should be the industry standard as these allow to build on top of the current state of the art and advance the use of ABMs for WR-DRM. Such standard can help also to have a more transparent model, where anyone can verify, analyse and reuse the code.

Table 6.3 presents a summary of the main characteristics of several ABM software that we believe have great potential to be used in DRM. We include both free-open source and commercial software, and we present a column with our recommendation on which cases each software has the potential to be used. We present two sections, one with generic ABM software to be used across different DRM categories, and another for ABMs specifically developed for evacuation purposes. We present these subcategorization given that evacuation is the DRM domain with most publications in our review.

To build this list, we took into account the current state-of-the-art of the software and learning curve, current uses of the software in WR-DRM, easiness to set up, use and re-use models, the scale to which the ABM can be used, the type of license and availability, and software support such as existing user manuals, tutorials, developer's or user's support. One commonality across the listed software is that all of them provide a friendly graphical user interface for model set up, for both the ABM modeller and for the potential end-users, built-in graphical capabilities and all have predesigned agent templates and built-in libraries and packages that can be re-used and adjusted to the modeller needs but avoiding to start from scratch.

Table 6.3. Comparison of the most suitable ABM tools for WR-DRM and its recommended use. Listed in alphabetical order and considering general-purpose ABM and ABMs for evacuation.

ABM Software	License / Availability	Coding Language IDE[8]	Model development effort	Modelling Strength Scalability Level	Recommended use in DRM
			General-purpose ABM software		
Anylogic	Proprietary Commercial	Java IDE: Own platform	Moderate	High Large Scale	If access to the software is not a limitation. When the model is already built in this software For large applications When transportation and evacuation is in the centre of the ABM for DRM When user-friendly IDE is preferred
GAMA	Open source, GNU GPLv2, Free	GAML modelling language IDE: Eclipse platform+ Plotting/ graphical editors	Moderate	Medium Large scale	To prototype models To be used in medium to large scale case studies When advance and sophisticated and built-in GIS capabilities are required When programming skills are intermediate or not a limitation
NetLogo	Open source, GPL, Free	Netlogo language IDE: Own interface	Simple /Easy	Medium Large scale	To prototype models To be used in small or medium-scale case studies To be used in academic settings When GIS and graphical representation is secondary When programming skills are not strong

[8] IDE: Integrated Development Environment is the software suite that consolidates basic tools required to write and test the ABM.

Table 6.3 (Continued)

ABM Software	License / Availability	Coding Language IDE	Model development effort	Modelling Strength Scalability Level	Recommended use in DRM
Repast-HPC	Open source, BSD, Free	C++ IDE: Eclipse platform	Complex / Hard	High Large Scale	When the system to model is very large or complex scale When GIS integration is a must. To be used in large scale parallel or distributed computing clusters When programming skills is not a limitation
Repast Simphony	Open source, BSD, Free	Java IDE: Eclipse platform	Complex	High Large Scale	When the system to model is very large or complex scale When GIS integration is a must. When programming skills is not a limitation
Traffic and evacuation focus ABM software					
MATsim	Open source, GPL, Free	Java IDE: Eclipse/Maven	Complex / Hard	From Medium to Very Large Scale	Use recommended when mostly focus on transportation problems such as evacuation When Transport (evacuation) networks are very large When GIS (in transport/evacuation) is desirable
TRANSIMS	Open source, NASA agreement, Free	Python IDE: Own interface	Moderate	Medium Large scale	Focus on transportation simulations. Evacuation planning has been tested at microsimulation level for large metropolitan areas. High 2D quality movie rendering and some pseudo-3D representation capabilities When the model is already built in this software

Table 6.3 (Continued)

ABM Software	License / Availability	Coding Language IDE	Model development effort	Modelling Strength Scalability Level	Recommended use in DRM
VISSIM	Proprietary, Commercial	Not needed, graphical interface. Scripts can also be used (Programming language independent, i.e. C++, Java, Python) IDE: Own interface	Easy	High Large Scale	It is a traffic simulation software If access to the software is not a limitation When the model is already built in this software No programming experience required Fast testing of evacuation scenarios, quick and simple model setup. Can be used with the default values of the proprietary software Driving and pedestrian modelling allowed

It should be highlighted that the aim of compiling the list of software is not to discourage the use of different software than those suggested in Table 6.3. However, we believe that using the same software will advance the use of ABMs for WR-DRM faster and more reliable, and will facilitate model replicability. A recommendation for the ABM modellers in WR-DRM is to investigate and analyse these and other tools to evaluate which is aligned with their specific needs.

In terms of software use, we cannot conclude which one is better to use to implement ABMs for WR-DRM, since that depends on modellers' preferences, abilities and also the scope of the model itself. Nevertheless, we encourage modellers to use Netlogo for basic and proof of concepts because it is easy to learn and use, and the ability to run prototype models. We recommend the use of Repast Symphony for more robust implementations because it can handle high computational demands, a larger number of agents and interactions and better representation of a more complex environment (space and time). In addition, with the sophistication of models and the increase in the scale of simulation, it is expected as well a surge in computational demand. In such cases, ABM tools that support parallel computing would be required, which Repast HPC, Swarm and MATSIM are the most promising tools (Abar et al., 2017).

Model and Code sharing

Regarding code sharing, it is advisable to use well-known data and code repositories rather than personal web pages or research institutes or university repositories as it will facilitate the maintenance and digital preservation of the code, software citation, as well as making it as standard procedure when coding ABM. Code and data sharing is being highly promoted recently and is gaining momentum from both academic and specialized journals, which are encouraging and sometimes making it a requirement to publish the data and codes alongside the articles. CoMSES-OpenABM and Github are the two repositories used by the papers we analysed that provide a good option to share the code.

Model Verification, Validation and Analysis

Model verification and validation play a central role in the acceptance of using ABMs for WR-DRM. ABMs should include in their reports a brief description of these two elements. Reporting explicitly validation and verification will make ABMs more robust and easy to verify for external reviewers as well to increase the credibility of the simulation results. Using standard protocols, such as the ones suggested in this review, and including them as supplementary material, alongside with a short section in the paper, should become the industry standard. Active involvement of final users, stakeholders or decision-makers of the ABM in the development and validation phases will help on the acceptance of ABMs as an appropriate (if not necessary) tool for DRM.

Sensitivity analysis or other types of model analysis is still vaguely applied in ABM for WR-DRM. We argue it should be present by default in the type of simulations done with ABM in order to make the outputs more robust, believable and gain the trust of authorities and stakeholders. In addition, identification of key parameters in an ABM, i.e. those to which the model is more sensitive, can benefit modellers in several ways. It will allow modellers to concentrate the data collection on those parameters and have more relevant data that will, in turn, reflect in more accuracy of model results. The modeller may also decide to remove the less influential parameters for model simplification, which requires less computational resources. An appropriate SA technique should be selected: some applications may find the use of simple SA techniques, such as OFAT, more appropriate and other models will necessarily make use of more robust techniques such as GSA.

Practical Recommendations and Future Directions

Space and time scale representation in WR-DRM is often used to represent as close as possible the real-world, here the use of vector data or high-resolution raster datasets to represent the environment of the ABM is the current trend. We believe this is a good approach in order to have more appealing results and engage stakeholders. This approach, in turn, compromises the computational time and needs for better computational resources such as parallel computing or other schemes that can handle large scale representations. However, no attempt was found in our review to explore the impacts of both space and time on the results of the ABM simulations; hence the ideal space and time representation for ABM with the sole purpose of DRM remains unanswered, and it is a line of research worth exploring.

Even though ABM is not a new tool as a computational modelling technique, its expansion and full potential is just started to be explored, and we will continue seeing its expansion and use. Parallel computing (PC) and high-performance computing (HPC) can be foreseen as the next breakthrough in the use of ABM since they allow to have faster and more robust implementations. Current implementations of ABMs using PC or HPC are still limited, and just a few could be found in our literature review. However, the results obtained are very promising to expand the use of ABM by tackling the computer capacity limitation, faster computing, bigger geographical extension of the models and more behavioural rules and number of agents in the simulations.

The use of mathematical or probabilistic models prove to be the dominant architecture to represent human decision making, and it seems from our results that it will continue to be the rule in ABMs for WR-DRM for the time being. However, the full potential of a more realistic human type of decision making continues to be unexplored, the underused of more sophisticated architectures deserves more exploration and implementation in ABM for WR-DRM. We believe that BDI type of implementations is the way to move forward in this direction due to the already usable extensions in ABM tools such as Netlogo and Gamma. However, modellers need to keep in mind whether or not questions can be answered with more simplistic approaches before moving to more complex model representations.

6.7 Conclusions

In this chapter, we address several common issues faced by modellers while implementing ABMs for DRM. We identified trends in model implementation and the main methodological issues that a modeller face, such as model purpose, data collection, validation and verification of the models and how to perform model analysis. Furthermore, we also draw attention to how human behaviour has been modelled, as this is a central issue to have a good and representative ABM for DRM.

Among those critical aspects of the use and implementation of ABMs is the purpose of the model. Predictive models are predominant in the field of DRM despite their focus on the theories rather than implementations in case studies. Similarly, validation and verification of models seem to be less important than presenting results from the ABM simulations. We believe this is not only a bad practice but is also limiting to some extent the credibility and acceptability of ABMs.

Sensitivity analysis is still not a common practice in ABM for DRM. The complexity of model design and parametrization requires modellers to be transparent with their implementations and performing any type of SA, can enhance the accuracy and robustness of the model. Therefore, the acceptability among users and policymakers can increase.

With the increasing acceptance of ABM for DRM, it is expected an increase in model complexity as well in the number of agents. This implies the need to have more robust ABM software to support the use of technologies such as parallel computing, artificial intelligence, genetic algorithms, to take advantage of the full potential of the ABM paradigm. Similarly, the need to explore the potential of ABM running in a cloud computing environment remains a challenge worth exploring.

In ABM for DRM, the central role of human behaviour as individuals or as institutions were identified as the main challenge to properly represent the beliefs, desires and behaviours in a set of rules. Qualitative data, such as interviews, group works, and field data collection can help to make these rules more robust and hence the output results. ABMs for DRM based only on quantitative data (i.e. census data, GIS data) can lack the trust from policymakers and stakeholders. Hence active stakeholder involvement is crucial to gain acceptance and use of ABM for DRM.

The use of protocols for model description and sharing the source code is slowly becoming the rule amongst ABM modellers for DRM. We believe that this good-practice, and is beneficial to advance faster the theories around DRM. Having a good model description, which is available will allow modellers to focus on refining existing models rather than having to start from scratch and focus the attention on model development. Modellers can have the time actually to learn from the models.

Case study applications are the dominant type of application for DRM, and we foresee a slow growth of generic models in the near future. We believe that ABM practice is still far from having the computational resources needed, and there are still theoretical gaps on how to generalize and upscale the finding offered by local or case study applications.

Adequate visualization and communication of ABM results to the general public or decision-makers are vital in order to gain acceptability of this modelling technique. In addition, having a friendly user interface with a strong core ABM model can facilitate testing and understanding of the model by the final user

7

DYNAMIC EXPOSURE ASSESSMENT USING ABM

In this chapter, we start presenting a framework to be used in the implementation of Agent-Based models for disaster risk assessment. The framework, in the form of an ontology, presents the system agent and environment identification. After the framework, we present an implementation in the case study of Sint Maarten to evaluate how information can be used to change the levels of exposure by promoting (or limiting) protective action behaviours in the households of the island. We present some experiments using the ABM as a simulation tool to assess how weather-related information and the trust in the information source may impact exposure. We end this chapter presenting some analysis and exposure maps based on the results of the simulations.

This chapter is partially based on:

Medina, N., Sanchez, A., Vojinovic, Z., 2016. The Potential of Agent Based Models for Testing City Evacuation Strategies Under a Flood Event. In Procedia Engineering, Volume 154, 2016, Pages 765-772, ISSN 1877-7058, https://doi.org/10.1016/j.proeng.2016.07.581

Medina, N., Sanchez, A., Vojinovic, Z., 2017. Planning Strategies for Flood Disaster Risk Prevention with Human Behaviour Models. Proceedings of the 37th IAHR World Congress, 13-17 August, 2017 Kuala Lumpur, Malaysia. ISSN 2521-716X

Medina N., Abebe Y.A., Sanchez A., Vojinović Z., Nikolic I. (2021) Dynamic Exposure Assessment to water related natural hazards. An Agent-Based Modelling Approach. Manuscript in preparation

7.1 INTRODUCTION

This chapter presents the assessment of exposure to water-related natural hazards events using an adaptive approach. Concepts from Complex Adaptive System (CAS) and Agent-Based Models (ABM) are the core of building such an adaptive approach. The choice was made according to the characteristic and the advantages of CAS, as it allows to explore inter-relationships, inter-actions and the inter-connectivity of elements within a system and between a system and its environment.. First, we present a framework with the relevant concepts and elements when implementing an ABM to measure disaster risk exposure. Then, a prototype Agent-Based Model is used to assess the exposure component of the disaster risk assessment using Sint Marten as a case study to evaluate the feasibility of the proposed framework.

7.2 ADAPTIVE EXPOSURE ASSESSMENT USING AN ABM

ABM has been widely used in DRA, covering the different phases of a disaster, including planning and policymaking, evacuation response, recovery and mitigation (Figure 6.7 in Chapter 6). Exposure to natural hazards using ABM can be evaluated using two distinctive approaches concerning the disaster phase. One approach is to evaluate the root causes of exposure, this is an evaluation of the hidden drivers that led to a community or society being exposed to a natural hazard (i.e. flood, hurricane), and these type of ABM are system focused. The second approach refers to direct measurement of a system's exposure to a particular hazard, and to evaluate the consequences in the system, categorization and examples of these approaches were presented in Chapter 6.

The first approach can be categorised within the policy and disaster management phase of a disaster. An ABM of this category is expected to include (mainly) institutional actors implementing norms and regulations, and it is used to evaluate how a system may evolve in the long term (i.e. years, decades), (some examples: Abebe et al., 2019b; Haer et al., 2019). The second approach is used mostly during the response and recovery phases of a disaster where the main purpose is related to measure the direct impact on the system due to a hazard and is typically configured in short terms (i.e. hours, days), these ABM can be categorized as operational ABMs (some examples: Chen, 2011; Takabatake et al., 2018).

The distinction presented above is important because the implementation in an ABM will be completely different, starting from the system identification, the type of agents to be used, and the rules governing the systems and interaction, the total simulation time, the time step among others. For the disaster risk assessment presented in this chapter, we are interested in the second type of approach; an operational ABM to evaluate the hazard's direct impact as a measurement of exposure.

7.2.1 System Identification and Formalisation

For an ABM to be useful, a proper system identification needs to be done; this is to define the system composition and boundaries. System identification consists of the inventory on the physical and social entities of interest for a particular system under study and the identification of the connections and interactions between them; this is the agents and environment characterisation. During the system identification, it is essential to define all the relevant concepts, actors, possible behaviours, interactions, and states of the agents and the definition of the agents' environment (Van Dam et al., 2012).

For disaster risk management of water-related disaster, as described in Chapter 6, it is crucial to identify and capture the complex interactions between individuals and their institutions with the hazards and other elements that characterise the environment in order to be able to discover the drivers of exposure to disasters triggered by natural hazards. Therefore, we have identified three major entities to characterise the system: the urban environment, an external stressor (water-related hazard) and the agents (Figure 7.1).

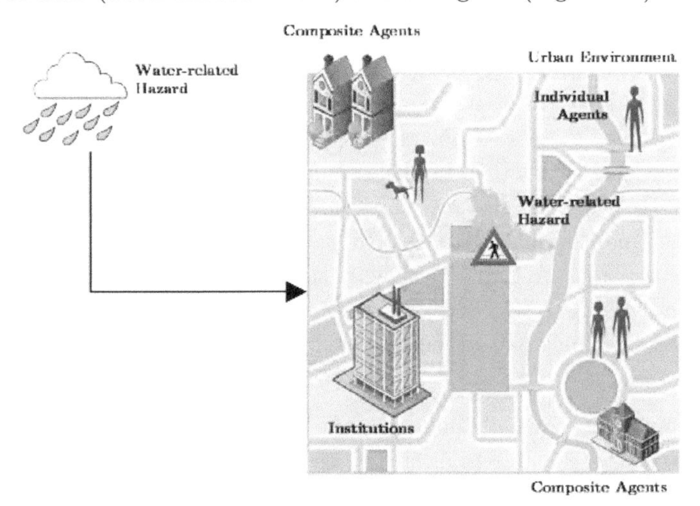

Figure 7.1 Overview of the composition of elements in an ABM for WR-DRM.

The characterisation of the properties of the three entities in an ABM for WR-DRM is presented in the form of an ontology (Figure 7.2). The following sections of this chapter will describe in detail each branch of the ontology.

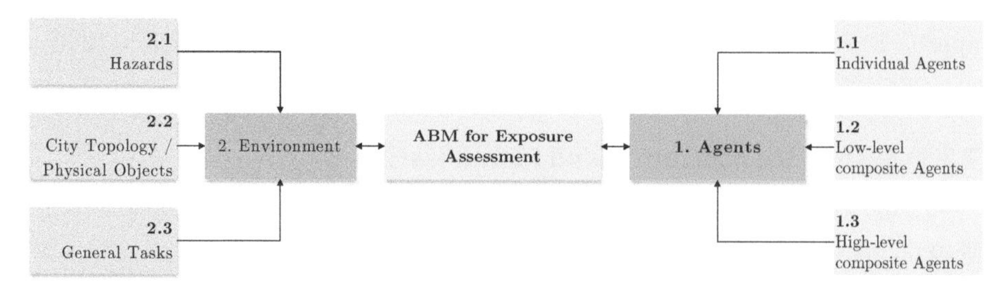

Figure 7.2 Ontology of the system formalisation for an ABM to assess exposure to natural hazards.

7.2.2 Agents identification and description

An agent's concept in an ABM model is just a simplification of the reality to represent the components of a system. Agents act as autonomous entities that are capable of performing actions according to a set of rules and without the direct user intervention, that is, an agent can interact with other agents and the environment and act (change state) based on these interactions (Abar et al., 2017; Nikolic and Dijkema, 2010).

Agents in the framework were categorised either as individuals or as composite agents based on an adjusted definition presented in Ghorbani (2013). We consider agents as individuals if they represent a single human, characterised by individual decision-making, such as pedestrians or tourists in evacuation models. In contrast, composite agents are those that represent a collection of individuals. We assumed two levels of these agents. Low-level composite agents represent a group of individuals who get together in the simulation and act as one unit, i.e. cars, households, housing developers, companies; and high-level composite agents representing institutional actors such as government entities, NGO's and insurance companies. Actions/decisions made by high-level composite agents can influence individuals or low-level composite agents directly and can change the environment (e.g. make an evacuation compulsory, implement flood protection, change land use).

The formalisation of individual or composite low-level agents in an ABM for WR-DRM needs to be able to discretise agents so that they can react differently to a potential hazard and a differential behaviour across the different phases of a disaster leading to different levels of exposure. In an ABM for DRM, high-level composite agents exist (or planned to exist) institutions whose "actions" will enhance or reduce the exposure of individual and low-level composite agents. Based on the analysis of vulnerability and evacuation behaviour presented in previous chapters, as well as the literature review of ABM for WR-DRM, we have compiled a set of variables that can be used to discretise agents and the possible set of actions to perform when agents' are faced to a disaster triggered by water-related natural hazards. Individual agents and low-level composite agents can be categorised using the same principles and variables, in the rest of this document they will be called Exp-Agent, recalling that these are the agents of interest to simulate its exposure, in Figure 7.3 we present the list of variables that can be used to discretise them.

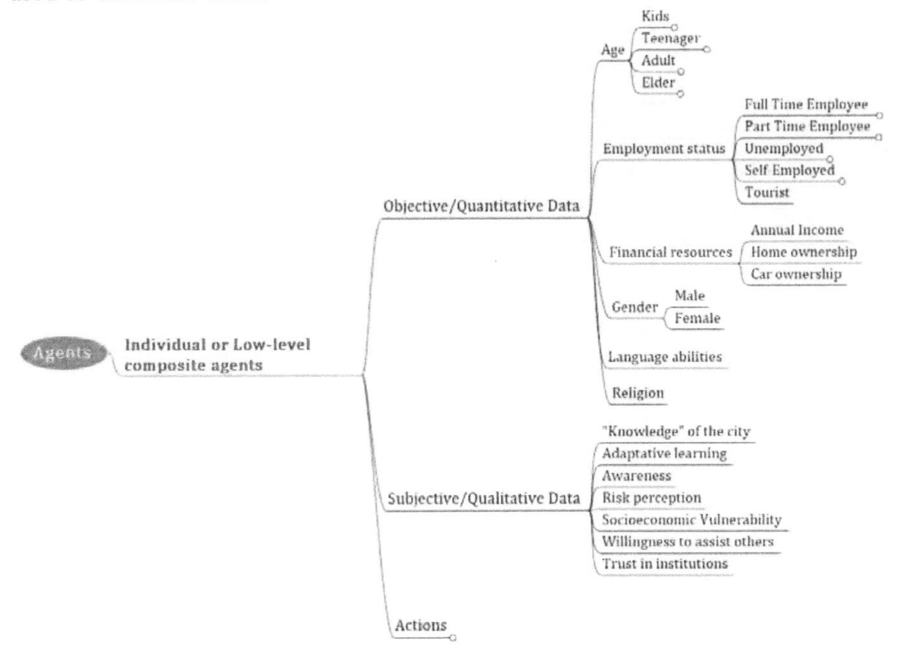

Figure 7.3 Variables characterising Exp-Agents.

The variables are divided into two categories: objective data, which corresponds to data of quantitative nature that can be typically measured using a numeric or categorical value, such as age, gender, financial resources, language ability, employment status, and subjective type of data, which is a more of qualitative type of data, that is used to measure individual's feelings or perceptions, such as risk perception, trust in institutions and vulnerability indexes. The list presented here includes the most typical variables used in the ABMs we review in Chapter 6, it is extensive, but we do not claim completeness. Hence, it should be used only as an illustration of possible variables; the final selection should be made based on the case study's characteristics and data availability.

The possible actions an agent may perform when face to a disaster is directly associated with the variables presented in Figure 7.2, for example, a female individual will be more likely to evacuate than a male one (Dash and Gladwin, 2007; Thompson et al., 2017), or the status of homeownership will likely determine the willingness to reinforce the house to withstand the associated winds of a hurricane (Huang et al., 2016; Thompson et al., 2017). An AMB modeller should create as many paths as possible to combine variables as many distinctive behaviours towards natural hazards may be of interest. However, without overcomplicating the system representation.

Based on the discretisation, a set of actions to execute needs to be provided for each possible distinct agent. Actions are a set of possible interaction of the agents with the environment or with other agents during the simulation. In Figure 7.4, we present a list of possible actions performed by Exp-Agents when faced with the impact of a natural hazard. The list is divided in two: (i) an agent can become a "pseudo" high-level composite agent, it can perform actions that are distinctive of high-level composite (e.g. assisting other peer-agents, be a source of information dissemination). (ii) Individual actions, an agent may choose to do nothing once they are aware of a possible threat based on its assessment of risk, they can choose to evacuate to a place that is perceived as a safer place than their own, or an agent can stay at its location but prepare to the hazard.

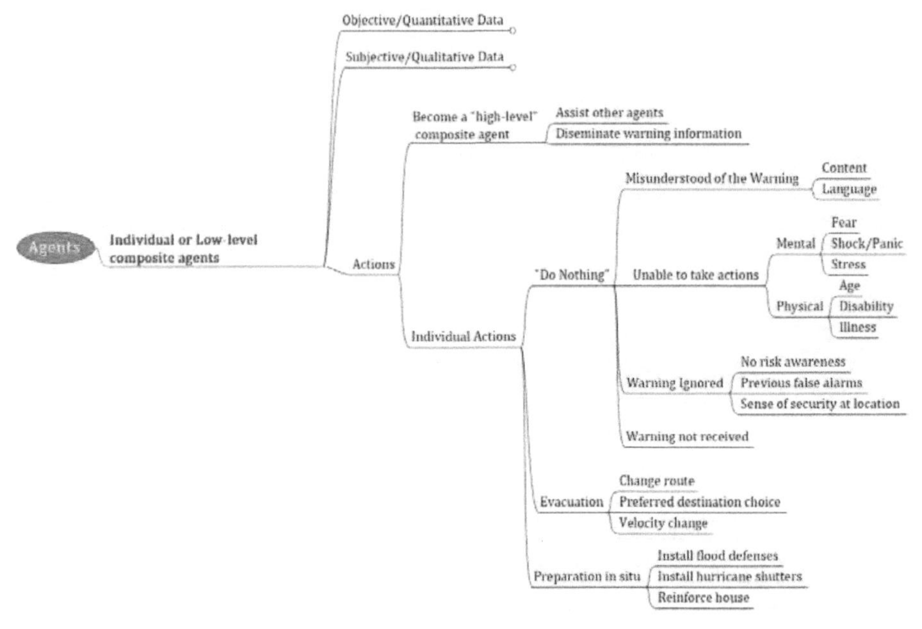

Figure 7.4 Set of possible actions for the individuals and low-level composite agents (Exp-Agents).

The last type of agents to be considered in ABM for operational DRM is the high-level composite agent. These usually are the institutions in charge of the DRM. For the exposure component of DRA, those organisations can be disaster departments, evacuation and shelter managers, forecasting and communication offices, among others. During the different phases of a potential disaster, they are in charge of forecasting possible threats. They also assimilate the forecast and decide what type of information and when it needs to be disseminated among Exp-Agents. They also need to define the way messages will be distributed, such as TV, radio, and emergency apps. High-level composite agents are also in charge of evacuation procedures, such as shelter preparation, traffic control, and mandatory evacuation orders, among others. The characteristics of high-level agents are presented in Figure 7.5.

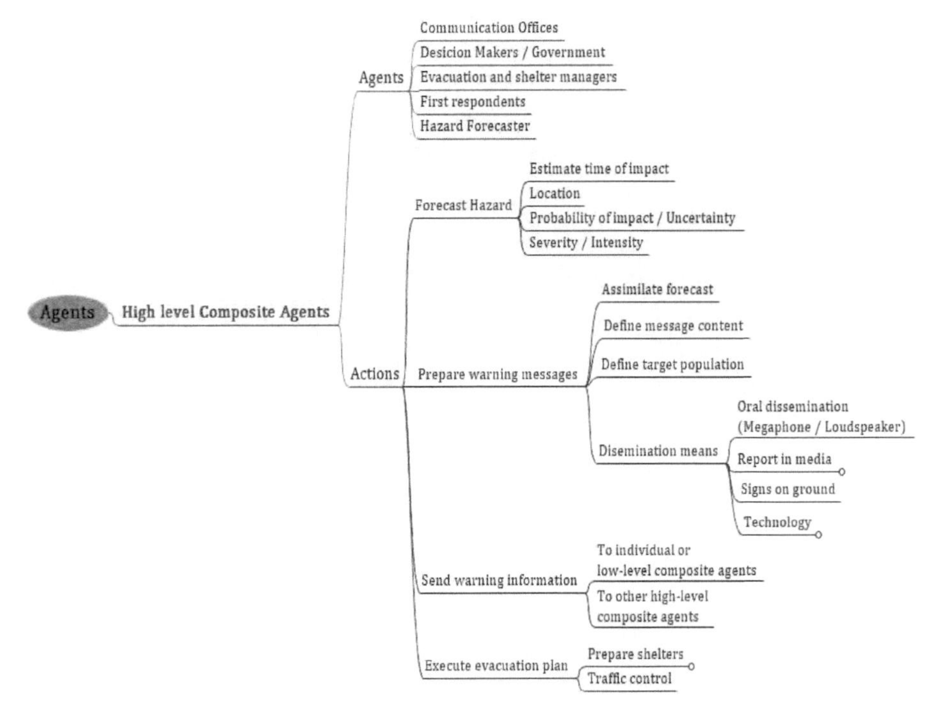

Figure 7.5 Formalisation of the high-level composite agents. Type of agents and actions.

7.2.3 Environment

The environment corresponds to the "physical world" where agents interact, perceive the "world" and act according to pre-defined rules. In the ABM model setup, the environment section describes all the external variables that under normal circumstances cannot be changed by the agents, either as a physical restriction or as a model decision. In the framework, the environment is composed of three main components. First, the hazard itself, for the scope of this thesis, the hazard is constrained to water-related disasters, such as floods, hurricanes and Tsunamis Figure 7.6. In addition to the type of hazard to be evaluated, it is necessary to include a geographic and time representation of the hazard. Different approaches can be used to represent a hazard in an operational ABM. One approach is considering a dynamic hazard that evolves as the simulation of the ABM advance (e.g. flood depth and extension), or it can be static over the simulation (i.e. flood-prone zones), or it can be represented as information in the agent's state (i.e. hurricane category, location and uncertainty cone). The approach to represent the hazard should be in the direction that allows answering a particular ABM's research questions. In an ABM for evaluating the impact of a flood event during an evacuation, a dynamic hazard can be more useful, as it will allow accounting the agents in direct contact (and potentially) with the hazard (Dawson et al., 2011). For an ABM where the intention is to evaluate the potential number of affected agents in a hard risk area, a representation of different hazard levels could be sufficient (Coates et al., 2014). Alternatively, in an ABM where the focus is on how information flow during a disaster, a pseudo-representation of the hazard could be enough (Rand et al., 2015).

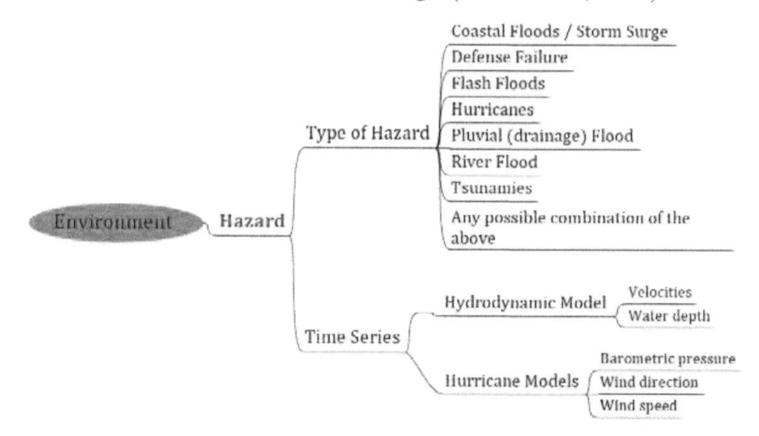

Figure 7.6 Formalisation of the environment, hazard component.

The second element of the environment corresponds to the physical objects or the infrastructure and elements that define a city topology. In our framework, the city topology is divided into four distinct features. Natural features such as water bodies and coastlines, Public infrastructure, in which it can be included, utilities networks, roads, flood defences and waterways. Geographic features of interest, such as elevation models and land use data. Furthermore, the buildings, which depending on the model complexity can be divided according to the type of building; also, the physical vulnerability of the building is usually represented as an attribute (Figure 7.7).

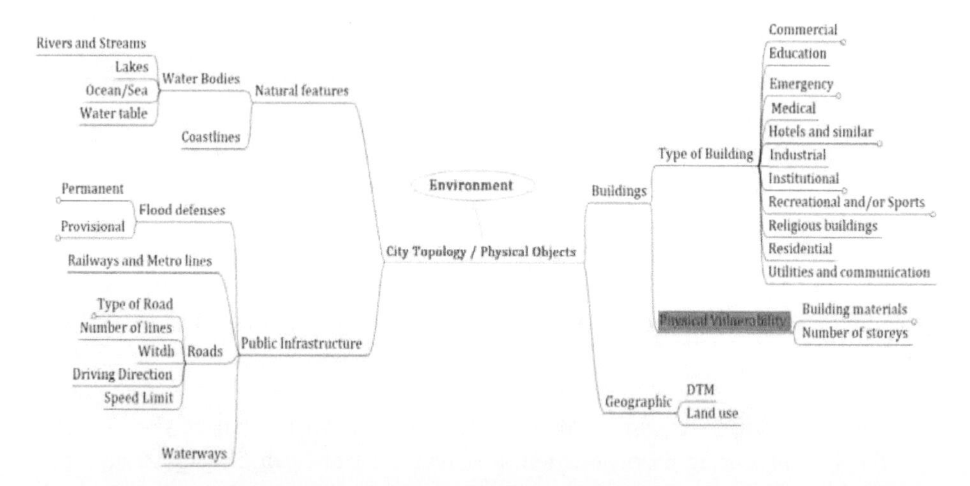

Figure 7.7 Formalisation of the environment, city topology.

The third component of the environment is defined as tasks; these are a set of rules that need a central control for the system. This component can include: distribution of agents in space and through time, to account for the output metrics, and to change some agents' role during the simulations (Figure 7.8). In the ABM categorized as an operational ABM, the simulation's impact measure is often associated with the impact of the hazard on the agents, such as loss of life, injured or in contact with the hazard.

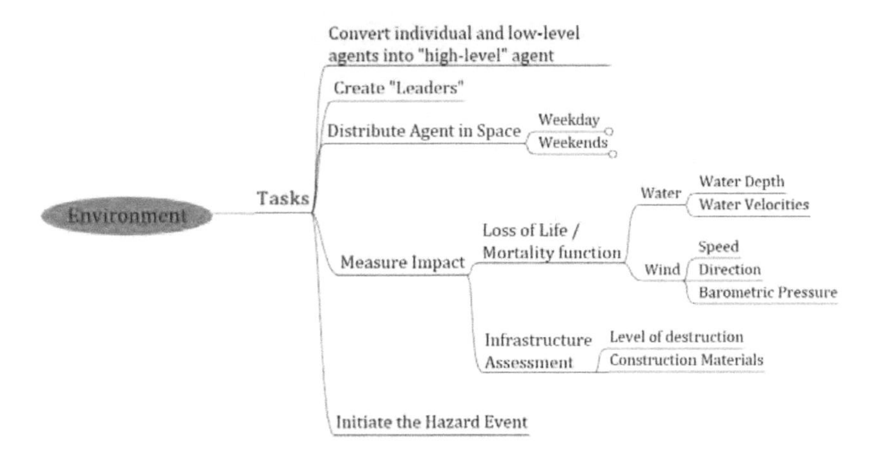

Figure 7.8 Formalisation of the environment, Tasks.

7.3 ABM Implementation. Sint Maarten Case Study

7.3.1 Scope and Conceptual model

In the case of Sint Maarten and based on the results of predictors of evacuation behaviour presented in Chapter 5, the ABM aims to evaluate exposure with a focus on the effects of information flow during different phases of a disaster and to assess how information can potentially motivate (or discourage) more individuals (or households) to perform evacuation or other types of protective behaviour. Our ABM implementation allows the different agents in which the system was decomposed to access, communicate, receive and process hazard-related information, and react accordingly to the agent's assessment of risk.

The ABM is focus and developed around the Exp-Agent that was defined in the sections above. The main goal is to evaluate the effect on evacuation behaviour of information content and the level of "trust" in the source of information. The ABM uses the Logit-iii- logistic regression model developed in Chapter 5 to evaluate each household's likelihood to evacuate given new content of information, and the source from where the agent received it.

Given the ABM's scope, the system decomposition was done based on meetings we held with several members in charge of the different components of DRM in the island during the fieldwork in the aftermath of Hurricane Irma, using the elements of presented in Figure 7.2. The conceptual model contains four blocks. The **first block** consists of the case study's geographic representation; this is the city topology/environment, which is the Dutch territory of Saint Martin. The **second block** accounts for the hazard, and it is represented by the winds and floods caused by Hurricane Irma. The **third block** corresponds to the information module, in which agents receive, interpret, and assess their own risk to different hazard-related information. The **fourth block** consists of the actions taken after an individual (or household) evaluate their risk. All the modules run in parallel, and they are connected. Figure 7.9 present the schematisation of the different elements of the ABM and how their connections.

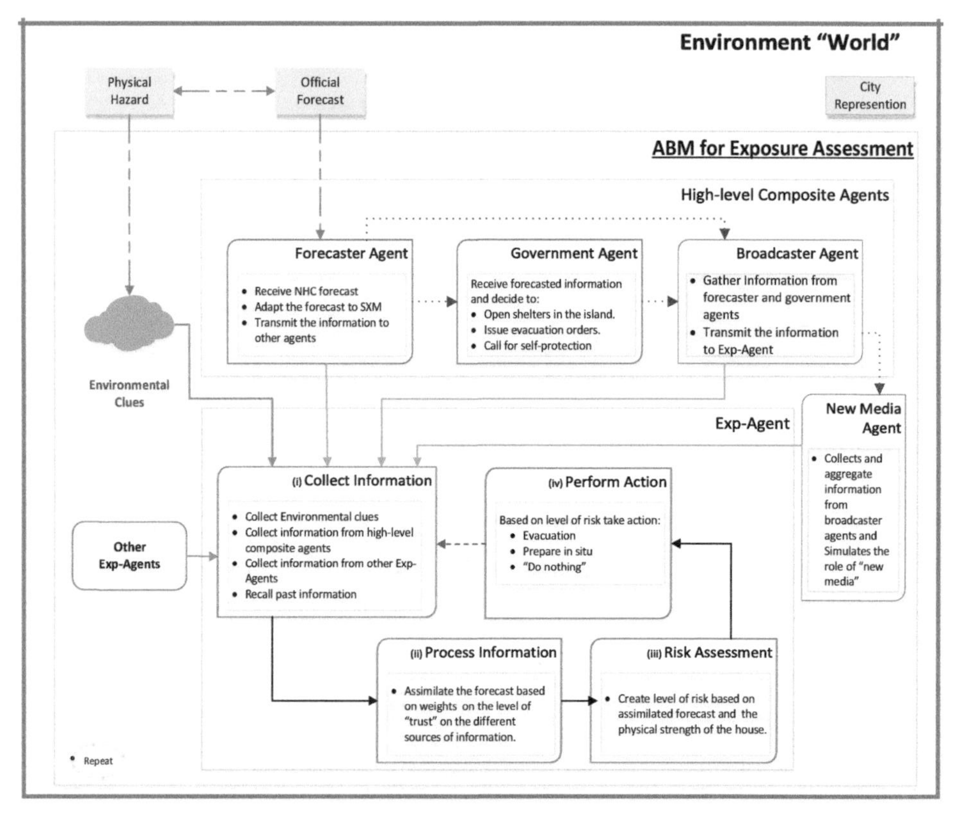

Figure 7.9 Schematisation of the main components of the ABM model and its interactions. Black continuous lines represent internal communication of Exp-Agents. Red dash line is the "remembered" information for the Exp-Agent after each time step of the model. Blue lines are external sources of information. Orange lines represent the communication between Exp-Agents through social connections. The green line corresponds to hazard information "senses" by the agent from the environment (e.g. seeing or in contact with the hazard). Purple dotted lines are information exchange between high-level composite agents. Grey dash line is information transmitted from the environment to the agents.

7.3.2 ABM Inputs and Setup

Based on the four modules of the ABM, a detailed description of model setup and inputs is presented in the following sections.

City Topology / Urban Environment

The Dutch part of the island was chosen as the environment that simulates the "real world" where the model's agents "exist in" and interact. Based on the elements of an urban environment presented in Figure 7.7, the elements that were considered of importance in Sint Maarten include the surface land represented with a digital elevation model, the surrounding ocean, coastline, the residential buildings, roads, inland water bodies and streams and the drainage channels (see Figure 7.10). In addition, the environment includes flood risk zones that correspond to flood inundation maps corresponding to a rainfall of 100-year recurrence interval. These areas are meant to capture different levels of risk in Sint Maarten. Conceptually, these risk areas are treated as potential evacuation zones, not as a dynamic hazard.

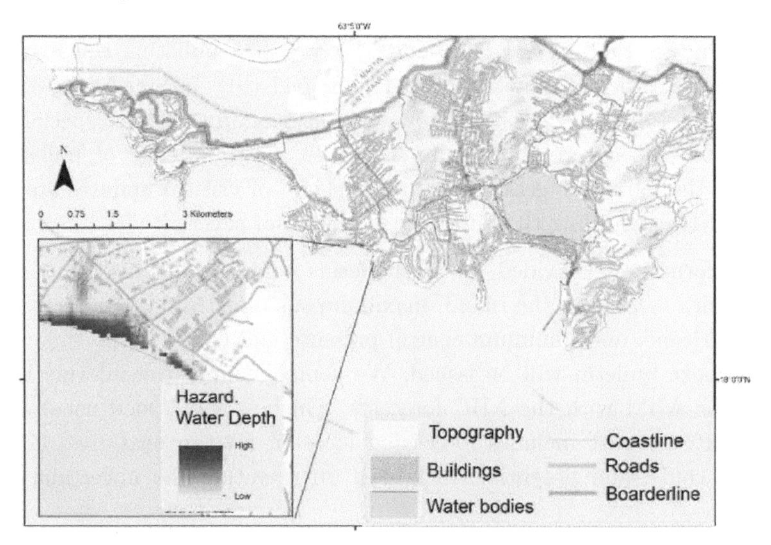

Figure 7.10 Map of Sint Maarten showing the urban elements (city topology), included as the ABM environment.

Hazards

The hazards that are within the scope of this thesis are those that can trigger a water-related disaster. We have chosen to include the effects of Hurricane Irma as the testing scenario for the ABM. The main hazard associated with Hurricane Irma in Sint Marten was the intensity of the winds associated with a hurricane category 5. Due to Hurricane Irma's magnitude compared to the island's size

(Figure 1.6), and the observed damage to the island's infrastructure, it can be safely assumed that the whole island faced potential disruptive winds of this hurricane (up to and above of 295 km/h) (Figure 4.1). The best track location of Hurricane Irma presented in Figure 7.11-(a), shows that Hurricane Irma eye crossed Sint Maarten.

Hurricane Irma did not bring extreme rainfall; hence inland flooding was not a significant issue during this hurricane. However, Irma caused localised flooding and storm surges in some areas, see detail in Figure 7.10. Furthermore, the flood hazard zones described in the city topology are also used to represent possible hazards. In the current development of the model, the awareness of an Exp-Agent knowing they live in a flood-prone area was assigned as a random probability to each household geographically located within a radius of 5 meters from the hazards inundation maps.

Information

In the ABM implementation, the hurricane wind and location is modelled as forecasted information. The forecasts about Hurricane Irma is extracted from the NHC archives (Figure 7.11-(a)), and the weather bulletins released by the meteorological office of Sint Maarten (Figure 7.11-(b)). In Sint Maarten, when a potential hurricane is forecasted, the meteorological office release special weather bulletins on average every 6 hours. However, the frequency of bulletins may change as the hurricane gets closer to the island, or critical updates are received from the NHC regarding the hurricane's location or severity.

The information provided in the bulletins contains the hurricane's current location with respect to the island, maximum sustained winds, the current speed of the hurricane, and minimum central pressure and contains the time and date that the next bulletin will be issued. We complemented the information to be used in the ABM with the NHC forecasts, which is also issued usually every 6 hours. NHC forecast includes forecasted location for the next five days, path, expected winds, and potential track area, representing the uncertainty on the information.

A key feature of the information "flowing" among the different agents is that not only the most updated information is available for recollection and use, past forecast is also possible to be circulating in the environment. This feature is important in the ABM design since in real life, not every individual update its information with the release of every new forecast (Lazrus et al., 2017).

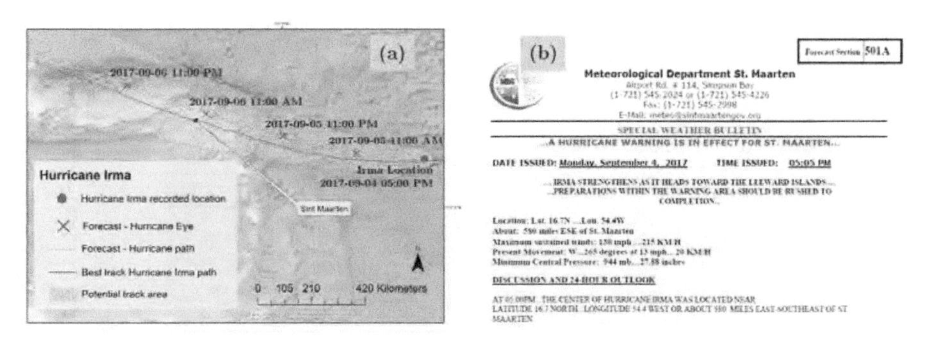

Figure 7.11 Hurricane Irma actual and forecasted information. (a) NHC 5-day forecasted location[9] and best track location of Hurricane Irma[10]. (b) Weather bulletin from the Meteorological Department of Sint Maarten for the corresponding NHC forecast.

Agents

There are two types of agents considered in the ABM. The high-level composite agents and the Exp-Agent. The high-level composite agents have four classes of agents: (i) a forecaster agent, (ii) broadcaster agents, (iii) "new media" agents and (vi) a government agent. High-level composite agents in our ABM are represented abstractly; they do not physically have a place in the city representation but can perform their actions through the environment. However, high-level agents are plotted randomly in the city representation for illustrative purposes.

The Exp-Agent in our ABM represents individual households. The Exp-Agent type of agents is represented explicitly in the ABM by using the geographic coordinates of the centroid of each residential building on the island using the updated database presented in chapter 2. Because our primary interest with the ABM, was to evaluate the flow of information and how it can change (or not) protective actions, we decided to implement the Exp-Agents in the ABM as static entities, this is, not moving agents in the road network and selecting a specific destination if they chose to evacuate. A detailed description of the agents and its actions is given in the ABM implementation section that follows.

7.3.3 ABM Implementation

The development of the ABM was implemented using NetLogo modelling tool. (Wilensky, 1999). An overview of the graphical user interface is shown in Figure 7.12. The ABM was built adapting the original CHIME ABM v1.4 (Watts, 2019) and initially presented in Watts et al. (2019). We have re-used the code and adapted it to reflect the local characteristics of information flow and institutions in charge of disaster management during a disaster in Sint Maarten. Besides the

[9] https://www.nhc.noaa.gov/gis/archive_forecast_results.php?id=al11&year=2017&name=Hurricane%20IRMA

[10] https://www.nhc.noaa.gov/gis/best_track/al112017_best_track.zip

understandable environment change, we have adapted some of the conceptual design and some rules governing agents' behaviour.

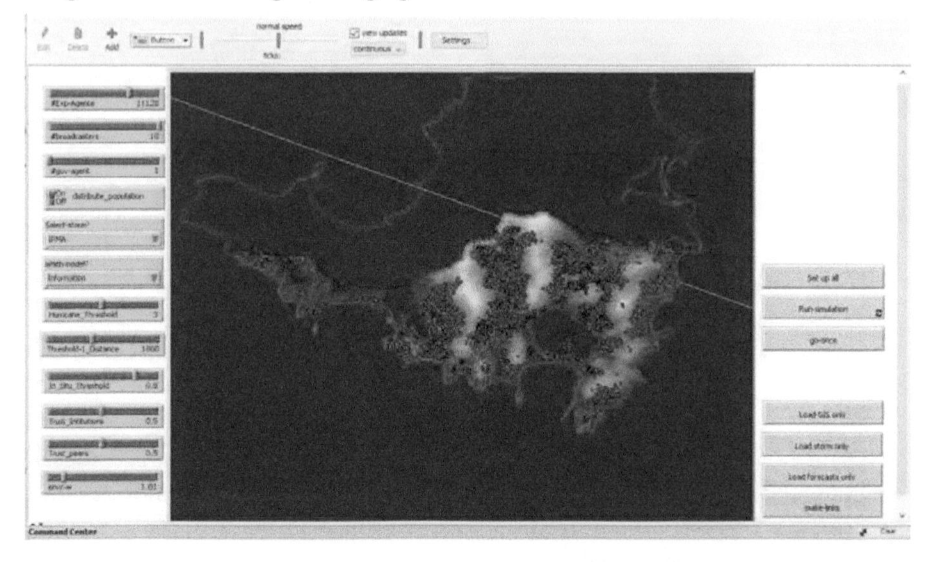

Figure 7.12 GUI of the Netlogo implementation of the ABM. To the left, user-adjustable factors. In the middle, the geographical representation of the environment, Sint Maarten is shown in full extension using the DTM to represent the inland, in dark blue the ocean and inland water bodies. Points here, represent the agents using the centroid of the building. The grey line represents the best track of Hurricane Irma. To the right side, the actions buttons to load data and run the simulation.

As we were interested in the effect of information flow before the hazard's impact, the model was simulated for five days, starting on 1 September 7:00 am local time and ending on 6 September at 7:00 am. Hurricane Irma made landfall on Sint Maarten on 6 September at 07:15 am local time. The eye of Irma crossed through the whole Saint Martin Island, and it took around two hours for the hurricane to cross the island and an estimated 45 minutes between the front and tail of Irma's eye according to information collected from residents in the field mission. Therefore, the information flow during the hurricane and the aftermath phases is not accounted for in the model's present version. In addition, the time step of the model was set to 1 hour, for a total of 120-time steps.

To adequately capture the effects of information flow on the protective behaviour of Exp-Agents before a disaster triggered by a hurricane, we have implemented five distinctive types of agents in the model. Each agent with a particular role in the DRM chain on the island, and in the model. The simplified conceptual design of the ABM consists of the interaction of the five types of agents, as depicted in Figure 7.13. The ABM starts by initialising the agents and the environment, followed by the acquisition of forecast information made by the forecaster agent, which sends forecast information to the other type of agents in the model. The information activates their respective modules. An overview of the agents' types and some model specifications are presented in Table 7.1, followed by a more detailed description.

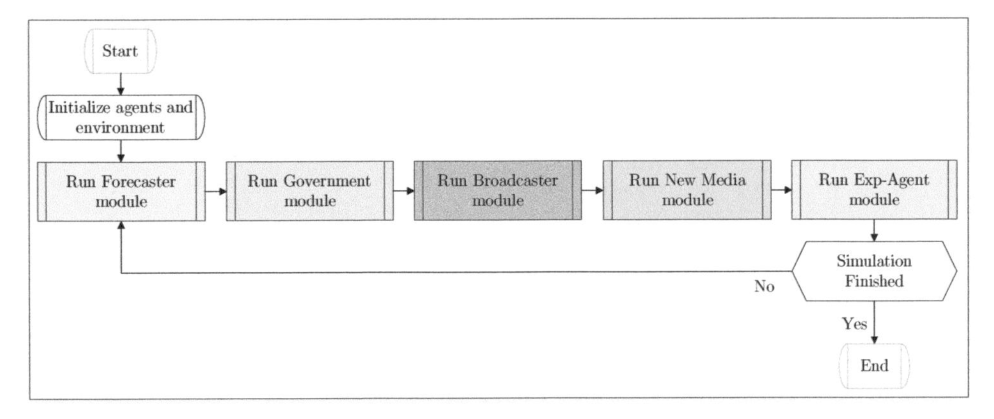

Figure 7.13 Simplified overview of the ABM implementation. Each module represents the behaviour of the different type of agents. The modules are better described later in the document.

Forecaster Agent

In Sint Maarten, the forecast of possible water-related hazards is in charge of the meteorological office. This agent is in charge of receiving the official weather bulletins issued by the NHC in the USA and adapt them to the local needs of Sint Maarten. Once the forecaster agent has the forecast ready for the island, it sends it to the broadcaster and government agents, using official communication channels.

Forecaster in Sint Maarten also sends information to the general public through the Meteorological office's Facebook page. We include this communication between the forecaster and the Exp-Agent in the model implementation. Forecaster agent information is in meteorological bulletins with all the technical components, including uncertainty and five days forecast (Figure 7.11). The flow of information from the forecaster agent is shown in Figure 7.14.

Table 7.1. Type of agents and actions and their implementation in the ABM.

Type of Agent	Number of Agents	Actions	Distribution in the model	Temporal scheduling
Forecaster	1	Forecast Hurricane. Transmission of information	Random[1]	At every time step
Broadcaster	8	Transmit information to Exp-Agents from the forecaster and the government agents using traditional dissemination mechanisms	Random[1]	At every time step
New media	8	It takes information from the broadcaster agent, aggregates it, and later transmits it to the Exp-Agents in a simplified form and using alternative communication channels.	Random[1]	An assumption was made to assign a probability of 33% chance of being active every time step
Government	1	Receives and interpret forecast and decide the type of actions expected from the Exp-Agent	Random[1]	At every time step
Exp-Agent	11128	It collects information from forecaster, broadcaster and new media agents. As well as from "own" memory and environmental clues. Once the message is assimilated the evaluation of risk is performed, and actions are taken accordingly	At the centroid of the house	The agent is active for the first 36 time steps. After that, random scheduling is assigned for the agent to collect (or not) new information

[1] Geographic location is only for illustration. No physical impact is assessed on this agent

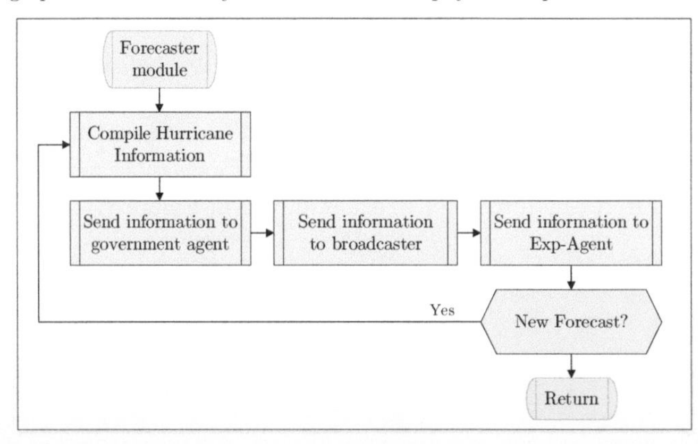

Figure 7.14 ABM implementation flowchart of forecaster agent module.

Broadcaster Agent

Broadcaster agents in Sint Maarten represent the traditional and official sources of warning dissemination and communication. In the list of this type of agents in Sint Maarten are radio, television and newspapers. This agent receives information directly from the forecaster agent and from the government agent and transmits it directly to the Exp-Agent using traditional warning dissemination sources and without changing its content. The broadcaster agent's information to the Exp-Agent differs from the one given for the forecaster in the content; it is simplified only containing the hurricane's current location, wind speed, and the 5-day forecast containing possible path and winds, but without the uncertainty band. Figure 7.15 shows the information flow for the broadcaster agent.

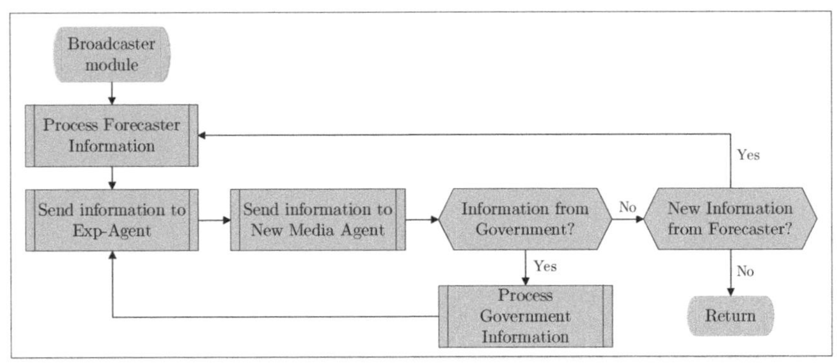

Figure 7.15 ABM implementation flowchart of broadcaster agents module.

From the government agent, the broadcaster agent receives official orders regarding disaster management, such as evacuation advise to official shelters, accompanied by a list of shelter availability, or a suggestion to protect their houses, or suggestion to evacuate to a stronger house if the own house is perceived weak, which if received at any time step is also communicated to the Exp-Agent.

An important fact for Sint Marten is that there are no mandatory evacuation orders in the DRM policies on the island, only the advice to evacuate to a safer place if seeming necessary. According to the interviews we had with DRM officials there are no intentions to change this in the future as they believe Sint Maarten inhabitants will never follow such orders given the idiosyncrasy of St. Maarten's residents and the island administration lack the resources to enforce it.

New Media Agent

New media agents are added into the ABM's conceptualisation to capture the role of "new media" information means (e.g. social media, designated apps). The intention is to represent a current trend on emergency-related information provided from non-official sources; thus, the new media agent's information is unverified. In our model representation, the new media agent does not receive information directly from the government agent in contrast to the broadcaster agent.

The new media agent does not run throughout the whole simulation; at every time step, a probability of 1/3 is assigned to each agent to allow it to become an active source of information for the Exp-Agent. The probability of transmitting information implemented in this type of agent aims to capture that new media information tends to be random without a defined time step for publication in real life. Information flowchart for the new media agent is depicted in Figure 7.16.

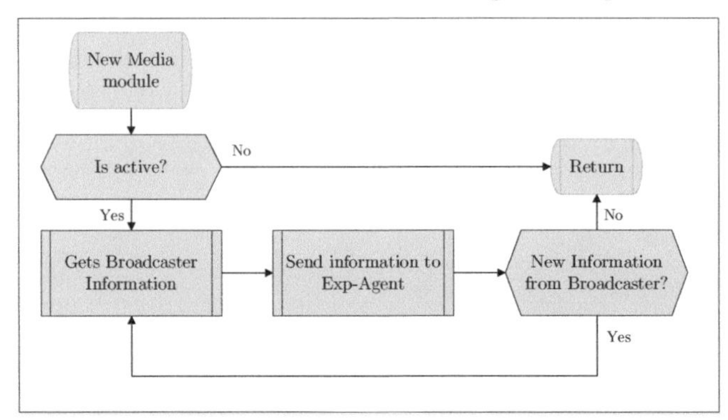

Figure 7.16 ABM implementation flowchart of new media agents module.

Government Agent

This agent represents the island's prime minister and supported by the different ESF (see Figure 5.1). This agent receives the forecast information directly from the forecaster agent. Once the forecasted information is received, it is processed. It is for the government agent responsibility to decide to open or not the shelters and inform Exp-Agents about protective actions such as evacuation or in-situ preparation. The information between the government agent and Exp-Agent uses the broadcaster agent as an intermediary. Figure 7.17 shows the information flow for the government type of agent.

Government agents receive information from the forecaster agent and update its risk assessment at every time step of the simulation. Government agent uses the hurricane category, the distance to Sint Maarten, and the probability of a direct hit, measured using the potential track area (uncertainty cone) to assess how risky a specific storm is when approaching the island. The uncertainty cone uses a weighted function to account for the higher uncertainty in the hurricane path associated with a more extended period ahead forecast.

As mentioned before, no direct evacuation orders are sent because it is not an actual scenario in Sint Maarten, and according to several officials, it is not likely to change in the DRM structure.

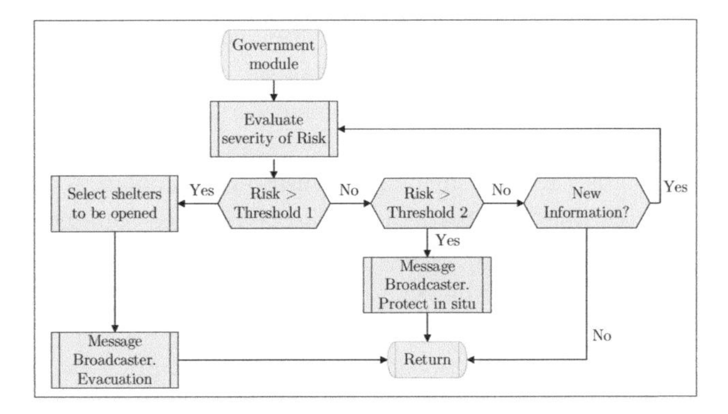

Figure 7.17 ABM implementation flowchart of government agent module.

Exp-Agents

In our model conceptualisation, each Exp-Agent represents a household in Sint Maarten at its geographic location. The modelling choice to evaluate the protective behaviour at household level rather than an individual level was made based on the fieldwork findings; in the island, when a hurricane is forecasted the protective behaviour is performed as a household unit rather than individually (Chapter 2). In addition, the model conceptualisation for this type of agent includes a particular behaviour detected in Sint Maarten. During the surveys, we detected that some households on the island always leave one person behind to protect the house from the storm or from looters even if they decide to evacuate their premises. The ABM allows capturing this behaviour so that Exp-Agent can be fragmented at any point in the simulation and measure the possible impact of such behaviour.

The Exp-Agents are the central concept of our ABM model, and as such, a more sophisticated design was done in order to be able to capture the complex protective decision-making process of households during emergencies. Two protective actions are implemented in the ABM based on the findings in Sint Maarten. We observed that households on the island either choose to evacuate to a place they consider safer than their own (i.e. friends or relative, hotels, shelters) or decide to stay in their residence and take in-situ protective measures. Also, if the assessment of risk results in low risk for a particular household, the "do nothing" behaviour is assigned.

To simulate the two protective behaviours mentioned above, we adopted the conceptual framework Protective Action Decision Model (PADM, Lindell and Perry, 2004, 2012), adapted in Watts et al. (2019). The implementation of PADM in our model considers the previous chapters' findings in this thesis, on vulnerability and evacuation behaviour, which allows us to find the variables that may lead to protective behaviour on Sint Maarten. The processes that agents undertake to evaluate the actions to perform during the simulation is shown in Figure 7.18.

At initialisation of the simulation, each household is assigned values for the six variables/predictors of evacuation in the Logit-iii regression model presented in Chapter 5. The variables are gender, homeownership, percentage of property damage, quality of the information, number of house storeys, and the vulnerability index. These variables remain unchanged during the simulation, except for the variable quality of the information, which is changed through the agent's evacuation assessment module (Figure 7.18 and Figure 7.19). Alongside the information for the Logit-iii model, each household (Exp-Agent) is assigned multiple variables that influence its behaviour towards the collection of information and the type of actions taken towards adapting or not protective behaviour (Table 7.2).

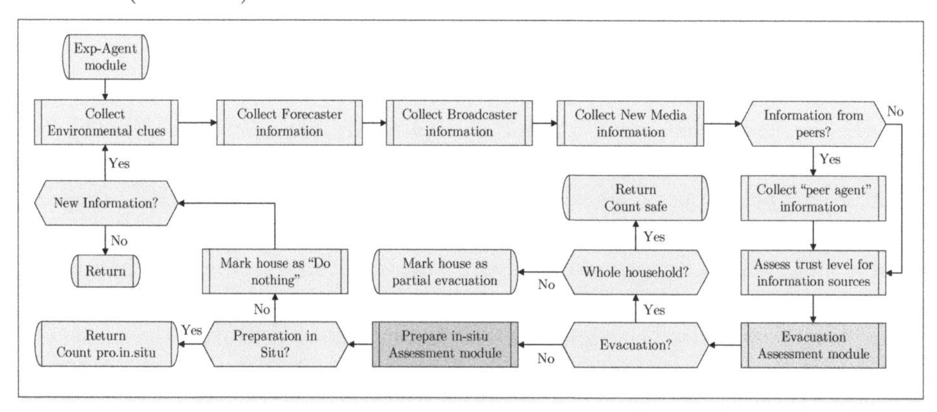

Figure 7.18 ABM implementation flowchart of Exp-Agent.

Table 7.2. Main variables for Exp-Agents, evaluated for each household independently.

Variable	Type	Observation	Value
Trust Score	Static	Assign the level of trust for each source of information (Forecaster, Broadcaster, New media, Government/Institutions, and peers)	Random [0 – 1]
Self-trust	Static	Indicates the level of trust in the previous assessment of risk (collected from Exp-Agent "memory")	Random [0.6 – 1]
Flood prone zone	Static	Indicates whether the Exp-Agent is located in flood-prone areas (see 7.3.2)	Binary. [0=No, 1=Yes]
Risk-level	Static	Evaluation of agent's risk, If above the threshold, evacuations protective action is assigned	Random-normal with mean 14, std. dev. 2
In-situ-risk	Static	If risk-level does not activate evacuation, a second threshold is evaluated to assign 9or not) protective action in-situ	0.75*risk-level

Table 7.2 (Continued)

Variable	Type	Observation	Value
Environ-clues	Dynamic	It indicates if a given time step an Exp-Agent is "sensing" Category 3 or bigger hurricane winds. It is based on the NHC best track reports	Binary. [0=No, 1=Yes]
"memory"	Dynamic	Previous interpretation of the information	Stored previous assessment
Partial-evac	Static	Used to assign a probability of partial evacuation to those household performing evacuation	Random [0 -0.35]*risk-level
Evac-dec	Dynamic	Result of the assessment of the evacuation decision	Categorical. 2 = Full evacuation 1= partial evacuation 0=No
In-situ dec	Dynamic	Result of the assessment of the in-situ protection	Binary. [0=No, 1=Yes]

In the current development of the ABM, Exp-Agent collects and process hazard-related information from all other agents in the ABM, and from environmental clues as represented in Figure 7.9. The agent first "look" in the environment and "sense" the wind speed and evaluate if it is in contact with flooding waters. From the forecaster and broadcaster agents, the Exp-Agents gets the latest official information regarding the hurricane. From the new media, the Exp-Agent gets what is called in this research as non-official information. Also, the agents start communicating with peers to access their assessment of present risk. Finally, the Exp-Agent remembers the last time-step information and uses it to evaluate the hazard's evolution (i.e. stronger winds or hurricane cone deviating).

The Exp-Agents are active through the entire simulation, and access to every new information release during the first 36 hours of the simulation. After hour 36, a probability function is applied to whether or not the agent will collect new information. The probability function is introduced to simulate that in real life, an individual (or household) may not check as actively as in the start of the simulation for information regarding the hurricane. Once an Exp-Agent has collected the information from the different sources, it proceeds to rank the information according to the source; a trust-score variable is used to rank each source of information. The trust-score gets a random value that ranges from 0 and 1 at initialisation of the model. The level of trust in the government agent and the information it gets from peers can be defined separately, given our research goals. In the model setup, for experimental purposes (see 7.3.4), it is possible to select a value for the level of trust for these two types of agents, ranging from 0 to 1.

Once the information is collected and weighted, the evacuation assessment module is used to evaluate whether a household will perform evacuation, given the new information level (Figure 7.19). The households' risk assessment includes the level of trust on the source of information, the severity of the hazard, the hurricane's distance to the island, and uncertainty on the forecasted information.

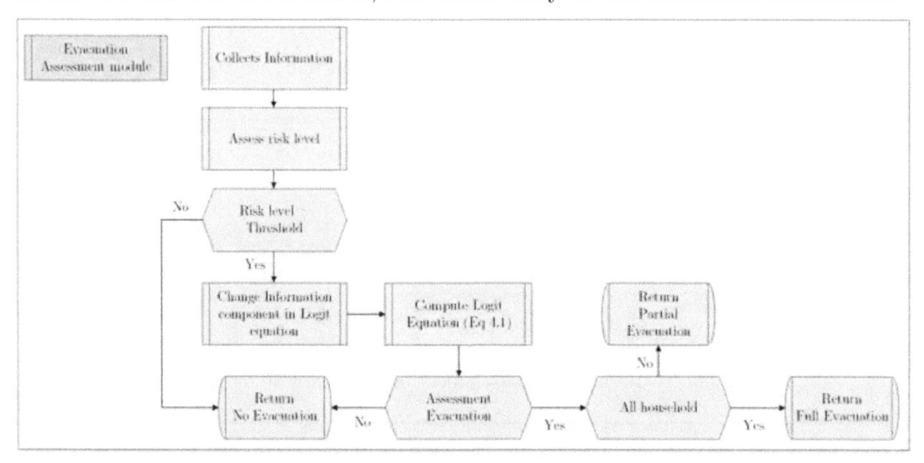

Figure 7.19 Exp-Agent decision-making process to evaluate evacuation as the protective action.

Under the current development, an assumption was made that only hurricanes with a category at least three will change the information component of the evacuation logistic regression. Hurricane category 3 was chosen based on the findings of Chapter 2-Figure 2.1 referring to intentions to evacuate according to the storm's intensity. This parameter can also be changed through the user interface for experimentation purposes. After the hurricane category, the agent evaluates how far the hurricane is in relation to Sint Maarten; a threshold is defined to assess if this parameter may trigger a change in information. The agent also uses the potential track area or cone of uncertainty of the 3-day NHC forecast to assess the uncertainty. If all three criteria described above are met, the agent evaluates the level of trust of the different sources and decides to which source gives the priority in the assessment and proceeds to compute the risk level.

The computed risk level is used to evaluate if the agent will increase or decrease the information component in the logit function (Eq 5.1-Chapter 5). Using the new value of evacuation probability, the new likelihood to evacuate is assessed. If a household evacuates, it is at this point on the agent's decision-making that it is decided if one member of the family is left behind to protect the house (return partial evacuation) or if they evacuate as a whole (return full evacuation) using a random probability. The probabilistic function is used because we did not collect actual information on the partial evacuation level; the value random value used is expected to produce low levels of partial evacuation.

If an Exp-Agent decide not to evacuate, a second conditional evaluates if the agent would take protective action in situ (highlighted purple in Figure 7.18). We did not collect information during the fieldwork regarding how many households undertake protective action in-situ, hence for the implementation of the in-situ protective behaviour we have included a random procedure to assign a probability of deciding to take protective actions in the house. Protection in situ is computed as a percentage of the risk assessment computed in the evacuation module using a predefined threshold (Figure 7.20). The Exp-Agent module ends by marking as "Do nothing" the households that did not evacuate or did not prepare in-situ. At every new time-step, the evacuation and prepare in situ models are only run using the "Do nothing" households.

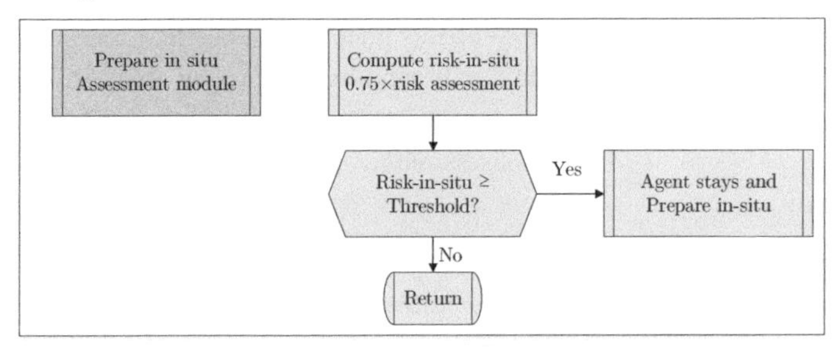

Figure 7.20 Exp-Agent decision-making process to evaluate in situ preparation as the protective action.

Exp-Agents information network

Based on the original code of the ABM (Watts, 2019), the information is shared between agents through a social and information network. In our model, one-way communication with Exp-Agents was set up for the types of agent forecaster, broadcaster, and new media. Exp-Agents can only receive information from them but cannot send information back. Also, all Exp-Agents can connect with all existing agents of these types. In addition, there is no direct communication between government agents and Exp-Agents; this is done through the broadcaster.

Information exchange between peer Exp-Agents is done using a simulated social network. The social network size and connections are created randomly at initialisation of the simulation. An Exp-Agent can connect to a maximum of other ten Exp-Agents for a given simulation. The agents remember the social network through the whole simulation, but it changes from simulation to simulation. Further development of the ABM should consider assigning an ID to the connections in relation to the agents to remain unchanged at every new simulation. The communication between Exp-Agents is bidirectional, allowing to exchange information between two Exp-Agents.

Key Assumption and model simplification

When an agent evacuates, it is assumed that the evacuation was performed to a safer place than its own; hence, the agent is accounted as safe. We assume that the self-assessment of risk is a good indicator to assume if the agent decided to evacuate they will be rational enough to choose a safer place and that the evacuation was done before the hurricane. In this version of the model, choosing a place that results in a more or equal exposure level is not included. Future implementation can simulate this effect by incorporating movement on the agents to account for agents exposed to winds or flooding waters and assign a specific destination.

Once an agent has decided to take protective action, either by performing evacuation or selecting protection in-situ, the model remembers the selection, and the Exp-Agent is no longer considered for information evaluation in the subsequent time-steps. Once an agent has adopted a specific protective behaviour in the model implementation, this will remain unchanged during the rest of the simulation. However, the agent remains in the simulation to still be able to pass information about its risk assessment to other Exp-Agents via peer communication.

The current implementation of the ABM focused on the exposure of households (Exp-Agents). High-level composite agents (government, forecaster, new media), were also included in the model, but with a lower degree of sophistication in its design. Further implementation should consider increasing the rules and connections of these type of agents interact with the hazard and with other agents, this may reveal new insights on the role of information during an emergency. Also, high-level composite agents' explicit location should be included in future model developments; as they too can be impacted by the hazard, and undermine their capabilities to communicate with other agents.

The ABM implementation focuses on the role of information; however, in Sint Maarten, evacuation behaviour was also dependent on other variables (i.e. homeownership, property damage). A more advance ABM should also evaluate the role of such predictors in the protective actions behaviours.

New media agents in the current model version only transmit in a simplified form what they receive from the broadcaster; we did not implement changing the value or content. Future implementations should include a more refined design of this type of agent to capture the complex flux of information over the internet. In reality, information from new media agents can be accessed almost all the time during a disaster, which is usually not filtered and validated, often presenting fake or partial news. Hence, Exp-Agents could be accessing more constant information, but more uncertain, which may lead to different results to the ones presented here.

Our ABM conceptualisation does not include feedback in the decision-making process regarding previous experiences that an agent may have had in the past, simplifying reality. In new implementations of the ABM, for a new hurricane season, the Exp-Agent should carry at the start of the simulation a "memory" regarding past experiences (i.e. previous evacuation behaviour, losses).

7.3.4 Experimental Setup

At initialisation of the model, all the required inputs are loaded; this includes base maps and the hurricane-related information (i.e. best track and forecast/bulletins). It is followed by the agents' distribution on the environment, and the peer network is created. Next, all system and agents variables are initialised based on the default or user assigned values. Once the simulation start, all the modules run synchronously performing their respective tasks in the model as conceptualised.

To assess the effects of information distribution on the exposure component of risk, we run a series of experiments varying key variables related to information flow and how they promote or restrict protective behaviour. The variables of interest to evaluate in this experimental set up can be divided into two groups. First, a variable directly related to the hurricane, we selected to test the effects on activating protective behaviour according to the hurricane's intensity. The second group of experiments is associated with the level of trust in the source of information. In our model, the effect of trust in institutions and trust in peers are evaluated.

The range values used in the modified setups of experiments are presented in Table 7.3. Experiment 1, is the based model run with the default values defined for the variables. Experiment 2, corresponds to the evaluation of varying the hurricane intensity as a precursor of protective action behaviour, while other variables remain constant. For experiment 3, the variable trust in institutions is evaluated, holding all other settings constant. Furthermore, in experiment 4, we evaluate the effects of trust in peers in protective action behaviour.

In addition, given the stochasticity associated with agent-based modelling, a broad variability on the results is possible from simulation-to-simulation. Hence, all experiments were run 100 times each to account for the stochastic effects, the number was selected based on literature review where it has been observed a stabilization on the main parameters after a few hundred replicates (Jenkins et al., 2017; ten Broeke et al., 2016; Yang et al., 2018). The results presented in this chapter corresponds to the aggregated values, using the average obtained in the multiple runs. Given the large numbers of, simulations, we used NetLogo *BehaviorSpace* tool to run them in parallel.

Table 7.3. Information related variables and their experimentation settings. In bold the modified values.

Variable	Experiment			
	1	2	3	4
Hurricane Category	As defined in Table 7.2	**1-3-5**	3	3
Trust in institutions		0.5	**0.10, 0.50, 0.75**	0.5
Trust in peers		0.5	0.5	**0.10, 0.50, 0.75**
Number of runs	100	100	100	100

7.3.5 Results and Discussion

Based on this research's primary goal, the output results from the multiple experiments and runs were analysed based on the percentage of Exp-Agents that decided to evacuate or protect in situ. In addition, because our agents are located in their geographical location, we could summarise the outputs as probability maps. The results of the experiments are presented below.

Experiment 1

First, we run a scenario where all the simulation variables were the default values given during the model conceptualisation. This base scenario serves as a validation of the model setup, by analysing that the agents are performing as intended in the model, the number of evacuees in this scenario were expected to evacuate at the rates we observed in the survey we performed after Hurricane Irma, this is around 31% of agents evacuating (Figure 2.13 and Table 4.1). Also, having a base case allows us to compare scenarios with the further experiments we run.

The base experiment results are shown in Figure 7.21-(a) and the results of the evacuation analysis performed in Chapter 5 are shown in Figure 7.21-(b) for comparison. The base experiment map is computed using the average value for each agent over the 100 simulations, and it is aggregated at the neighbourhood scale using the VROMI categorization (see Appendix E).

As observed, a similar pattern of protective action was achieved in the base model compared to the probabilistic evacuation map. The base experiment preserves higher protective action patches over the same neighbourhoods as the actual observed behaviour associated with Hurricane Irma, such as Philipsburg (ID=39), Dutch Quarter (36), Middle Region (37), Cay Bay (17). Changes in the maps' similarity in these neighbourhoods might be associated with the inclusion in the exposure prediction of in-situ protection's protective behaviour. Agents in this areas assess their risk as higher due to the associated socio-economic vulnerability and perception of living in higher risk areas, which for the model increases the sensitivity to risk assessment, hence increasing the probability of performing a type of protective behaviour.

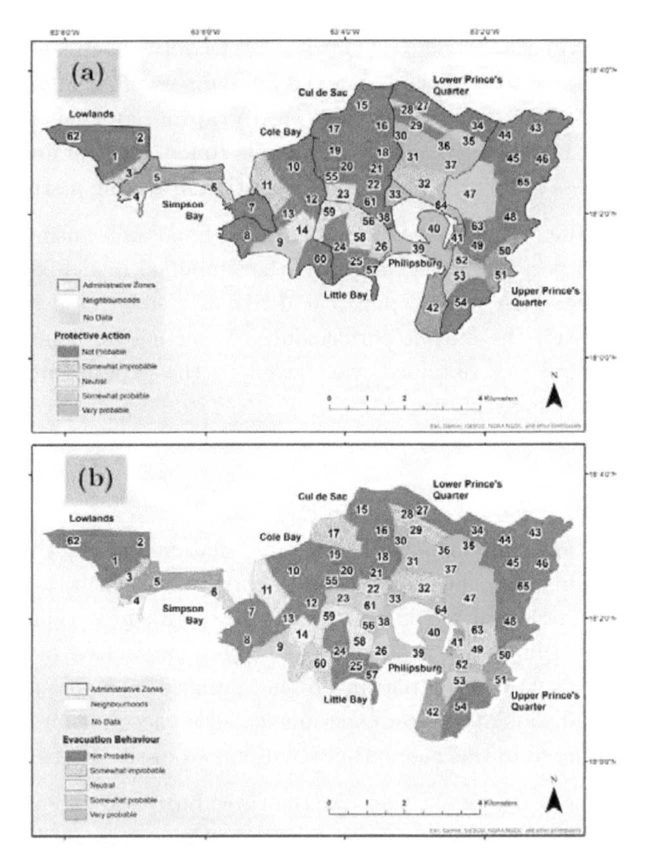

Figure 7.21 Evacuation Probability Maps. (a) Experiment 1- Base simulation. Based on the average response of households over the 100 simulations and aggregated at the neighbourhood scale. (b) Predicted Evacuation behaviour in Chapter 5-Figure 5.3-(b). Numbers represent the identification (ID) of each neighbourhood, as presented in Appendix E.

The base experiment also preserves the regions where a low probability of performing any type of protective actions such as Cul de sac, Lowlands and Upper Prince's Quarter administrative zones. The experiment 1 results show that overall it is capturing the protective action behaviour, but in those neighbourhoods where not a match was achieved it is observed a decrease in the intentions to perform protective action, see St Peters (17), Little Cape Bay (60) and Cockpit (11) neighbourhoods. The simulations we set up are meant to replicate low evacuation rates in the zones mentioned above to replicate the observed patterns. Setting up the model to achieve low evacuation rates may influence having more neighbourhoods categorized as not probable or somewhat improbable categorie.

One crucial validation of the model is that not all Exp-Agents in high-risk areas are taking protective behaviour, as it was observed in the case of Sint Maarten, see Simpsons Bay Village (6), Union Farm (29) and Little Bay Village (57). The opposite also applies, there are some agents in no so-hazardous areas taking protective behaviour, see Sucker Garden (47) and Bloomingdale (63). This behaviour is possible to achieve due to the randomness of some variables used to set up the model. The randomness allows us to capture an agent deciding not to evacuate from a high-risk area due for example to low trust in authorities or the threshold to assess risk is too high (e.g. perception of having a strong house).

Replicating the same behaviour is nearly impossible using the type of simulation ABM perform, especially given the simplified model conceptualisation used in this research. A more sophisticated risk assessment and decision making model may increase the output performance. Nevertheless, being satisfied with the degree of similarity obtained, we moved to the experimentation with the model in the following sections.

Experiment 2

This experiment varies when agents start assessing the risk associated with the forecasted hurricane's intensity. Collecting information and reacting earlier in the hurricane's timeline associated with lower hurricane intensities or later in the hurricane's timeline for higher intensities. This experiment's results are shown in Figure 7.22, describing the probability of taking protective behaviour at the neighbourhood level. It is important to note that the maps show the result averaging the 100 simulations per experiment and its account for both protective behaviours evaluated in this chapter; evacuation and in-situ protective behaviour.

Comparing the protective actions for the three hurricane intensities evaluated in experiment 2, shows a clear trend to decrease the number of households (and hence neighbourhoods), as the hurricane intensity increases as a precursor of protective behaviour. In other words, fewer agents will take protective actions if they consider a high category hurricane is not a threat. Even though this behaviour was expected, it serves as another validation that our code is performing as intended.

In addition, interestingly, even decreasing the activation of protective action to the lowest in the hurricane scale (Category 1), still produces results in which some neighbourhoods will not take action. Cul de Sac, Upper Prince's Quarter and Lowlands administrative zones contain most of the neighbourhoods that will not perform protective action even when a category 1 hurricane is used to activate the evaluation of risk. The behaviour of not taking actions in these areas can be explained in these areas hosting some of the wealthiest neighbourhoods of the island.

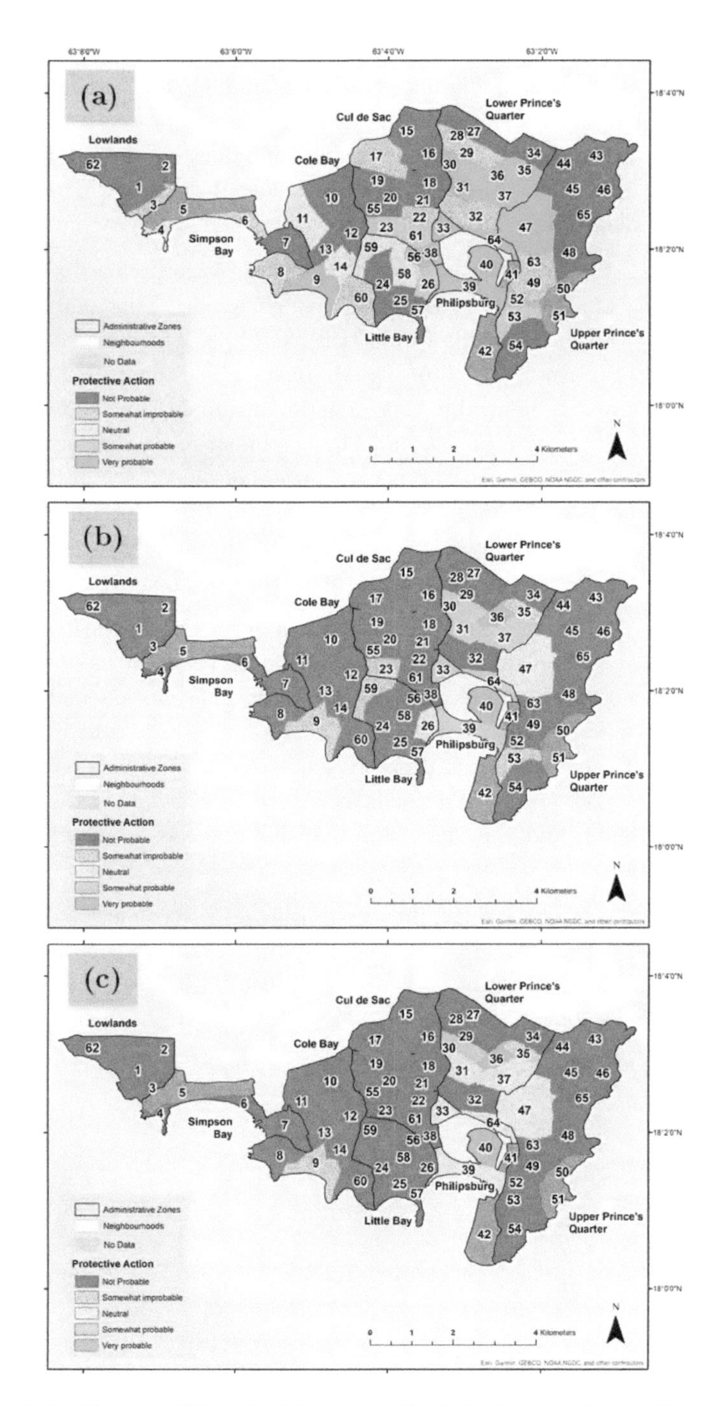

Figure 7.22 The probability of taking protective behaviour at the neighbourhood level in experiment 1. (a) Hurricane category 1. (b) Hurricane category 3. (c) Hurricane category 5. Numbers represent the identification (ID) of each neighbourhood, as presented in Appendix E.

A perception of strong houses associated with (perception) of good construction materials and methods is highly likely to be explaining why the confidence of not taking protective action in these areas. Also, not taking protective action can be explained that lower hurricane intensities for Sint Maarten do not create a sense of urgency to take precautionary actions. A long history with hurricanes has promoted the island's construction to be resilient to relative low hurricanes

In contrast, the scenario where a category 5 hurricane is used, most neighbourhoods would not perform protective behaviour. This analysis allows us to detect which neighbourhoods would still likely take protective actions in this scenario: Zorg En Rust (ID=30), Dutch quarter (36) and Pond Island. Six neighbourhoods remain neutral in this scenario, meaning there is a 50-50% chance of taking or not protective behaviour. The neighbourhoods that are probable to perform protective action and the neutral neighbourhoods are either located in the poorest areas and/or in flood-prone areas. This is a promising result for disaster risk managers in the island, as some of the more dangerous zones are presenting good chances of performing protective actions.

Not taking protective behaviour associated with the hurricane Category 5 scenario can also be explained in the lack of time to react to the potential threat leading to fewer agents protecting in situ or evacuating, associated with less frequent recollection of information. More constant collection of information at earlier phases of the hazards can promote protective actions associated with more risk awareness. Also, fewer agents will be willing to evacuate if the hurricane is closer to the island, and they may perceive that there is no enough time for a secure evacuation.

In addition, the results are also presented using a three colour code representing the decision taken at the household level (Figure 7.23). It is complemented with an analysis of what type of actions are taken based on the multi-hazard assessment in Chapter 4.

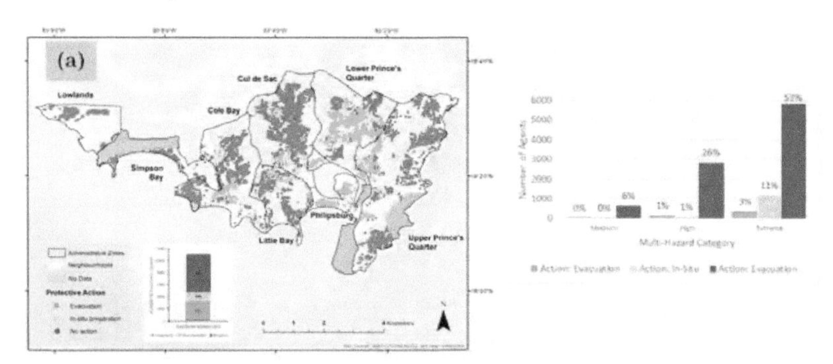

Figure 7.23 Protective action decision at household level in experiment 1. (a) Hurricane category 1. (b) Hurricane category 3. (c) Hurricane category 5. In the right side, the number of agents performing each type of action and grouped by the multi-hazard assessment.

Figure 7.23 (continuation)

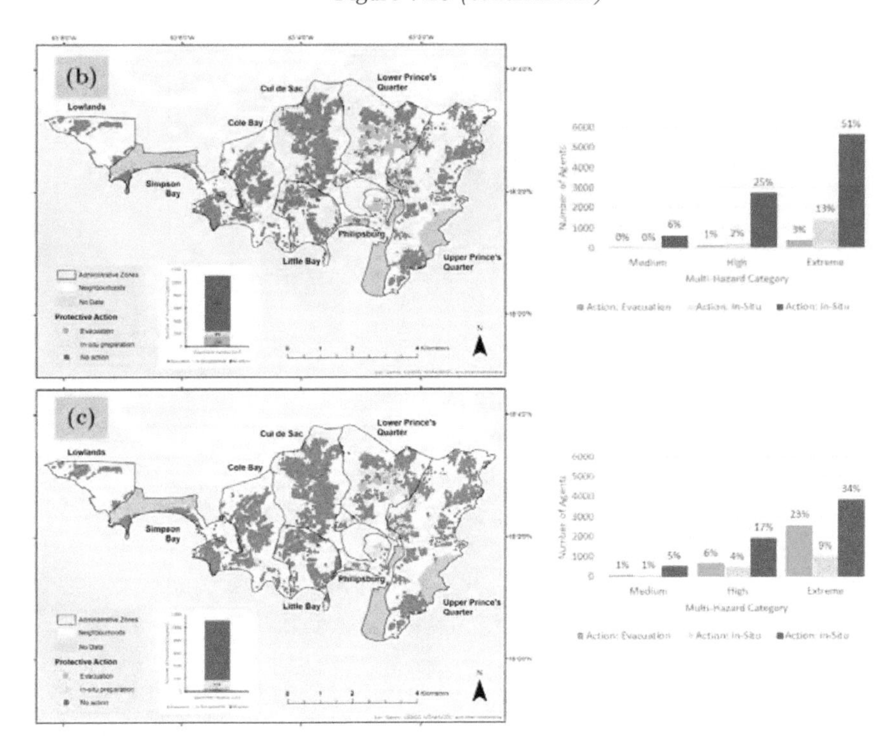

Analysing the distribution of the type of action performed using the household representation allows us to infer that actions towards protective behaviour (green and light yellow in the maps and bars) form clusters around those neighbourhoods where more actual evacuation was reported during Hurricane Irma. The most defined cluster centres are Philipsburg (ID= 39) and Dutch Quarter (36). In situ protection seems to turn in the periphery of the cluster of evacuation. Our results suggest that a type of social contagion phenomenon may be occurring in our simulations; it seems that the decisions taken by neighbours are influencing households. Our result seems reasonable as houses in a particular neighbourhood share some socio-economic conditions and similar housing infrastructure.

Experiment 3

The importance of trusting official sources of information was tested in this experiment (Figure 7.24). Currently, observed trends of protective behaviour, especially those associated with evacuation, are relatively low in Sint Maarten. This behaviour is partially explained in the low trust in institutions and the type of messages distributed during emergencies on the island. In our simulations, the more trust in institutions is assigned, the more neighbourhoods increase the probability of taking protective action. Trust in official sources of information may trigger the collection of information more frequently and create a sense of urgency in the households to perform any type of protective action.

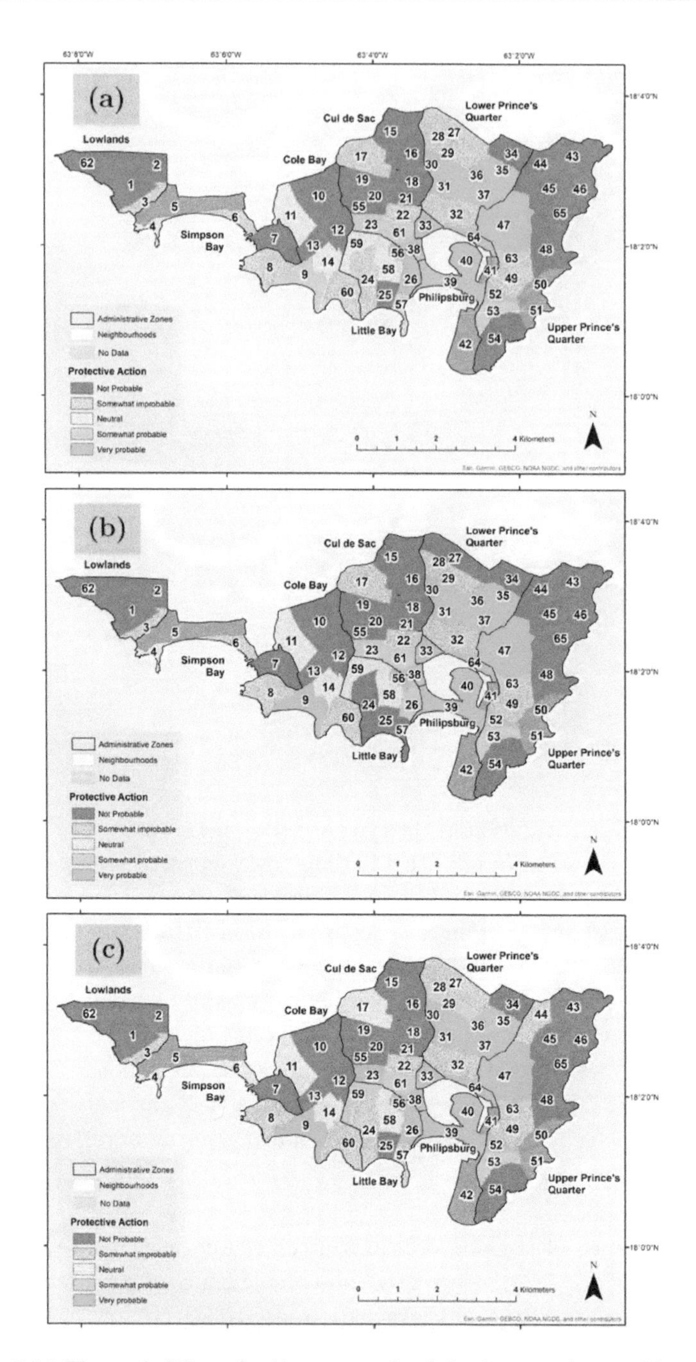

Figure 7.24 The probability of taking protective behaviour at the neighbourhood level in experiment 2. (a) Institutions 10%. (b) Institutions 50%. (c) Institutions 75%. Numbers represent the identification (ID) of each neighbourhood, as presented in Appendix E.

The distribution of neighbourhoods that are not likely to perform any type of protection is again distributed mainly among Cul de Sac, Upper Prince's Quarter and Lowlands administrative zones, and those more likely to evacuate in Lower Prince's Quarter and Philipsburg zones. Two new zones are presenting a positive change toward adopting measures, Little Bay, in the neighbourhoods Cayhill (24) and Little Bay Village (57), as well as in and those neighbourhoods located in the coastline of Cole Bay, Billy Folly (ID=8) and Little Cape Bay (60). These findings highlight the importance of increasing trust in institutions on the island to reduce exposure to natural hazards. Currently, low trust in institutions creates that agents rely more on other information sources to make their decisions regarding protective behaviour or to trust their "instincts" (our own assessment of risk).

In addition, for this experiment, the results are also presented using a three colour code representing the decision taken at the household level and in reference to the hazard zone as presented in Experiment 3 (Figure 7.25).

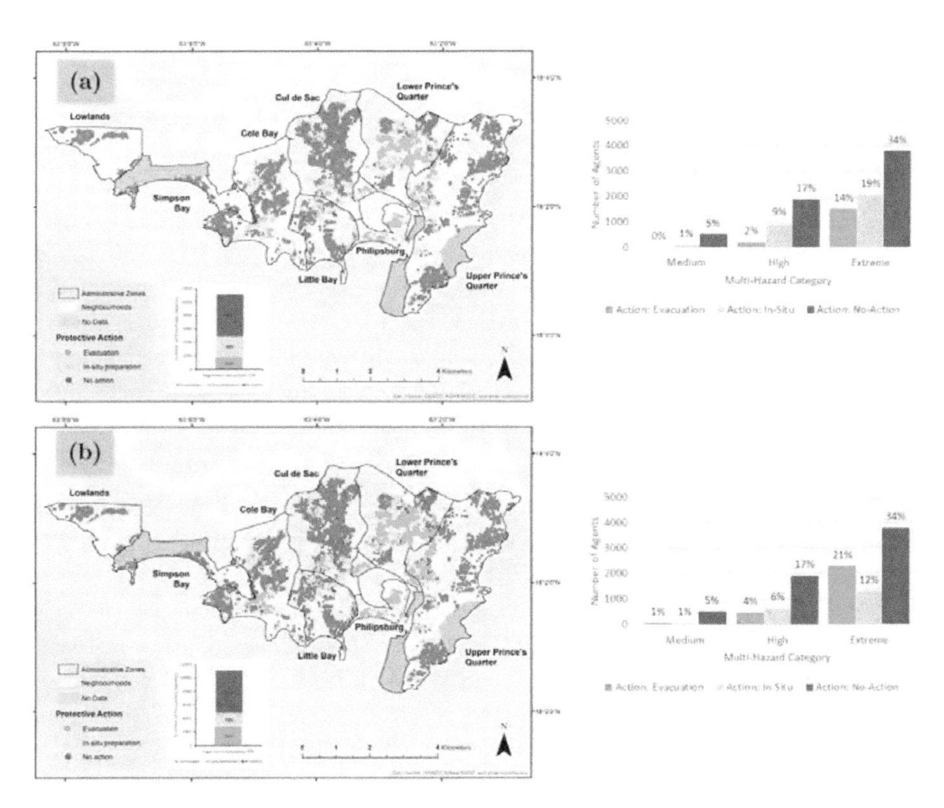

Figure 7.25 Protective action decision at household level in experiment 2. (a) Institutions 10%. (b) Institutions 50%. (c) Institutions 75%. In the right side, the number of agents performing each type of action and grouped by the multi-hazard assessment.

Figure 7.25 (Continuation)

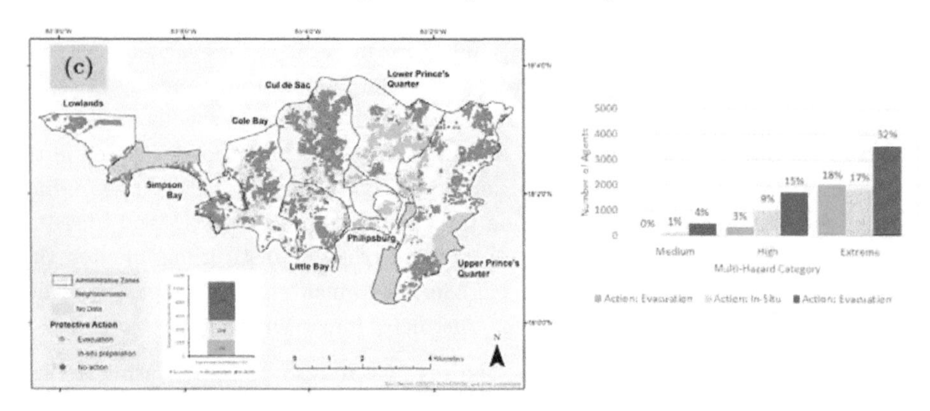

The protective behaviour is observed scattered across the island, but some clusters of protective behaviour are observed in Lower Prince's Quarter and Philipsburg. Two new clusters are observed, one near Cay Bay (ID=9) and other around Zaeger Gut (61). The positive effect of information from official sources in promoting protective behaviour in Sint Maarten seems clear, but the effect has more influence in those areas prone to pluvial flooding or storm surges.

Experiment 4

This experiment was built on the fieldwork findings, where there is low trust in institutions. As such, we wanted to examine what could be the role of information peers on protective behaviour. The averaged results aggregated at the neighbourhood scale are presented in Figure 7.26. A similar pattern as the one observed in Experiment 3 is observed in the role of information through peers, increasing the probability of protective behaviour. However, the contribution seems to be milder for peers' effect, with more agents remaining in the No action behaviour.

Results from our simulations do not show a significant change between the three scenarios of experiment 4. Between scenario, with 10% peers (Figure 7.26-(a)) and 50% peers (Figure 7.26-(b)) four neighbourhoods presented positive change towards performing protective actions, St Peters (17), Vineyard (53), Cay Hill Village (59) and Little Cape Bay (60); also two neighbourhoods have a negative change, Mary's Estate (22) and Western Fresh Pond. Between scenario with 50% peers (Figure 7.26-(b)) and 75% peers (Figure 7.26-(c)) six neighbourhoods present positive change (ID= 3, 56, 23, 17, 22, and 63).

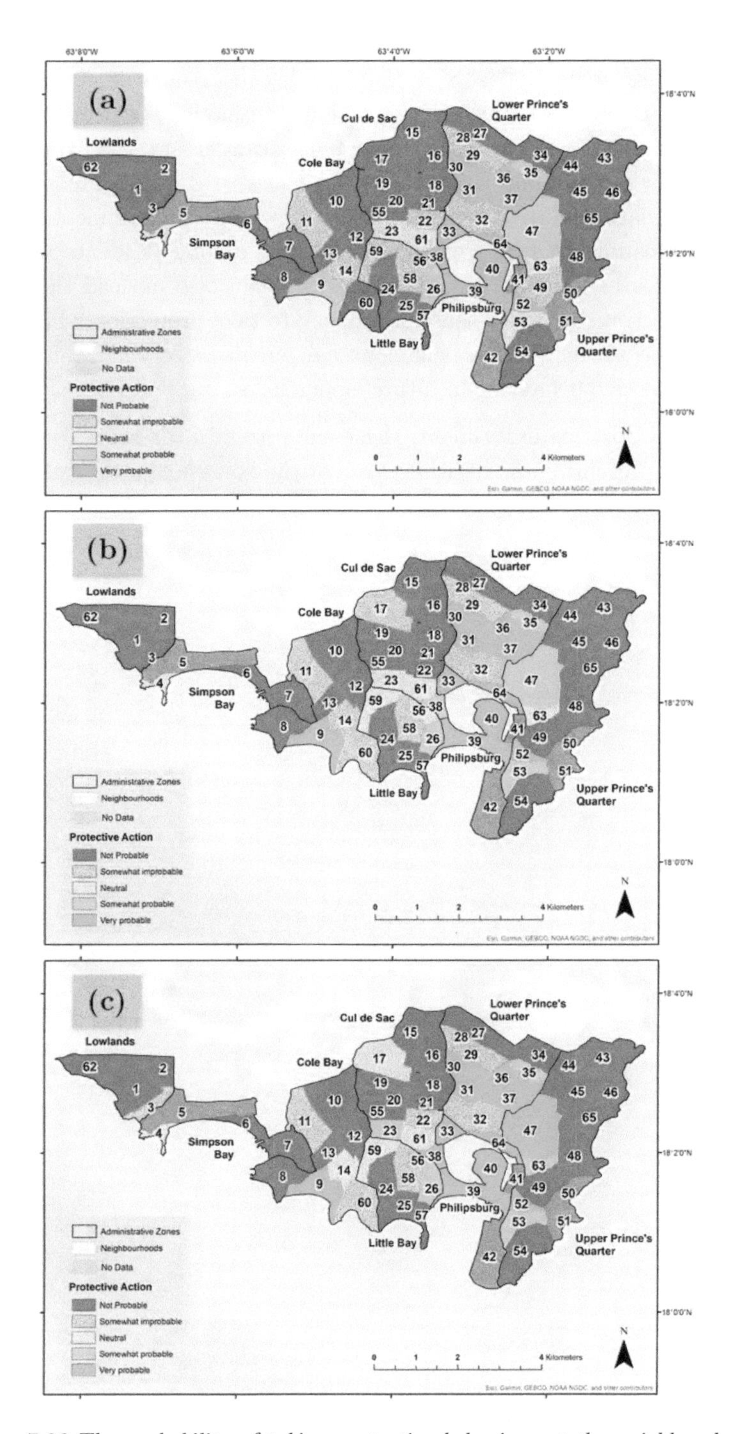

Figure 7.26 The probability of taking protective behaviour at the neighbourhood level in experiment 3. (a) Peers 10%. (b) Peers 50%. (c) Peers 75%. Numbers represent the identification (ID) of each neighbourhood, as presented in Appendix E.

Our result suggests that trust in peer's information can be counterproductive for disaster risk reduction in Sint Maarten because some neighbourhoods are changing negatively towards protective action. Not having a central control of the information that flows during an emergency can create the spread of false information or perceptions, rumours or fake news regarding the severity of a potential threat. Information through peers can undermine the threat's importance, leading to less protective behaviour among residents and creating more exposed areas. Information through peers can also mislead the severity of the hurricane, transmitting a sense of urgency to take protection while in reality, it may not be the case, this behaviour can create a "crying wolf" effect and reducing protective behaviour in future emergencies.

In addition, for this experiment, the results are also presented using a three colour code representing the decision taken at the household level and in reference to the hazard zone as presented in Experiment 4 (Figure 7.27).

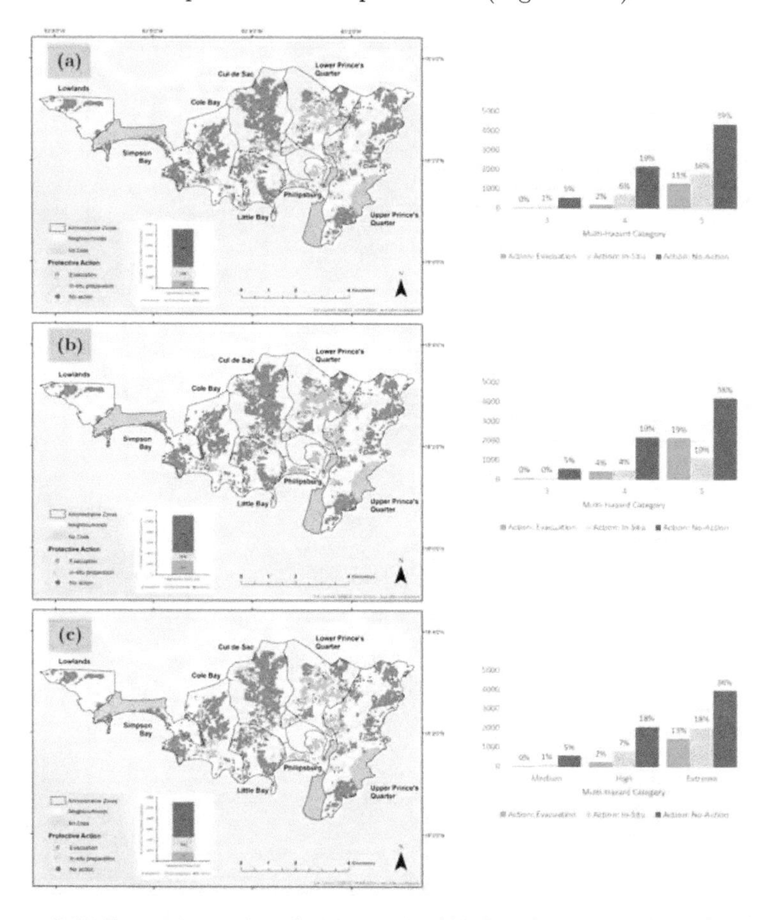

Figure 7.27 Protective action decision at household level in experiment 3. (a) Peers 10%. (b) Peers 50%. (c) Peers 75%. In the right side, the number of agents performing each type of action and grouped by the multi-hazard assessment.

Peers influencing behaviour tends to decrease the number of households slightly not performing any type of protective action, but only in those areas located in category 5 of the multi-hazard assessment. The number of households performing protective actions increases as the percentage of peers influencing increases, but a variation in the type of action is observed between scenarios. An increase on the number of evacuees from 11% to 19% is observed in the simulation from the scenario of peers 10% (Figure 7.27-(a)) to peers 50% (Figure 7.27-(b)), but a decrease from peers 50% (Figure 7.27-(b)) to peers 75% (Figure 7.27-(c)) from 19% to 13%. These variabilities in behaviour can be associated with the transmission of information of peers decreasing the severity of the potential threat.

7.3.6 Conclusions

The ABM developed and presented in this chapter can be seen as an operational simulation tool that integrates behavioural models, hazard models and information during emergencies. The results obtained demonstrate the usefulness of this type of simulations to understand and discover emergent patterns involved in information flow during emergencies. The model simulates how information is collected, interpreted, and actions are taken towards adaptive protective behaviour during an evolving hurricane hazard.

One of our ABM's main conclusions is that forecasting a threat is not sufficient to promote protective behaviour during emergencies triggered by natural hazards. It can be derived from our results that in parallel to a good forecast, the message content, the (trust in the) message provider, and the way how the message is delivered, accepted and understood by the community is also of utmost importance to promote timely protective behaviour in the island of Sint Maarten.

Our findings also suggest the need to understand better the effects of social networks on evacuation processes and other protective behaviours and how they can be used to achieve better evacuations rates or promote more in-situ actions. The work also shows that having a more informed community, where the trust in institutions is relatively high will increase the number of people performing protective actions when faced with a potentially disruptive hazard.

Given advances in technology it is almost impossible to control the flow of information through a network of peers, hence the importance of trust in institutions as the primary source from where inhabitants are collecting (and ultimately distributing) the warning information, so the spread of fake or misleading news is limited.

An ABM as the one we developed here, can be a useful tool for operational risk management activities, such as identifying critical areas where adoption of protective actions are limited, and exposure (of people) is usually more critical. The new knowledge of the system could be used to help prioritise areas for disaster risk reduction measures. Similarly, the ABM can also help identify evacuation patterns, identify critical infrastructure, and identify the needs for improvement of existing emergency locations (i.e. the number of beds, food storage).

With the results obtained from our ABM, we were able to show the dynamic and adaptive nature of exposure of risk to people explicitly. We conclude that the simple infrastructure (e.g. house) is not a precursor of exposure, which offers just a partial view. The actions that individuals may undertake to lower their exposure levels can significantly affect the reduction of exposure to water-related hazards, either by reinforcing infrastructure (protect in-situ), or by getting away from the potential risk area (evacuation).

Exp-Agents in the current version of the model do not carry information regarding past disasters, such as previous evacuation decisions and trust in institutions based on previous emergency performance. This simplification limits the interpretation of the model's outputs since past information may shape future emergencies' actions and behaviour. Still, we believe the outputs are useful to evaluate the impact of information flow among different DRM actors, as our modelling exercise reveals some system dynamic otherwise hidden, regarding the role of the different actors included in the model and the role of different warning information on risk assessment and protective actions.

We evaluate the potentiality of using the ABM with the real and uncertain NHC and Meteorological office bulletins. Further experimentation should consider simulations where the best track of the hurricane is tested and see the overall impact on protective behaviours of using what could be considered an "ideal" forecast. Furthermore, testing different hurricanes could help understanding better key drivers of protective behaviour in Sint Maarten.

8

ADRA -ADAPTIVE DISASTER RISK ASSESSMENT

This chapter presents the proposed framework for including the adaptive and dynamic nature of the exposure component to natural hazards in disaster risk assessment. The framework, called ADRA, aims to be a comprehensive and flexible methodology reflecting the local needs when implemented and based on the data availability. The proposed methodology is then illustrated using as a case study Sint Maarten during the emergency and disaster caused by Hurricane Irma on 6 September 2017. With the ABM results, we compute a set of exposure scenarios, which are then used to compute and map adaptive risk assessment for the island. The results obtained using ADRA are then compared against those obtained using the traditional approach for mapping risk.

This chapter is partially based on:

Medina, N., Sánchez, A., Vojinovic, Z., (2021). Adaptive Disaster Risk Assessment – ADRA-. Incorporating dynamic Exposure into Disaster Risk Assessment. Manuscript in preparation.

8.1 INTRODUCTION

This chapter presents the Adaptive Disaster Risk Assessment framework (ADRA). ADRA aims to explicitly integrate the three DRA elements (Figure 1.3) o support a holistic disaster risk management approach. The framework integrates the three elements that composes risk using a map-based approach. As stated in the introductory Chapter 1, the hazard component was identified as the most studied element of the three components of disaster risk, hence not in this thesis's scope, the multi-hazard map produced in Chapter 4 is used to assess risk. Vulnerability is used in the form of an index-based, using the findings of Chapters 2 and 3. The adaptive exposure component uses the chapter results on evacuation in Chapter 6, and the exposure maps produced in Chapter 7. The workflow of the ADRA methodology is shown in Figure 8.1.

The following sections in this chapter will illustrate the proposed framework's implementation to assess risk from an adaptive approach. We start by presenting the hazard computation and mapping. Next, we present the traditional disaster risk assessment to have a base scenario for comparing the ADRA outputs, followed by the adaptive risk assessment approach. We conclude the chapter by comparing the methods that allow reflecting on the new methodology.

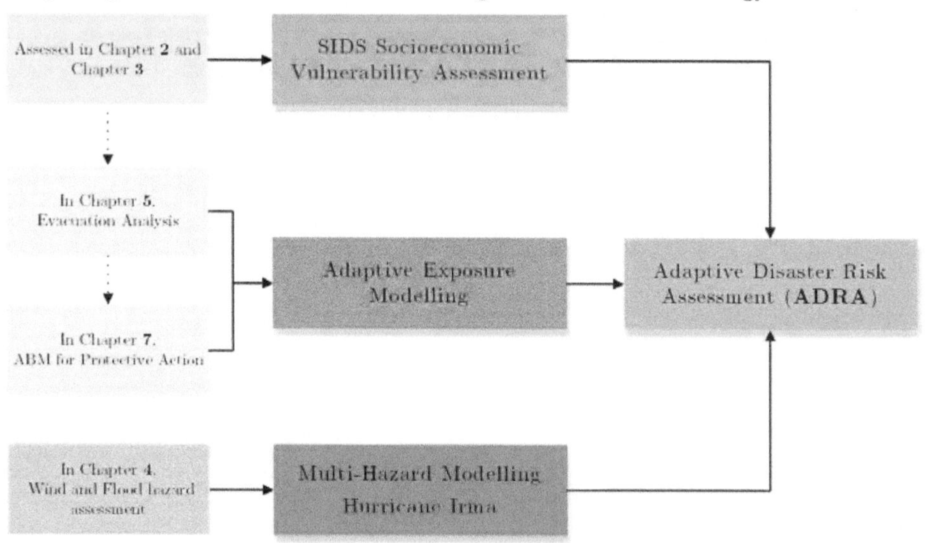

Figure 8.1 Overview of the methodological approach of the ADRA framework. Dot lines indicate findings of one chapter are passed as inputs to be used in another chapter. Solid lines represent the final outputs of one chapter are used directly on the components of ADRA.

8.2 TRADITIONAL RISK ASSESSMENT

Risk in traditional assessment methods is often represented as the probability or likelihood of occurrence of hazardous events multiplied by these events' impacts. Using GIS, traditional approaches compute risk as to the occurrence of hazards and vulnerability (Eq 8.1). Accordingly, reclassified raster maps of the multi-hazard and the vulnerability index are used to produce the final disaster risk maps. The risk maps produced in this section are used as the base scenario to compare those that we produce using the ADRA framework.

$$Risk = Hazard \times Vulnerability \tag{8.1}$$

As the hazard map for evaluating risk using the traditional approach, we selected the multi-hazard map computed in Chapter 4 and shown again in Figure 8.2-(a). We used the PeVI - vulnerability index for the vulnerability map presented in section 3 (Figure 8.2-(b)). The resulting (traditional) risk assessment is presented in Figure 8.3. The values resulting from multiplying the hazard and vulnerability raster maps can vary from 0, representing no risk in a particular neighbourhood, up to a maximum value of 25, representing a neighbourhood with both hazards and vulnerability being extreme. To reclassify the resulting aggregation, we used the values presented in Table 8.1. The reclassification values are multiplied; this is done to balance the weight of each component of risk into the result.

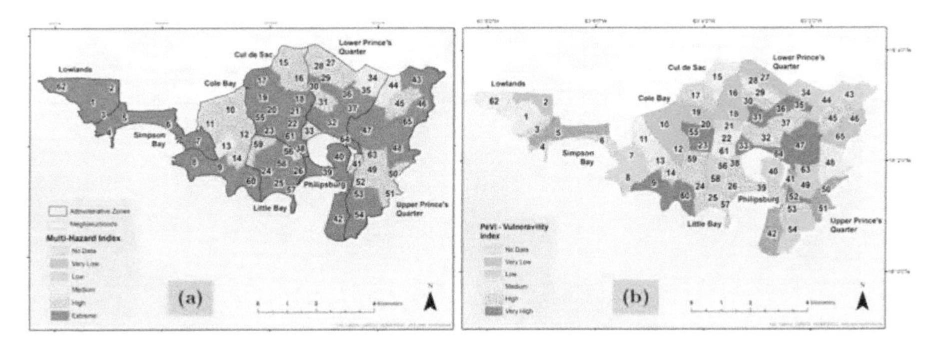

Figure 8.2 PeVI vulnerability index produced in Chapter 3 (a). Multi-hazard assessment produced in Chapter 4 (b).

The results obtained from the traditional risk assessment present a very similar result as those reflected by the vulnerability index (Figure 8.3). The possible explanation for this result is that vulnerability, not the hazard is the leading factor contributing to risk in the island of Sint Maarten, partially because the hazard used to compute the risk is almost affecting in the same proportions all the neighbourhoods. Only some neighbourhoods suffer in less proportion the effects of the extreme wind.

Table 8.1. Values used to reclassify the product of hazard and vulnerability to produce the Risk Map (Traditional approach).

Hazard × Vulnerability	Reclassified Value
0 - 5	Very Low = 1
6 – 10	Low = 2
11 – 15	Medium = 3
16 – 20	High = 4
21 - 25	Extreme = 5

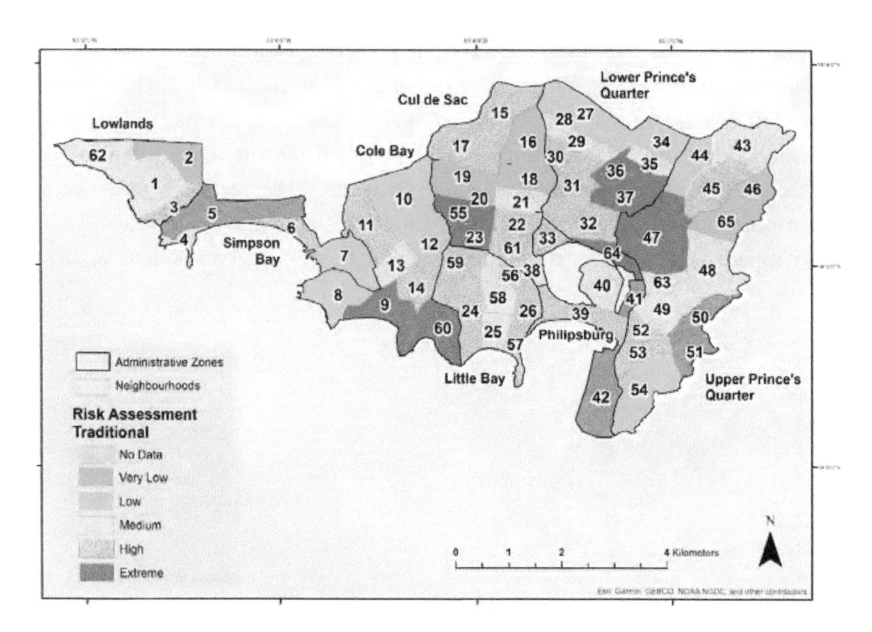

Figure 8.3 Disaster Risk Map for a multi-hazard risk assessment using the traditional approach (Risk= hazard × vulnerability).

Six neighbourhoods presented a reduction in the risk level compared to the vulnerability assessment result, Belvedere, Mount William Hill, Orange Grove, Over the Bank and Over the Pond. These neighbourhoods share some attributes; they are located inland, away from the shoreline, partially protected from wind due to the surrounding topography and not floods are reported in those areas. In the analysis presented, no neighbourhood increases the categorisation of risk level as the one obtained through the vulnerability analysis.

8.3 Adaptive Risk Mapping and Discussion

The ADRA framework's assessment of risk is built on the traditional risk assessment by incorporating the adaptive exposure component into the computation and mapping. ADRA is computed using Eq 8.2. for the hazard and the vulnerability components, we use the same inputs of the traditional approach (see section 8.2). For the exposure component, four exposure maps are used as a proof of concept.

$$Risk = (Hazard \times Vulnerability) \overset{+/-}{\Longrightarrow} Exposure \tag{8.2}$$

8.3.1 Exposure Mapping in ADRA

The first exposure map to be tested corresponds to the probability of evacuation derived in Chapter 5 based on the logit regression model (Figure 5.3-(b)). The other three exposure maps are from the experiments with the ABM in Chapter 7. The second exposure map corresponds to the experiment 2-c, which corresponds to the testing scenario where a category 5 hurricane intensity is used as the threshold to promote protective actions (Figure 7.22-(c)). The third map is from the information from institutions, testing the value of 10% (Figure 7.24-(a)). The fourth map is information from peers with 50% value (Figure 7.26-(b)). The four exposure maps are shown again in Figure 8.4

Differences in probabilities of exposure are easy to observe by comparing the probability of exposure based on the actual behaviour during Hurricane Irma (Figure 8.4-(a)) and the results from the ABM simulations. When compared to the exposure map associated with the hurricane experiment (Figure 8.4-(b)), changes are mostly in the direction of fewer neighbourhoods performing protective action and hence increasing exposure to natural hazards, only two neighbourhoods preserve the protective action when these two scenarios are compared, Dutch Quarter (36) and Pond Island (40).

The differences between actual evacuation behaviour map (Figure 8.4-(a)) and the information from institutions experiment (Figure 8.4-(c)) are more noticeable to the south of the island in those neighbourhoods directly in the coastline, where neighbourhoods increase the probability to perform any type of protective action: Billy Folly (8), Cay Hill (24), Little Bay Village (57), also to the north Bethlehem (27) changes from not probable behaviour to somewhat improbable.

Comparing the information from peers experiment (Figure 8.4-(d)) with the probabilistic evacuation map (Figure 8.4-(a)) differences are promoting protective behaviour, and also some neighbourhoods reflect now less probability to perform protective action. Three neighbourhoods increase the probability of performing protective action: Cockpit (11), Cay Bay (9) and Welegelegen (58). In contrast, eight neighbourhoods present a reduction in the probability of taking protection: Maho (8), Wind Sor (14), Industrie (23), Zaeger gut (61), Bishop Hill (35), Sucker Garden (47), Bloomingdale (63), Hope State (49).

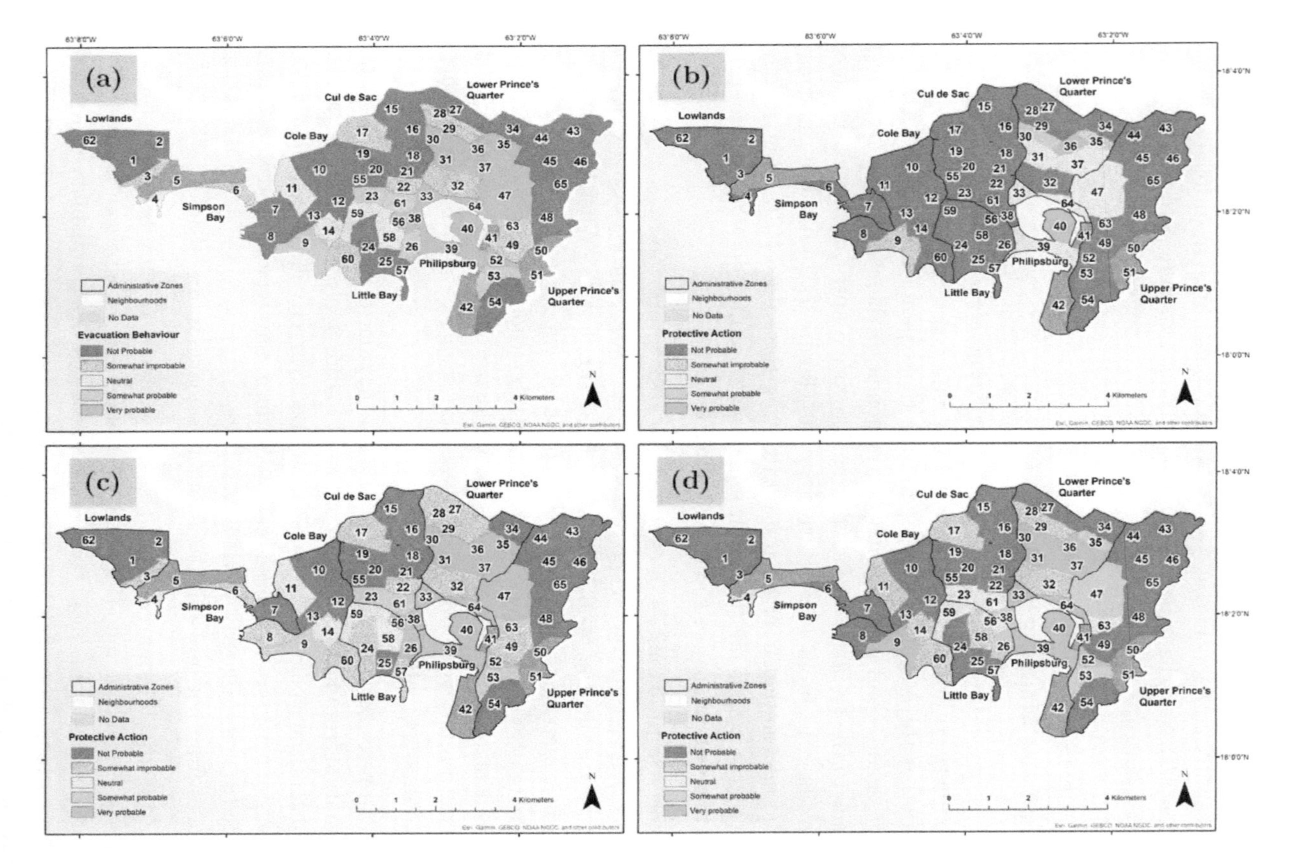

Figure 8.4 Exposure maps used in ADRA.(a) probabilistic evacuation associated with Hurricane Irma's actual behaviour (Chapter 5). (b) Exposure assessment ABM-Experiment 2 (Hurricane Cat-5)). (c) Exposure assessment ABM-Experiment 3-(Intitutions-10%). Exposure assessment ABM-Experiment 4-(Peers(50%).

Some of the probable reasons for the changes in probabilities of protective actions were discussed more extensively in Chapter 7. However, mainly, those neighbourhoods with observed positive changes are associated with the most vulnerable in the island according to our PeVI analysis, due among other factors to low-income and inferior construction materials or are located in flood/storm surge prone areas. Hence, promoting that peers communicate the urgency to take protective actions. In contrast, neighbourhoods changing negatively toward protective actions are located in the wealthiest neighbourhoods and located in areas where less damage from Irma was observed due to stronger houses or less exposed to the hazards.

8.3.2 ADRA Computation

The disaster risk assessment that is offered by the ADRA framework can be categorised as exposure of people, in contrast to exposure to infrastructure or assets, as such the dynamic associated with actions taken by households towards reducing its exposure to a natural hazard should be taken towards reflecting reduction or increase in the final computation of the disaster risk (to life). The selected method to reflect the potential change to disaster risk, associated with a dynamic exposure is based on the values presented in Table 8.2. The method aims to change the risk level to a higher level if the neighbourhood is rated in the not probable or in the somewhat probable categories on the exposure maps, or to lower the associated risk level in those neighbourhoods in which a protective action from households is somewhat or very probable to happen. The adaptive disaster risk maps are presented in Figure 8.5.

Table 8.2. Values used to compute the Adaptive Risk Assessment.

		Protective Action / Exposure Assessment				
		Very Probable	Somewhat probable	Neutral	Somewhat Improbable	Not Probable
Risk Assessment	Very Low	Very Low	Very Low	Very Low	Very Low	Low
	Low	Very Low	Low	Low	Medium	Medium
	Medium	Low	Medium	Medium	High	High
	High	Medium	Medium	High	Extreme	Extreme
	Extreme	High	High	Extreme	Extreme	Extreme

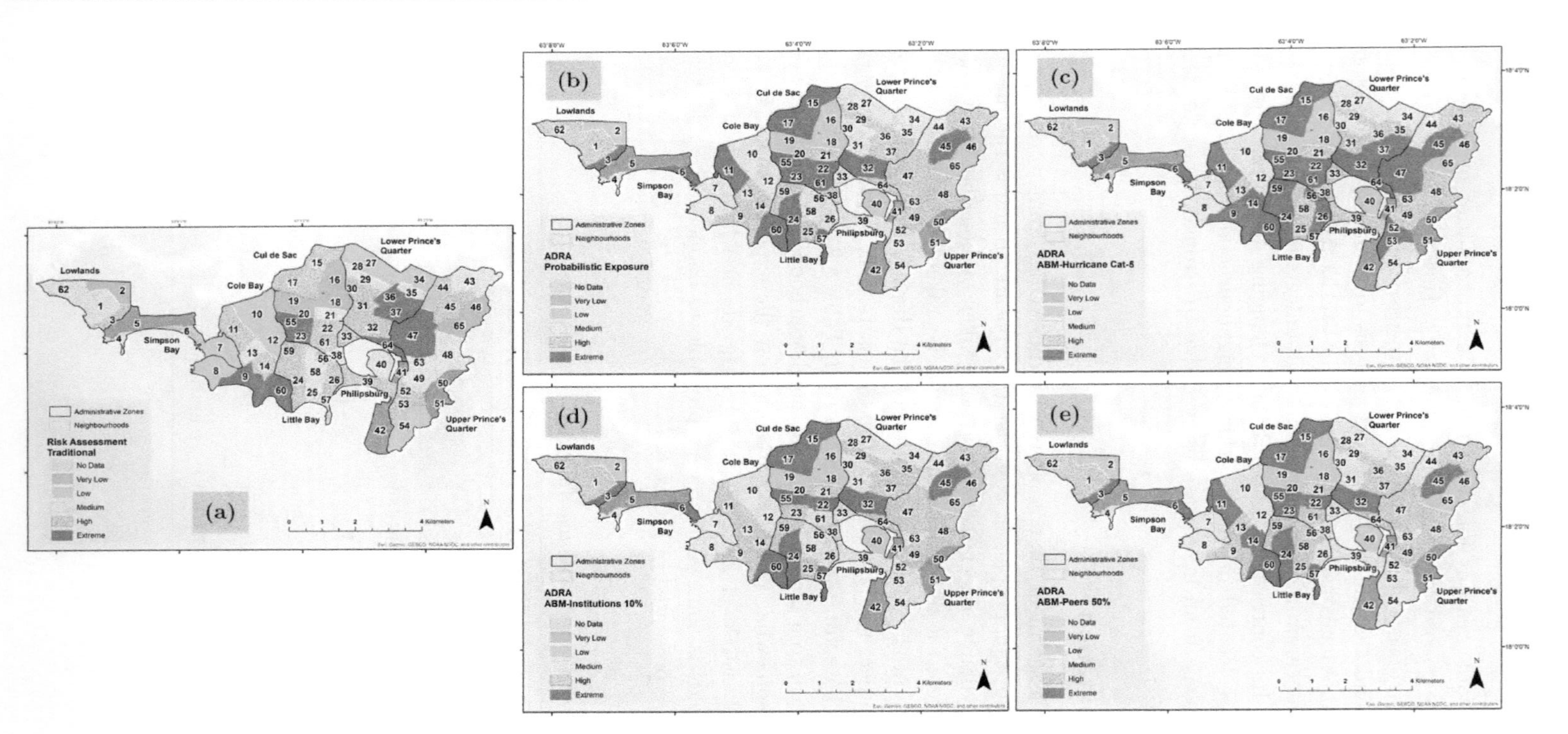

Figure 8.5 Traditional DRA vs Disaster Risk Map using ADRA methodology. (a) Traditional DRA (b) Exposure from evacuation probabilistic map. (c) Exposure from evacuation ABM Hurricane (Category 5) experiment. (d) Exposure from evacuation ABM Institutions (10%) experiment. (e) Exposure from evacuation ABM peers (50%) experiment.

The disaster risk assessment changes between the traditional approach and those obtained using ADRA shows more neighbourhoods where the risk assessment is amplified when the dynamic exposure is included. A lower proportion of neighbourhoods tend to lower the level of risk. Moreover, a handful of neighbours do not vary the risk level across our exposure experiments. The changes in risk level assessment based on the ADRA framework, are better shown in Figure 8.6. In blue, those neighbourhoods whose risk assessment was lower using exposure. In light-red, those neighbourhoods whose risk assessment was increased using exposure. Light-yellow corresponds to neighbourhoods, whose risk value was not changed.

The overall trend towards increasing risk level is predominant in the administrative zones lowlands, Simpson Bay, the northern neighbours of Cole bay and Cul de Sac and the north-west neighbourhoods of Upper Prince's Quarter. The neighbourhoods belonging to these zones were found to perform low evacuation, as presented in Chapter 7. The results are associated with the actual behaviour observed in our survey after Hurricane Irma. As mentioned in previous chapters, in Sint Maarten, the tendency is not to evacuate (68.6% of respondents did not evacuate). ADRA penalised this behaviour by proportionally increasing the risk assessment level. The behaviour was supported by wealthy households, better construction materials, lower losses during Irma and higher house ownership.

In Contrast, some neighbourhood tends to have a positive change towards protective behaviour across the experiments. The neighbourhoods are located in the central portion of Lower Prince's Quarter administrative zone, one neighbourhood in Cole Bay (cay bay, ID=9), three neighbourhoods in Upper Prince's Quarter (Sucker Garden (47), Over the Bank (52), and Vineyard (53)), Fort hill (26) in Little Bay and Philipsburg. These areas were identified as those where more evacuation is expected associated with poor housing condition and high socioeconomic vulnerability indexes and may promote agents to activate their protective measures when a hazard is forecasted. In addition, in these areas floods are a constant threat, either from pluvial flooding or storm surge, its residents showed recognition of the flood risk area they are located, and it seems that the ABM manage to capture such behaviour by promoting risk reduction measures.

A few neighbourhoods show a trend to remain unchanged on the risk level assessment across the experiments; these are Rockland (55) and Industrie (23) in Cul de Sac, Little Cape Bay (60) in Cole Bay, Cay Hill Village (59) in Little Bay. It seems that the socioeconomic conditions in these neighbourhoods are such that the new assessments of information are not enough to change the conditions of protective behaviour dictated by the logistic regression model used in our simulations. A further investigation is required to understand what can be done to improve these neighbourhoods' protective behaviour.

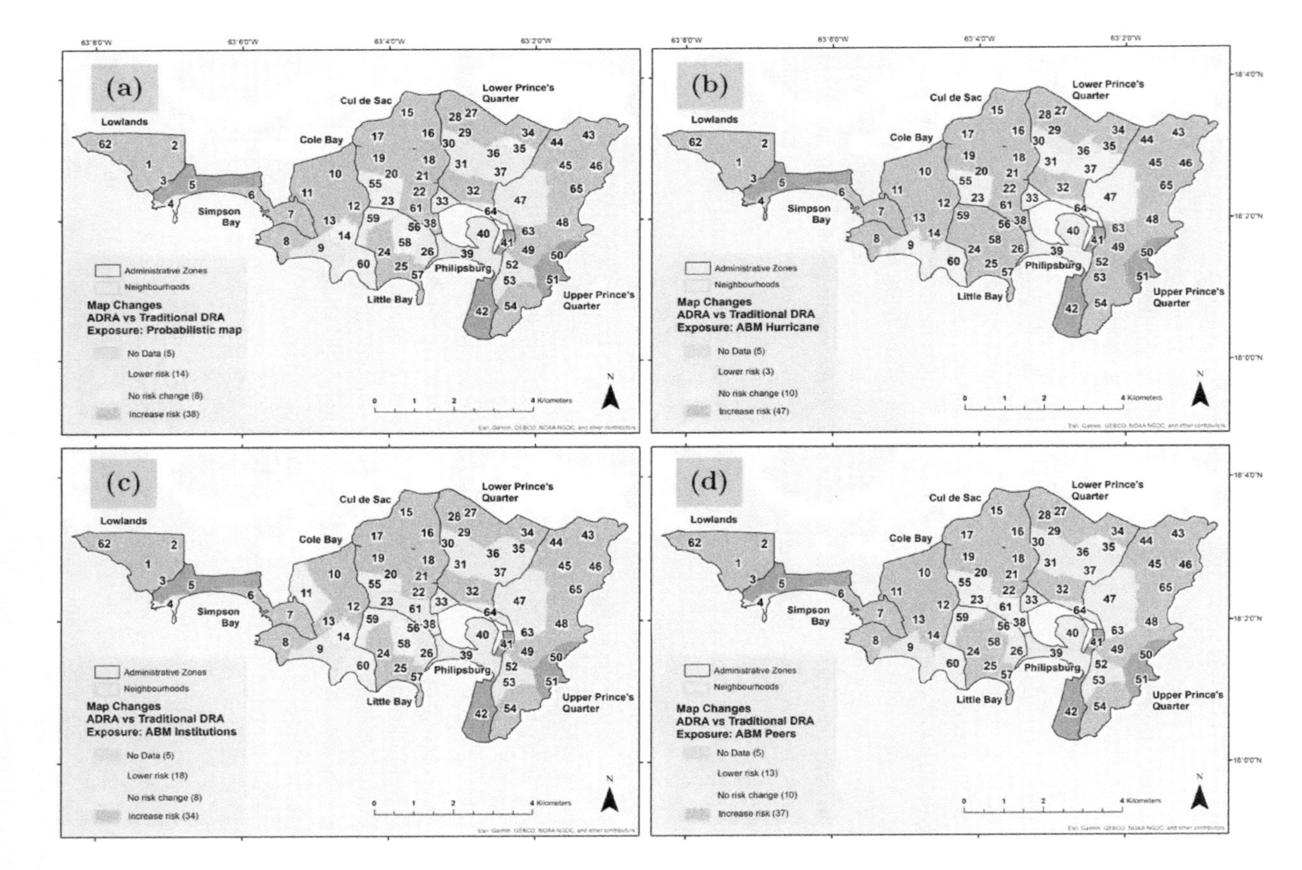

Figure 8.6 Changes in the Disaster Risk Map by incorporating ADRA, adaptive exposure. (a) Exposure from evacuation probabilistic map. (b) Exposure from evacuation ABM Hurricane (Category 5) experiment. (c) Exposure from evacuation ABM Institutions (10%) experiment. (d) Exposure from evacuation ABM peers (50%) experiment.

In terms of overall performance to lower risk, the experiment corresponding to Institutions (10%) is the best with 18neighbourhoods lowering its risk assessment level (Figure 8.5-(d) and Figure 8.6-(c)). For the institutions' experiment, the number of neighbourhoods showing an increase in its risk assessment is still high, accounting for 34 neighbourhoods. The critical scenario is the hurricane Category 5, in which only three neighbourhoods would decrease its risk level compared to the traditional approach, and 47 neighbourhoods will be assessed as riskier than using the traditional approach.

An insight for Disaster risk managers in Sint Maarten arises when comparing the traditional risk assessment with the four maps produced using ADRA methodology. Some neighbourhoods are always assessed as extreme risk; Little Cape Bay (60), Rockland (55). Furthermore, those neighbourhoods that were evaluated as extreme in all four ADRA assessments are: Maho (3), Simpson Bay Village (6), Cay Hill (24), Little Bay Village (57), Mary's Estate (22), St Peters (17), Reward (15), Madame's Estate (32), and Ocean Terrace (45). Particular attention should be put into these neighbourhoods towards DRR action plans or contingency plans for post-disaster relief.

8.4 Conclusions

The holistic framework proposed in this chapter, ADRA, assessed disaster risk from an adaptive approach, in which the exposure component is explicitly quantified and mapped. The exposure component has a human life-centred approach, in which protective actions performed by households is taking into account to measure the effects on the increase or decrease of the exposure level. Two protective actions are evaluated under the current framework. First, a probabilistic map derived from actual evacuation behaviour observed after an extreme event. Second, three maps from the ABM approach in which the intensity of the potential hurricane threat and the role of information flow from different sources are evaluated.

We believe that disaster risk assessment should be people-centred, and human lives should be the top priority of disaster risk management, regardless of the economic and infrastructure impact. As such, a human-centred approach, such as ADRA is necessary for a more effective DRM strategy. ADRA allows disaster risk managers to gain more knowledge of their systems and focus their (often) limited resources more efficiently.

The inclusion of the ABM results into ADRA shows how practical bottom-up approaches can be for disaster risk management. Effective communication of risk, increasing the trust in authorities, and information broadcast using the advantages of the internet and smartphones can effectively be used to reduce exposure (and risk). However, understanding the effect of peers and emergency information flow from non-official sources has not yet been fully explored, and more research in this field is not only desired but necessary.

Our model results show that evaluating the risk of a hurricane using only the hurricane intensity can lead to dangerous and catastrophic situations in Sint Marten. Activating protective behaviour only to high-intensity hurricanes may create high levels of exposure across the island leaving most of the population at risk exposed to other hazards associated with a multi-hazard assessment, such as floods, storm surges or landslides.

If actions and decisions on where to focus the resources for disaster risk reduction are taken based on methods that focus on risk-to-life, such as ADRA, additional measures need to be undertaken at government and household level that allows for compensation of the losses to infrastructure. Insurance schemes, increase the protection level of flood infrastructure, improve building regulations, are amongst the complementary measures that can be used.

Identification of areas that do not change (or increase) the risk assessment with ADRA requires a more in-depth analysis of which mechanisms can be used to reduce risk in those areas. The research method used in ADRA, suggest that these are neighbourhoods that probably will not evacuate under any level of hazard, as such the government of Sint Maarten should undertake an exhaustive inspection of infrastructure to warranty they are built to withstand a Category 5 hurricane. Additionally, if they are located in flood-prone zones, they have elevated houses. Awareness campaigns to promote protective actions can also be used, but our findings suggest that the impact of such measures may be limited.

The risk assessment used the neighbourhood as a scale of representation based on the lower resolution of the inputs used to compute ADRA maps; the scale was restricted from the information available to perform our analysis. This aggregation method is useful to observe the general picture of the effects of exposure on risk assessment. However, it may also hide the real protective behaviour of bigger resolutions (e.g. block scale or household level).

The aggregation method also provides an average value of hazards from which more detailed models are available. The peaks of wind and flood are then averaged in this analysis. Caution is then advice when interpreting and using the risk assessment results presented in this chapter. If a neighbourhood is selected from our analysis to implement actions towards reducing impacts from the hazard, the use of individual risk maps (flood and wind), should accompany the decision where to implement them more effectively.

9

A WEB-BASED APPLICATION FOR EMERGENCY MANAGEMENT

Emergencies and disasters caused by weather-related events are increasing in severity and intensity. There is also an increase in the number of people and assets exposed to weather-related disasters due to unplanned urbanisation. The combination of climate change and urbanisation growth is causing more disasters every year around the globe. As a consequence, more people are being left homeless, displaced or requiring emergency help. Better communication in times of crisis has been proved to be an effective way for disaster-risk reduction.

Consequently, in this chapter, we present an innovative web application that aims for better information flow during the different phases of a disaster. EvacuAPP is a web-based application built using the insight we collected during several interviews with stakeholders on the island of Sint Maarten in the aftermath of Hurricane Irma. EvacuAPP functionalities include shelter identification and administration, wayfinding directions, critical infrastructure management, emergency assistance requesting, incident reporting, and real-time hurricane tracking.

This chapter is partially based on:

Medina, N., Sánchez, A., Vojinovic, Z., (2021). EvacuAPP. A web-based application for emergency management for Hurricane disasters. International Journal of Disaster risk Science (IJDRS). Under Review

9.1 INTRODUCTION

It is estimated that in the last 20 years more than 1.3 million people have been killed and 4.4 billion have been injured, with millions being left homeless, displaced or requiring emergency help as a consequence of such disasters (CRED-UNISDR, 2015; Gu, 2019). The amount of people in need of assistance due to disasters triggered by natural hazards makes it clear that a more precise understanding of the role of communication during the different phases of a disaster is needed, as well as the development of communication tools that allows the number of casualties to be reduced and the assistance for those in need to be faster and more accessible (Eiser et al., 2012; Paton, 2008).

Emergency management cycle consists of four phases; prior to an emergency are the prevention and preparedness phases (pre-disaster), during an emergency is the response phase, and in the aftermath of the emergency is the recovery phase (Benjamin et al., 2011). C Communication across all the phases of a disaster triggered by natural hazards is of utmost importance for evacuees, survivors, rescue forces and disaster risk managers (Boulos et al., 2011; Mythili and Shalini, 2016). Each phase in a disaster requires a different type of information, influencing the type of communication tool required and its functionalities (Zlatanova and Fabbri, 2009).

In the pre-disaster phases, information can be used to increase risk awareness and enable proper communication of potential hazards, such as the hurricane path, strength, lead time, and dissemination of warnings or evacuation orders. During a disaster, it can help to keep people updated about the potential threat and developments regarding the hazard and if new measures are needed to be taken by the community, as well as to request assistance and rescue. In the aftermath of a disaster and in the recovery phases, communication is crucial for the survivors and the relief teams; it enables people in need (e.g. trapped, injured) to be identified, and better planning for the rescue and response teams in the allocation of resources. Also information collected in this phase can be useful to improve future disaster risk management (i.e. photos of damaged infrastructure) (Sebastien and Harivelo, 2015; Zlatanova and Fabbri, 2009).

Information and Communication Technology (ICT) during a crisis has been proved to be not only useful but necessary in accordance with modern disaster management challenges (Kavanaugh et al., 2013). After the disaster caused by Hurricane Irma on the island of Sint Maarten in September 2017, a post-disaster fact-finding mission revealed the importance that the role of communication had in the overall impact of the hurricane on the island and its inhabitants (PEARL, 2018). Furthermore, a study on evacuation behaviour during Hurricane Irma found that one of the main predictors of evacuation behaviour on Sint Maarten is the quality of the content of the information they receive during a forecasted hurricane. (Medina et al., Under Review).

Given the importance of information during a disaster, we have developed a web application (web-app) that combines several elements identified as important or necessary from different stakeholders groups we interviewed, and it also incorporates key elements of disaster-risk communication recommended in the literature. Following the identification of crucial functionalities, the web-app aims to provide a two-way communication tool that provides fast, reliable and personalised access to emergency-related evacuation.

This chapter presents a web application that has been developed to be used as a tool for disaster risk management (DRM) and evacuation purposes on the island of Sint Maarten. We start by presenting the justification for the web-app followed by application development, including the conceptual design and each one of the functionalities incorporated in the web-app. Then we present the discussion of the main advantages and expected uses of the application in future emergencies, as well as some identified future developments.

9.2 JUSTIFICATION

The number of smartphones, tablets and other devices that allow communication using the Internet is increasing at an accelerated rate (Sebastien and Harivelo, 2015), and their use during emergency situations has been increasing notably over the last decade (Maryam et al., 2016; Omaier et al., 2019). This increase in the number of ICT tools gives the potential to reach a large proportion of the population in high-risk areas during emergencies, and has been proven to be an important factor in disaster risk management. Nowadays, with the advancement of the Internet and mobile technologies, disaster risk managers have the potential to reach out directly to the general public for whom their messages are ultimately meant, and to engage citizens in disaster risk management; this is with no intermediaries, and in (almost) real time when it is needed (Boulos et al., 2011; Zlatanova and Fabbri, 2009).

ICT tools can enable end users and disaster risk managers to communicate back and forth during the whole lifecycle of an emergency and potential disaster (Zlatanova and Fabbri, 2009). As a consequence, disaster communication using ICT technologies has the potential to save many lives through better communication of the severity of an impending threat, and the transmission of warnings and evacuation orders. It can also be used to send timely requests for help during or in the aftermath of a disaster and as a tool for better planning for rescue and relief teams (Maitland et al., 2006).

Accordingly, after the disaster caused by Hurricane Irma on Sint Maarten, we developed a web-based application (web-app). A web-app is a website that is designed to be flexible, responding to being viewed on several devices and platforms; it functions like a mobile app but from the device browser (Stevens, 2018). With this development, we aim to reach a large proportion of the population of the island, providing the end users and the disaster risk managers with some of the most critical functionalities identified for Sint Maarten's needs.

A web-based application was favoured over a mobile app (smartphone users), because it is a solution that works on most user devices (Sebastien and Harivelo, 2015). Web developments are more flexible, they do not have to be custom-built for a platform or hardware, they can be accessed from multiple devices and platforms, the user does not need to download and install it on their devices, and a web-app does not need app store approval; hence they are faster to be launched and operational. In contrast, native mobile apps are built for a specific platform (i.e. Android or iOS) and are limited to one brand/platform or having to do multiple deployments, each platform needs to approve the app before it is accessible to the general public, a native mobile app needs to be installed, and it uses the mobile phone memory.

Furthermore, there is an increasing trend in the use of ICT technologies to gather and transmit information during crises. An online survey by the American Red Cross shows that during an emergency, one in five American adults would try to contact first responders through a digital means such as e-mail, websites or social media, and about 69% of respondents expect a fast response to their request through social media (Walter, 2010). This trend was also observed on Sint Maarten. Based on the results of the survey we conducted in the aftermath of Irma, we were able to determine that after the radio, the Internet is the preferred method to get the latest information during a hurricane; over 23% of respondents access warning information using the Internet (Figure 9.1).

In addition, on Sint Maarten we detected that social media platforms such as WhatsApp and Facebook were also widely used during and after Hurricane Irma, which shows a substantial shift towards mobile and Internet-based technologies and shows the vital role that mobile applications could have in disaster management on Sint Maarten (Medina et al., 2019).

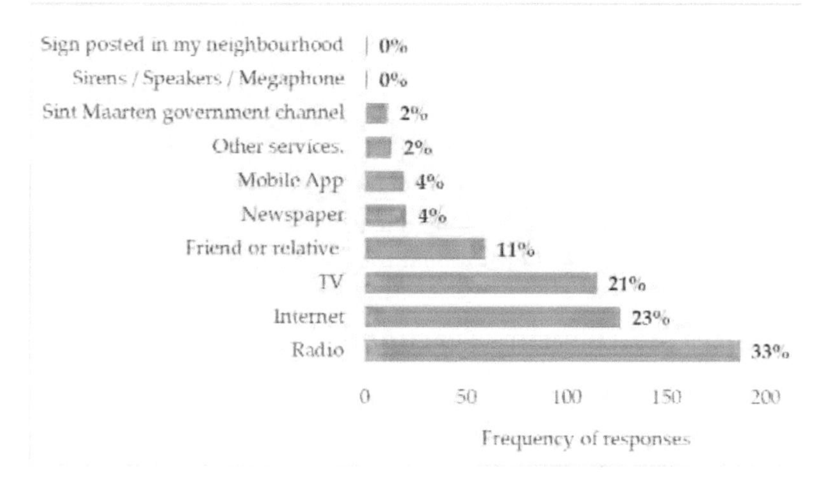

Figure 9.1. Sources of information on Sint Maarten according to a survey question. 'From where do you get the latest updates on warnings or evacuation information? -- Mark all that apply.'.

Additional justification is the number of handheld devices in Sint Maarten households. We found that the number of smartphones or tablets on the island is relatively high. From our sample (N=255) only 6.2% of respondents answered that they do not have one in their households, 73.7% have between one and four devices and the remaining 20.1% having five or more (Figure 9.2). The number of devices shows that communication using mobile technology has the potential to reach across the whole island regardless of socio-economic status. Furthermore, as presented in Kremer (2017), mobile devices have been shown to be viable platforms for information transmission, as they have become the number one way of retrieving information in various situations.

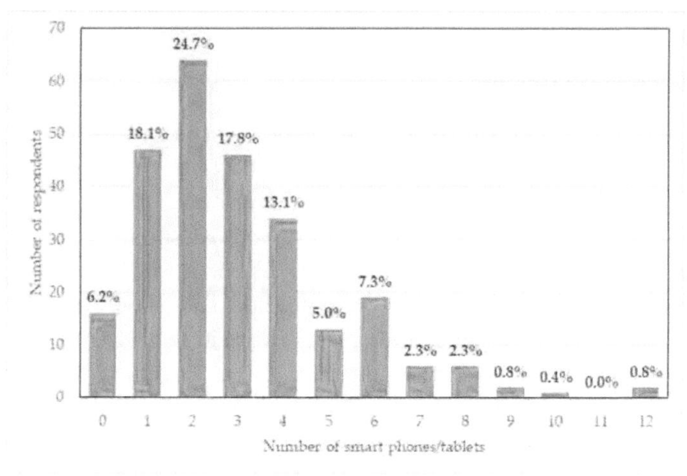

Figure 9.2. Total number of smartphones or tablets in the household.

Finally, regarding communication disruptions caused by the devastation of Irma, severe damage to the island telecommunication infrastructure was reported, with damage of approximately 50% because of the hurricane (PEARL, 2018). Broadband internet was able to remain operational in most sectors of the island as these systems are mostly underground. Wireless communication was provided only in some regions because only 11 of 31 cell towers withstood Irma's disruptive wind, and the hilly topography presented a challenge to provide wireless Internet due to the limited number of operational towers (ECLAC, 2017; PEARL, 2018).

9.3 Web-app development – EvacuAPP

EvacuAPP was initially developed and adapted to be used on Sint Maarten on the Dutch side of the Caribbean island, as part of an EU-funded research project called PEARL (PEARL, 2013). It was designed based on the findings of different needs identified in the disaster caused by Hurricane Irma on the island. The web-app was initially launched as a beta version on 13 October 2017, just one month after Hurricane Irma struck Sint Maarten, but due to the impact of the hurricane and the extensive damage to the island infrastructure, it was only possible to travel to the island in February 2018. We intended to disseminate EvacuAPP first to the government agencies in charge of disaster risk management, such as the section head for disaster risk management and a representative of the emergency support group ESF-7 as the one responsible for evacuation, shelter and relief, and subsequently to the general public. During the fieldwork, we interviewed several stakeholders, including government representatives, as well as the residents, to identify different users' needs. The web-app was designed to include several capabilities identified as crucial to reduce disaster risk in the island by promoting evacuation or protective behaviour on the island based on these identified needs.

9.3.1 Conceptual Design

When designing successful ICT tools to be used in DRM, several principles should be followed. First, end users should be able to request help, report observations, and receive critical information (Bahk et al., 2017). Second, in order for the tool to be useful, the design needs to be context-specific and should include individual identified needs (Kremer, 2017). Third, it should provide real-time information on how the emergency is unfolding and should provide two-way communication between users and managers (Mythili and Shalini, 2016). Fourth, an intuitive, easy to use, and appealing user graphic interface has been reported to increase the likelihoods of adoption of new technologies for emergency management (Zlatanova and Fabbri, 2009). Consequently, EvacuAPP incorporates the above mentioned principles in its design.

The design of EvacuAPP, as a web-app, was intended for two main and distinct types of users. On the one hand, it is targeted to be used for residents, visitors and tourists on the island of Sint Maarten (end users). On the other hand, manager-users are represented by the app administrators; these are emergency managers and shelter managers on the island. The type of user determines the content that is accessible to them. Manager-users can access the control panel of the web-app through authorisation login. They are in charge of maintenance of the system and update information regarding emergency management, such as status of a shelter, or add barriers in the road network (i.e. floods, fallen objects). They have editor capabilities and can add or remove functionalities on the web-app. They can control and monitor the use of the web-app by the end users, for example, they can see who requests emergency assistance or a looting incident, and coordinate and send the appropriate help.

End users, on the other hand, have access to all the functionalities of EvacuAPP at their disposal, and accordingly, they can access emergency information and send requests, but do not have editing privileges; end users do not need to register or authenticate to access the web app. Figure 9.3 shows the home screen as it will appear on a smartphone, with all the functionalities available for the end user. In the upper left corner, the user can use the *zoom in* and *zoom out* buttons, the *home button* which is configured to display the full extent of Sint Maarten if the user zooms outside of the area of interest, and *my location*, a button to centre the screen at the current location of the user, by using the GPS built-in functionalities of the device or by using the network information. In the upper right corner, there are two buttons.

The first button, by default, is the *Legend button*, which is a visual explanation of the symbols seen on the map, and the second button allows the user to navigate through the different functionalities of EvacuAPP. The functionalities are distributed into three sections. Those are the *Edit* section, the *Layer List* section and the *Directions* section. In the *Edit section*, the user accesses the emergency assistance and the incident report functionalities. The *Layer List* allows the end user to select which operational layers will be displayed on the screen (by activating the checklist next to the name of each layer). The *Directions* assistance provides wayfinding directions from a starting point to a selected evacuation destination.

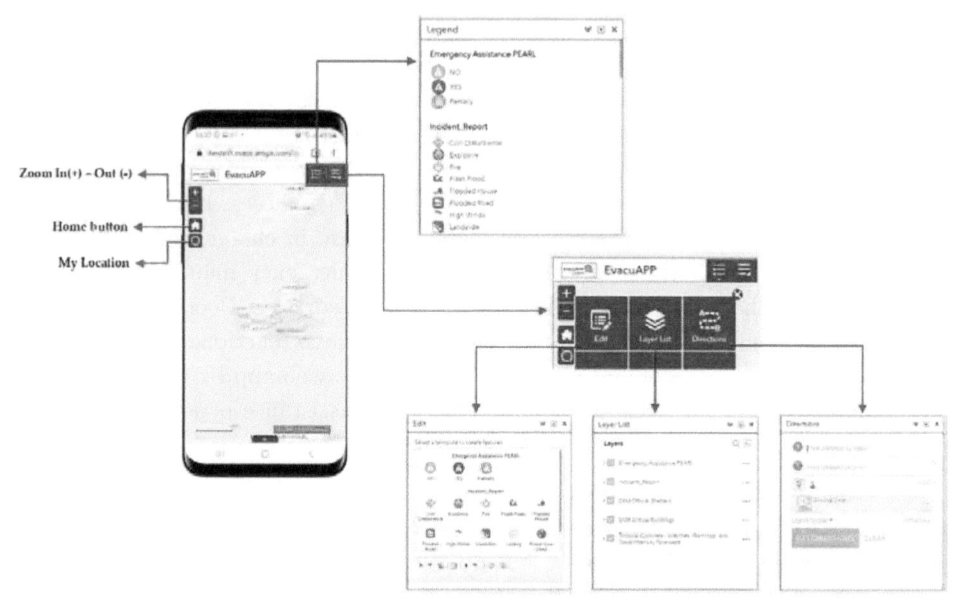

Figure 9.3. Home screen and functionalities access of EvacuAPP on a mobile phone view.

The design language of the web-app is English, and icons and layers are presented in this language. However, when used, the web-app detects the default language of the device being used and changes the titles of the different functions accordingly (Figure 9.4).

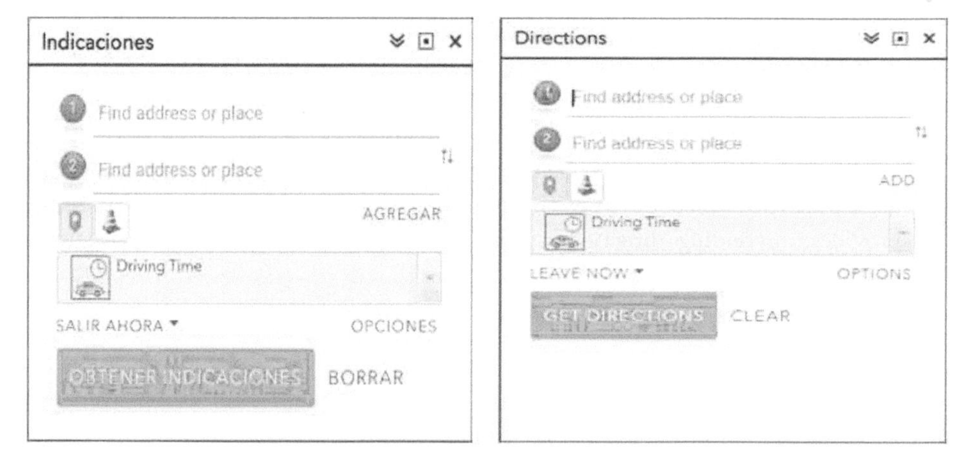

Figure 9.4. Example of language adaptability in EvacuAPP. Wayfinding direction: left: Spanish and right: English.

EvacuAPP graphics were created following design principles for emergencies. Palette colours across the web-app were chosen to transmit tranquillity and calmness to the user. In contrast, primary and secondary colours (red, yellow, orange, green, and blue) are used in situations where it is necessary to redirect the attention of the user, such as emergency assistance or the hurricane path. Also, icons were chosen to be visually appealing and with an intuitive symbol for acceptability and ease of use.

The current version of the web-application was developed using the ESRI server as a repository. It uses ESRI services for the base map and the traffic information. The hazard forecast layer is retrieved directly from are connected from the National Hurricane Center (NHC) (Medina, 2017).

9.3.2 EvacuApp Functionalities

In this section, we present the main features and functionalities of EvacuAPP. It includes shelter identification and administration, wayfinding directions, critical infrastructure management, emergency assistance, incident reporting, and hurricane tracking.

Shelter identification

EvacuAPP is divided into different layers/sections. First, it is possible to use the web-app to see the information related to a specific shelter in the case of an evacuation. The web-app enables end users to see on a map the location and main attributes of the official shelters that are available on the island for a particular hurricane season (Figure 9.5).

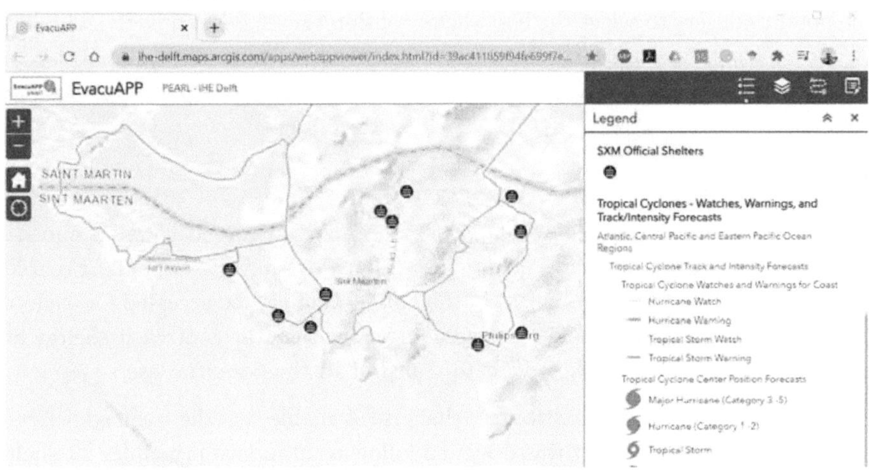

Figure 9.5. Location of all official shelters listed on Sint Maarten as seen by end users of EvacuAPP.

A potential evacuee using the web-app will be able to see if a particular shelter is open at the moment of a query as well as the level of occupancy to determine if it still has any capacity left. A shelter manager can use the web-app to inform potential evacuees what it is necessary to bring with them to the selected shelter, including food, water, blankets or medicine. It also provides a photo of the shelter, which is particularly useful for non-residents of the island that may not be familiar with the surroundings (Figure 9.6).

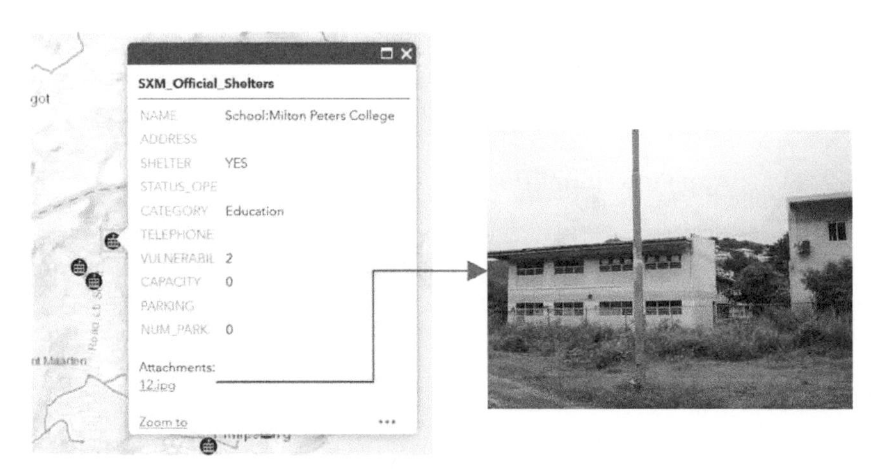

Figure 9.6. S helter Information available to the end users of EvacuAPP.

The information that is displayed regarding the shelters is intended to allow the potential users to select the best shelter option they have at any given moment of an evacuation. Information concerning shelters can be updated by the administrators of the shelters using the administrator rights to the web application.

Shelter administration

The second section of the web-app, which is hidden to end users, is aimed at shelter administrators or evacuation authorities in the island. Complete information and editor capabilities in the application can be accessed for each one of the shelters allowing them to have control on the status of each shelter and any relevant information they want to transmit to the potential users.

The complete list of attributes that is available to the manager-user is presented in Appendix F. It was designed following standard principles for shelter management (FEMA, 2006; USAID, 2003). Amongst the most relevant attributes are the name of the building, operational status of the shelter (i.e. open or closed), contact information, capacity and some physical characteristics such as parking capacity, as well as resources at the shelter (e.g. water, food).

Figure 9.7. Shelter information in manager-user mode in EvacuAPP.

Wayfinding evacuation directions

Once the user has selected the shelter that they want to evacuate to, they can use the built-in directions developed within EvacuAPP. The application allows the user to select any point on the map as the starting point (as default the current location), and the preferred/selected shelter as the destination. Step-by-step instructions to reach the destination are displayed on the screen, and the route is shown on the map (Figure 9.8). This functionality offers the estimated time of the trip, computed based on current traffic flow and the displacement method chosen by the user; at the moment the EvacuAPP allows driving and walking to be selected as the displacement method. Once a route has been selected, it can be saved, shared or printed, in case of a possible internet disruption.

Figure 9.8. Wayfinding functionality within EvacuAPP

Even though Sint Maarten is relatively small and long-term residents may know their way around the island, driving functionalities are particularly useful for users who are not familiar with the area, as is the case for tourists. Furthermore, EvacuAPP allows manager-users to add barriers to the roads in the map if they identify road disruptions due to flooding, fallen objects, or traffic has been diverted (Figure 9.9).

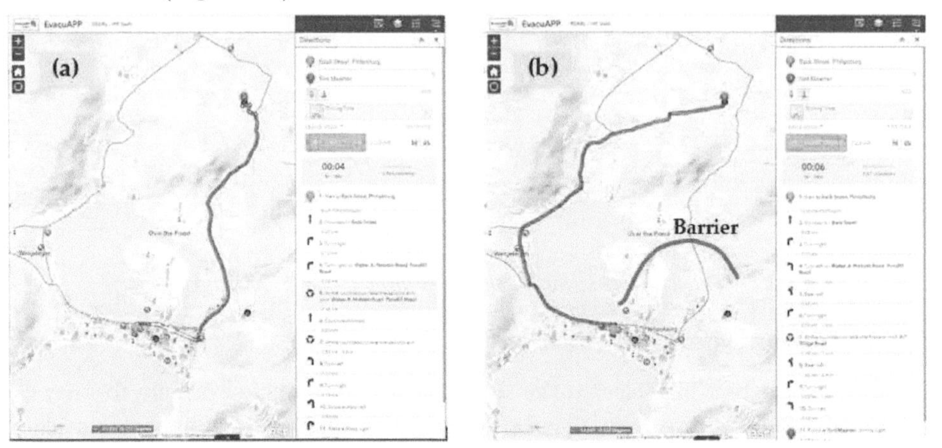

Figure 9.9. Wayfinding functionality within EvacuAPP; (a) normal flow and (b) barrier added.

Wayfinding functionality was important to include in EvacuAPP despite having well-known applications such as Google maps or Waze for this type of service. The advantage of a built-in wayfinding functionality lies in the possibility to re-route at any given point of the evacuation if a shelter changes its status to Closed or the managers-users have added an identified obstacle (i.e. e.g. flood, fallen tree) to a segment of a road on the map.

Critical Infrastructure / Buildings

Critical infrastructure in terms of disaster risk management is defined as the physical structures or assets that support services that are essential to the functioning of a society or community (Etinay et al., 2018). For Sint Maarten, a list of critical buildings was previously identified (UNDP, 2012), which includes facilities regarding transport, fuel and transport, banking, government and public services, lodging, healthcare, insurance, religion, and utilities.

It was decided to include this layer for the manager-users to be able to take specific actions over this infrastructure in times of evacuation. Also, in large-scale emergencies, some of these buildings can be used as shelters and managers can update their status to be reflected as a new shelter option for the end users. Figure 9.10 shows the different types of critical buildings for Sint Maarten.

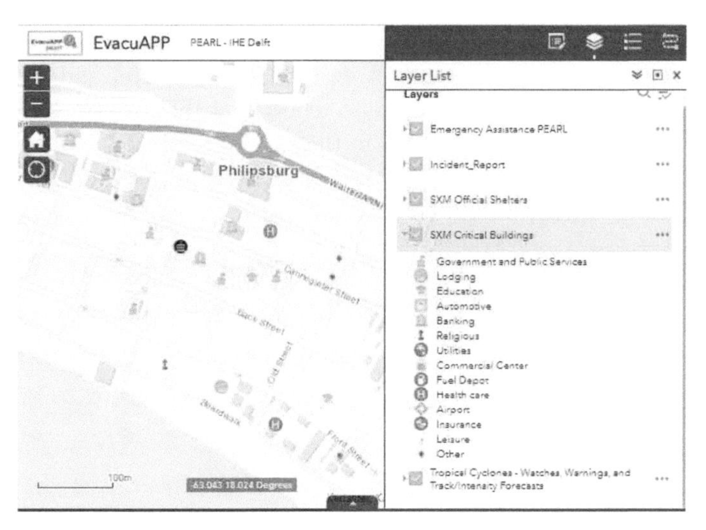

Figure 9.10. Operational layer of critical buildings within EvacuAPP for Sint Maarten.

Regarding critical buildings in the web-app, the end user will access similar information to that presented for shelters (Figure 9.6), whereas the administrator again has full access and editor control as in Table F.1. At any given moment administrators can change the status of any critical building to function as a shelter.

Emergency Assistance

One of the most valuables functionalities in an emergency app is the possibility to allow users to request assistance at any moment of the emergency. This functionality aims to provide a link between crisis responders to end users that might require assistance. With this functionality, any end user can report their need for assistance at their current geolocation at three levels (YES, NO and partially) as shown in Figure 9.11. Once a user has requested assistance, EvacuAPP will send a notification of their request with an attached geolocation tag to the disaster risk managers/app administrators. Having the geolocation of all possible residents in need of assistance will allow the authorities to have a better and more coordinated assistance plan. The functionality can be accessed during all the phases of an emergency, e.g. in the pre-disaster phase an end user can request assistance in transportation, or report him/herself trapped in the aftermath of a hurricane.

Through the editing section of the web-app, an end user can request full (YES), partial or NO assistance. We include the option to respond NO in this functionality in order to allow the user to report him/herself safe, because even if a user is physically safe they can be in need of other resources (i.e. water, food, security, or cover). Once the user has placed an emergency assistance 'pin' on the map (Figure 9.11), The information that can be provided through the text box includes the user's address and information about the house (i.e. the number of floors, access through a back alley), phone and mobile phone contact numbers, request for transportation and/or pet evacuation assistance, and provide some emergency contact details. They can also report that they are trapped and add a photo if they think it could be useful in the assistance process. Only the contact name and the need to evacuate fields are mandatory on this form

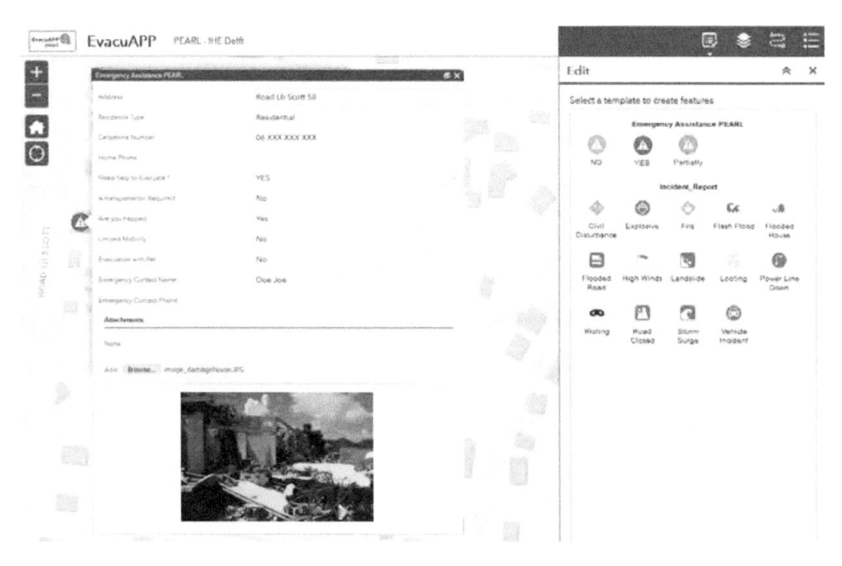

Figure 9.11. Emergency assistance functionality within EvacuAPP for Sint Maarten.

Also, a text box appears for the end-user to fill in details to guide the assistance. The information that can be provided through the text box include the user's address and information about the house (i.e. the number of floors, access through a back alley), phone and cell phone contact numbers, request for transportation and or pet evacuation assistance and provide some emergency contact details. They also can report as trapped and add a photo if they think it can be useful on the assistance process. Only contact name and need to evacuate fields are currently mandatory in this form.

The 'pin' when a request has been placed is by default located in the current location of the user based on the GPS functionality of the device, but it is possible for the user to manually locate the pin for assistance in another part of the map; this is part of the design, so that a user can request help for another person who they know needs it but does not have access to EvacuAPP. For the current version of the web-app only the need for help has been made compulsory to fill in, but future versions need to make other fields in the form compulsory to allow verification of the veracity of the request (i.e. name, mobile phone number)(Sebastien and Harivelo, 2015).

Incident Report

Another functionality that was reported as needed/useful during our field campaign is the ability for the users to act as a crowdsource of information regarding incidents related to the evacuation or emergency situation. The type of incidents that can be reported as useful by the end user include looting, floods, storm surges, road closure or vehicle incident, rioting, fire, explosions, civil disturbance, landslides, high winds, and power lines down (Figure 9.12). Incident reports are also accessed through the editor section as illustrated in the emergency assistance functionality.

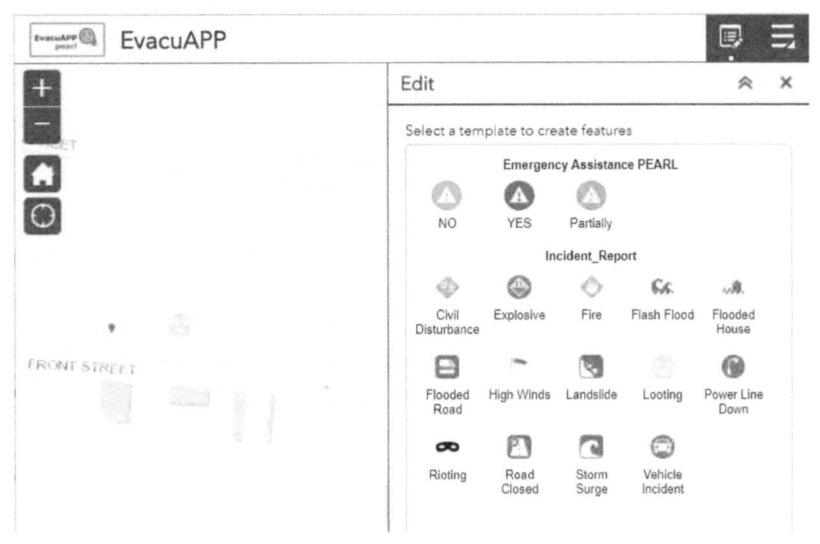

Figure 9.12. Information that can be provided when using the incident Report layer in EvacuAPP.

The information from this functionality has two main purposes. First, the information can be of use for the authorities when deploying ground assistance or planning rescue missions during the evacuation. Second, users may be more willing to evacuate when they know they can report different types of incidents that we found to discourage evacuation on Sint Maarten (Medina et al., Under Review), such as looting and rioting or when they can minimise the chances of being trapped in a traffic jam (road closed) when the potential hazard arrives.

Hurricane Monitoring

The final functionality of EvacuAPP allows users to activate the layer of current tropical cyclones in the Atlantic Ocean using data directly retrieved from the National Hurricane Center (NHC). This functionality allows the users to watch for warnings and track the intensity and location of hurricanes and tropical cyclones once they are forecast. This function allows an end user to be aware of a major threat, its possible path, and arrival time to its location.

Having warnings, evacuation and assistance request functionalities integrated into one application will make it easier for the final user to find all the information without the need to jump from one application to another. As illustrative of the track and warning functionality, Figure 9.13 shows Hurricane Ophelia in the Caribbean.

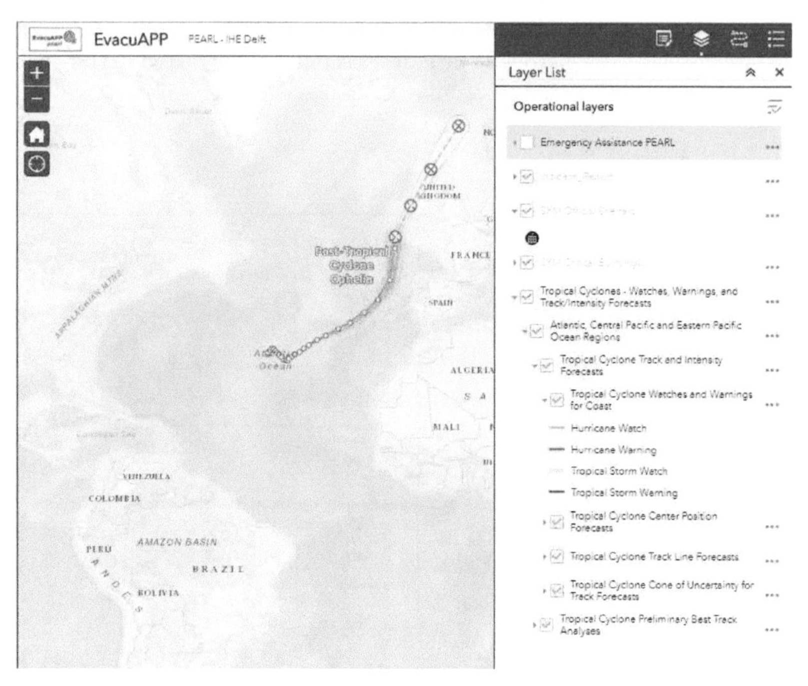

Figure 9.13. The National Hurricane Center's forecast layer of hurricanes within EvacuAPP.

EvacuApp on different screens

One of the main advantages of a web application such as EvacuAPP is its ability to adapt and fit different devices and screen sizes. This requirement was expressed in the fieldwork for the acceptability and usefulness of the application on the island. We present in Figure 9.14, Figure 9.15, and in Figure 9.16, how adaptable the EvacuAPP is to a laptop/computer browser, tablet and smartphone, respectively.

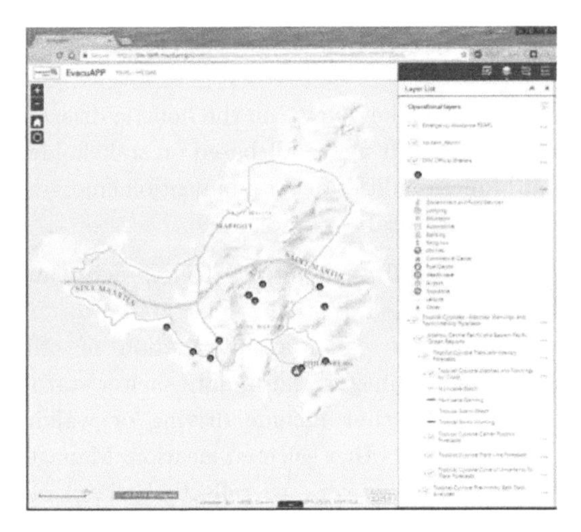

Figure 9.14. EvacuAPP on a laptop web browser.

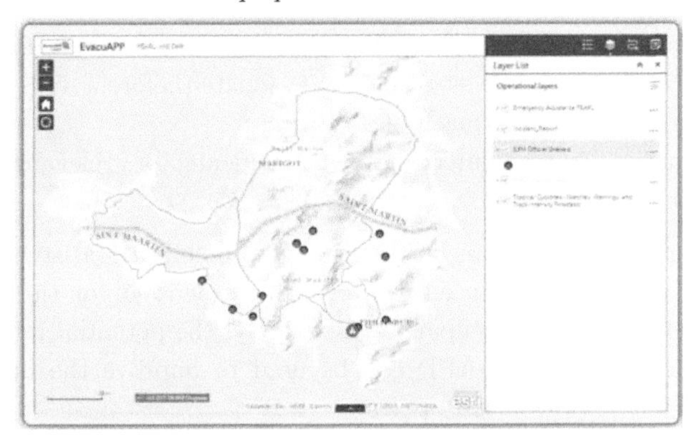

Figure 9.15. EvacuAPP on a tablet web-browser.

Figure 9.16. EvacuAPP on a smartphone: (a) iPhone X and (b) Samsung Galaxy S8.

9.4 DISCUSSION AND FUTURE DEVELOPMENTS

EvacuAPP is aimed to be a key player in the holistic disaster risk management on the island of Sint Maarten. It was built based on stakeholders' identified needs and as some standard functionalities of communication emergency apps. The main characteristics of EvacuAPP are:

- Access to the web-app from multiple devices, without logging on or any installation required.
- End users can request essential information of official shelters, and manager-users can administer and update shelter status.
- Wayfinding directions that include driving or walking directions to a preferred shelter, or any other selected location. Manager-users can impose barriers in the traffic network to reflect observed or reported limiting conditions on the roads.
- Direct access to the forecast information from the NHC about current tropical cyclones - Watches, Warnings, and Track/Intensity.
- End users can request assistance to evacuate (before a disaster) or request assistance in the aftermath of a disaster.
- Information crowdsourced about incidents of relevance during an emergency evacuation.

In addition, EvacuAPP can be further used to collect statistical data on the usage of the app and compliance rates regarding evacuation on the island, as well as shelter preferences and occupancy. Ultimately, the potential information that can be gathered with EvacuAPP can be used to improve the mechanism and strategies for future evacuation events.

However, despite the interest of several stakeholders to use and implement the app, due to time and economic constraints, no real testing and adoption of the system have been done in the following hurricane seasons after Hurricane Irma. At the time of writing the app is fully functional and still accessible, with more than 300 uses. Hence, even though the app is fully functional, and at the time of writing still accessible, no real testing and adoption of the system have been carried out in the following hurricane seasons, and its usability and limitations for a real emergency therefore remain unknown.

If EvacuAPP, or any other emergency app, were to be adopted on Sint Maarten some drills and training on their use would be necessary. As reported in Zlatanova and Fabbri (2009), people would be reluctant to use any new technology in times of crisis if they are not familiar enough with it. In addition, if EvacuAPP is used or tested on Sint Maarten, the disaster risk managers should measure the web-app's ability to transmit messages during the disaster, measure the ability of use in terms of cognitive capacity, and assess the acceptability and trust of users towards the presented development (Tan et al., 2020).

One foreseeable challenge is how to promote the use of the EvacuAPP for tourists and foreigners who might be unfamiliar with the island and its vulnerability to disasters triggered by natural hazards. Promoting the benefits among the hotel sector may increase the chance of tourists using it. Currently, the hotels are organised in an association called Sint Maarten Hospitality and Trade Association. They receive bulletins from the meteorological office and decide what actions to take. By also incorporating the use of EvacuAPP, more risk awareness and reaction not only from hotel managers but from tourists might be expected.

One of the most crucial elements in disaster risk management is the location of potentially affected individuals. Knowing the location of the hazard and the location of possible evacuees can allow personalised warnings or evacuation orders to be sent as the emergency is unfolding. On Sint Maarten, the quality of the message that people receive, as well as a more direct message, will lead to better evacuation rates and compliance with instruction, as reported in an unpublished work by the research team (Medina et al., Under Review). Currently, on Sint Maarten, where the most used source of information continues to be the radio, its effectiveness in promoting evacuation or protective behaviour is questionable as the results seen after Irma illustrate. Its lack of effectiveness might be explained due to the generic and non-individual-specific information it provides (Kremer, 2017). EvacuApp has the potential to be that missing tool for DRM on Sint Maarten by providing people with more personalised and in-depth information.

Concentrating emergency information in one single app such as EvacuAPP can focus the attention of the user on getting information from a verified source of information and potentially avoid the spread of false information that usually circulates during an emergency, especially with the advancement and wide-spread use of social media (Lovari and Bowen, 2019).

We can foresee some limitations in the current development of the app. First, it relies on an internet connection to have access to the main functionalities, and therefore in the case of a total failure of the Internet in the area where a user is connected they cannot use the web-app. Hence, alternatives to solve this issue need to be implemented in future developments of EvacuAPP. One alternative could be to use the so-called Device-to-device solution (D2D), which allows devices close to each other to communicate directly without using external networks (Sebastien and Harivelo, 2015). An implementation of D2D solution is presented in (Bahk et al., 2017), Bahk's solution is called a Mesh Network, where a device that has lost connectivity will look for other devices using the app in the surroundings and use them as a router to send the request; another example is the use of Wi-Fi direct technologies to allow direct communication for two or more devices (Camps-Mur et al., 2013).

Second, the EvacuAPP design assumed that the mental state and abilities of the end user remain unchanged during use in a disastrous situation, which sometimes might not be accurate for some users due to panic or some physical constraints due to the emergency. Further developments of EvacuAPP should include elements of human-centred design principles that account for this kind of possible impairment. Improvements include simplified maps when a disaster occurs, by using primary colours to highlight an escape route while keeping the background colour neutral. There should also be a simplified button to ask for assistance without the need to fill in a form to take into account users with limited mobility or high distress levels. The app could include a module that allows the user to send pre-configured messages not only to the administrators of the web-app but also to predetermined groups such as family or/and neighbours.

Third, in the current version of EvacuAPP, end users that have requested assistance cannot follow updates on their request. Therefore, to enhance the usability of the app, it should contain a function that allows users to verify the status of their request (i.e. the manager-users have seen it or help is on the way).

Fourth, a single audible alarm should play at critical moments of an emergency, such as every time a hurricane increases on the Saffir-Simpson scale or when a hurricane is located within a certain distance from the user. However, this should not be overused to avoid increasing the levels of stress.

Finally, further developments should include the near-real-time tracking of all users' locations and status as primary information for the disaster risk managers on the island. Location can help risk managers to visualise distribution of the population density and identify areas of potential need, such as the concentration of non-evacuees in a high-risk area, or detect traffic jams or traffic volume to a specific shelter, among others. This information can be used to prioritize resource allocation.

9.5 CONCLUSIONS

In this paper, we present the development of a web application that aims to be used during emergencies related to weather-related events such as hurricanes and floods on the island of Sint Maarten. It was designed and implemented based on the identification of users' needs as well as some standard functionalities of emergency management. Having included some of the identified users' needs might increase the chance of future adoption and use of the web-app on the island. However, the choice of web-app does not limit the applicability of the functionalities behind EvacuAPP in other kinds of development (i.e. smartphone apps).

Even though the thorough design process of EvacuAPP incorporates those needs identified by several stakeholders on the island and from ICT standards during emergencies, the web-app we developed can be considered as in development. We have already identified some room for improvement as presented in the discussion section and those can be incorporated in future developments. On the other hand, it could also be determined after some trials that some functionalities will no longer be needed for Sint Maarten. Only when it has been used and applied in drills and in real emergency situations can its full potential and pitfalls be determined

Social media platforms such as WhatsApp and Facebook were broadly used during and after the passing of Hurricane Irma (PEARL 2018), which shows a shift in technologies being used on the island to communicate during a crisis, and shows the important role of mobile applications for disaster management nowadays. Even officials from the government used these platforms to communicate and plan actions before, during and after Irma. However, while social media platforms are useful in an emergency, they are risky due to the possibility of spreading fake news and users have reported their frustration about not finding emergency communication that is fit for the specific purpose (Kremer 2017). Sint Maarten emergency managers should utilise this opportunity to invest in developing specific applications to communicate the latest and official news in one centralised place and promotes its use among residents and tourists.

Compared to traditional sources of warning and disaster information communication, web developments and mobile applications have a greater potential to be used in disaster risk management by taking advantage of the inherent sensors in current smartphones and tablets that enable more timely, directed and personalised information to be sent to the end users.

Despite the interest shown by the stakeholders that we interviewed, concerning the use of the app and the foreseen potential, at the time we presented the app, all the public workers of the administration were still dealing with the recovery phase of the disaster caused by Irma, and they expressed a lack of time to further explore the app and use within DRM in the island. It is essential to try to push forward the benefits of adopting this (or other developments) in order to increase emergency communication on Sint Maarten.

Regardless of the importance of new methods to be used in the disaster management cycle, such as is the case of web applications, emergency management organisations should always have another plan to communicate before, during and after a disaster given the fragility of communication infrastructure during disasters triggered by natural hazards. In the case of Sint Maarten, radio communication has been historically the most reliable and used source in this regard.

At the moment, the number of people that reported evacuation to shelters on Sint Maarten is very low, only 3% of those who evacuated during Hurricane Irma choose public shelters (Medina et al. 2019); this is a phenomenon not exclusive to the island, but is also common in other evacuation studies. However, due to the extended damage caused by Irma to the whole island, we expect that the number of evacuees requiring shelter allocation will increase in future evacuation scenarios. The present web-app can help new evacuees to shelters by guiding them in an unknown procedure. Nevertheless, even if people do not evacuate to shelters, some functionalities of EvacuApp are still useful, such as hurricane forecasting and tracking, wayfinding directions and the possibility to report safe even if they evacuate to a different place (e.g. friends' or relatives' houses).

Finally, we hope that EvacuAPP will become a standard to be use by the disaster risk managers on Sint Maarten, as well for residents of the island. We expect that it will be able to reach a large percentage of the population to create hurricane and flood awareness among the public when a threat is forecast and help guide residents to reach safety in the case of a possible evacuation. We also believe it can help to promote evacuation because it has functionalities that account for some of the limiting factors for households to not evacuate, such as reporting looting. Also, EvacuAPP has the potential to become a tool for rescue and relief teams in the aftermath of a disaster in the case that some residents require help because they can request help using his/her current location (GPS-based). If a sufficient number of users is reached in the island, it would be possible to develop (to a certain extent) real-time evacuation plans based on the evolving threat and reported location of residents of Sint Maarten.

10

Outlook

10.1 Bringing it all together

The previous chapters focus on developing models that aim to capture the adaptive nature of risk into an assessment framework (ADRA). They do this by proposing new methods to assess vulnerability and exposure in order to capture the complex interrelation of humans with the environment more accurately when faced with a disaster. This chapter summarises the previous chapters' work concerning the three research questions presented in the introductory chapter. Then the chapter continues by presenting reflections on the results of this dissertation. The last part of this chapter offers an outlook on the remaining open challenges, practical recommendations, and future or complementary research opportunities.

10.2 Main contributions of this thesis

On 27 August 2017, Hurricane Irma originated on the west coast of Cabo Verde, and after some days of turning into a catastrophic storm, it finally reached the SIDS of Sint Maarten on 6 September 2017 when it was at its peak of intensity. The result for Sint Maarten was catastrophic, with billions in economic losses and the loss of life of some vulnerable residents on the island. The catastrophe posed a challenge to the research I had already been carrying out for over three years, and so a decision was made, that it was not only necessary, but also an ethical responsibility to reshape the focus of the research, and feed our theories with ground data related to disaster risk in the aftermath of a disaster. After Hurricane Irma, the aim was that this dissertation's outputs could be used to help the reconstruction phase and the Building Back Better plan for the island.

To answer the new challenges, we started by asking ourselves: *"What elements of socioeconomic vulnerability are important in an adaptive risk framework in the context of a SIDS?"* This question was addressed in chapters 2 and 3. Chapter 2 presents the main results and analysis of our fact-finding mission on the island regarding vulnerability and risk assessment. In Chapter 3, using an index-based approach, we performed a socioeconomic vulnerability assessment with the findings and data collected in the aftermath of Irma.

Hurricane Irma was catastrophic for Sint Maarten but offered an excellent opportunity to perform an in-depth analysis of some of the root causes of vulnerability and to incorporate new variables into the computation of vulnerability indexes that are only possible to observe and detect after a disaster has unfolded. Travelling to the island gave insights into factors that shape risk and vulnerability on Sint Maarten that would otherwise be impossible to bring to light.

The interviews we carried out allowed respondents to give us insights into their assessment of risk and hidden drivers that a desk study would not have provided. One of the most important insights we gained was that despite the intensity of a forecasted hurricane, and certainty of a direct hit, most Sint Maarten residents would rather stay in their own houses. Two main reasons were exposed for this behaviour: to protect the house from looters and from the storm itself, and because they perceived their houses to be stronger than public shelters and to avoid the discomfort associated with an evacuation. This finding was later used to shape the agent-based model implemented in this research as a fundamental behaviour characteristic of St. Maarten's residents.

The fieldwork camping allowed us to collect relevant information that influences vulnerability in a post-disaster environment, which was directly incorporated into the index approach: level of trust in institutions, risk awareness, risk perception, the role of information, damage extent, and recovery speed, among others.

The indexes and associated maps produced in Chapter 3 are the first of this kind for Sint Maarten despite the potential hazards they encounter each year during the hurricane season. Our assessment identifies the neighbourhoods on Sint Maarten most vulnerable to natural hazards with insights into each neighbourhood's most important variables. We hope that the findings on vulnerability can be used to guide policymakers on where to focus the limited resources available to mitigate (or eliminate) the impact of a potential hazard.

To our knowledge, this is the most integrative study of this type in the context of SIDS that also incorporates post-disaster findings, and it offers a framework flexible enough to assess vulnerability in other similar areas with similar potential hazards and geographic characteristics.

Once we had answered the question on vulnerability, we moved on to the second disaster risk assessment component. We asked: *"What are the main predictors of adaptive behaviour to reduce exposure in a SIDS?"* To answer this question, again, the data we managed to collect through surveys in the aftermath of Hurricane Irma, combined with the findings on socioeconomic vulnerability, was the basis to assess the exposure component of risk. We answer this question in Chapter 5 and Chapter 7.

When humans are faced with a potential threat to life or damage to infrastructure, they react and adapt, to the best of their capacities (or intentions), to mitigate the possible impact of such a threat by performing reduction of exposure measures; this stresses the importance of capturing the dynamics by which exposure evolves over time for a more realistic DRA. Evacuation from the potential threat zone and adoption of in-situ protective measures are listed among the most effective (and performed) actions to undertake to reduce exposure to a threat. Evacuation behaviour is analysed in Chapter 5 and Chapter 7, whereas in-situ protection is assessed in Chapter 7.

Evacuation during disasters triggered by natural hazards is perhaps one of the most studied disaster risk reduction strategies. Still, despite the number of studies on the topic, we found a lack of agreement on which variables can be used as predictors for individuals (or households) to evacuate, which can be partially explained by local, environmental and cultural differences. Hence, a study leading to the local predictors that could explain the observed evacuation behaviour on Sint Maarten was needed. In Chapter 5, using binomial logistic regression analysis, we explored a set of variables or predictors to explain the behaviour to evacuate (or not) on Sint Maarten. For Sint Maarten, we found that six variables help to explain the observed evacuation behaviour. Gender, homeownership, percentage of property damage, quality of the information, number of house storeys, and the vulnerability index were all found statistically significant as predictors of evacuation. We formulated a regression equation model to assess the chances of evacuating a particular household on Sint Marten probabilistically, and the results are presented in probabilistic distribution maps per neighbourhood.

Despite the amount of studies on evacuation behaviour, and SIDS being listed as one of the most vulnerable and exposed territories to natural disasters, there was a lack of literature applied to evacuation in the context of SIDS. Our findings are amongst the first of this kind. Table 5.1 in Chapter 5 offers a guide to which parameters could be potentially used as predictors of evacuation in the context of SIDS.

To further explore the role of exposure on disaster risk, in Chapter 7 we used an ABM as a virtual laboratory that offers a new tool to explore how a system may evolve to changes in key components that may promote or restrict protective actions during a disaster. Based on the finding of Chapter 5, we focused our analysis on how the quality of the information received during an emergency may shape the reaction towards protective actions during an emergency; the quality of information being defined in this dissertation as the trust of the source and the content of the information received.

Despite exposure to natural hazards being a distinctive component of disaster risk, it is common practice to assess this component in DRA studies as part of the physical component of vulnerability, or it is studied assuming that the sole presence of infrastructure determines the degree of exposure of a system (i.e. land use maps, a layer of buildings). This conceptualisation of exposure fails to capture the adaptive nature of humans to their changing environment. It is an oversimplification to assume that (all) humans will not adapt or prepare (e.g. evacuate, in-situ protection) to a certain extent when a known threat is approaching, regardless of the uncertainty associated with forecast information.

Our ABM methodology builds on this gap in the exposure assessment, by studying the complex interaction between humans and their environment (i.e. the urban fabric and the hazard itself). We were able to capture the adaptive human behaviour into the exposure component. We have explored how evacuation and in-situ preparation can change the exposure level due to different fluxes of information and source credibility. What differentiates our exposure methodology is that it is built as a human-centred approach compared to infrastructure or physical vulnerability approaches that we found as standard practice in DRA studies.

In addition, in Chapter 6, we present an extensive systematic literature review on the use of ABM in water-related disasters. A comprehensive literature review on ABM used for water-related DRM was done because this modelling tool was used to incorporate (in Chapter 7) the adaptive behaviour of humans into the exposure component of the proposed adaptive disaster risk assessment of this dissertation. Despite the increasing popularity of ABM for DRM, a literature review in the intersection of both disciplines was found to be lacking. We believe that our findings manage to identify current (best) practices, research gaps and ways to move forward the field of ABM for WR-DRM and that our review can provide a starting point for further research in both ABM and DRM.

We asked the final question: *"How beneficial is an Adaptive Disaster Risk Assessment over traditional Disaster Risk Assessment methodologies?"* This question was answered in Chapter 8, in which we present a methodology to assess risk, including the adaptive nature of exposure. The proposed methodology, called ADRA, is a holistic framework that combines all previous disaster risk assessment findings in this dissertation. ADRA uses the vulnerability assessment results in Chapter 3, the probabilistic evacuation maps produced in Chapter 5, and the protective behaviour simulations of Chapter 7.

ADRA's major novelty is incorporating adaptive exposure maps that reflect the protective actions that households on Sint Maarten may undertake to decrease exposure levels when faced with an imminent disaster. We account for two protective measures, evacuation and in-situ preparation. ADRA is a human-life-centred approach in which the risk that it aims to assess is the loss of human lives by reducing contact with the hazard. ADRA visualisation maps allow the impacts of a particular strategy in reducing (or not) the exposure and risk levels at the neighbourhood scale to be evaluated. Understanding differences in response to different strategies for exposure reduction gives DRM managers the possibility to target different strategies that better fit a particular neighbourhood of interest.

Finally, in this dissertation, we wanted to offer a practical tool to answer some of the specific needs that this dissertation found significant for DRR on Sint Maarten. We proposed and created a web application tool that aims for better information flow during the different phases of a disaster. EvacuAPP is a web-based application built using the insight we collected during several interviews with stakeholders on the island of Sint Maarten in the aftermath of Hurricane Irma.

A communication tool was selected as a practical tool for DRR on Sint Maarten, based on the findings of Chapter 5, where information was found to play a significant role in reducing exposure by promoting evacuation and risk awareness. In Chapter 7, and in Chapter 8, the importance of information on Sint Maarten was again proved to be important when simulating how institutions and peers' role in communicating risk can completely change the island's exposure and risk maps

10.3 REFLECTIONS

10.3.1 On Vulnerability Assessment

SIDS socioeconomic and geographical characteristics make them very distinct to other territories. These characteristics are their undiversified economies, limited natural resources, multi-ethnic cultures, geographical isolation, and constant threat to extreme weather events. These all highlight the need to consider a selection of variables and multiple dimensions that may encompass the vulnerability to disasters triggered by natural hazards in such regions. In this regard, our index-based approach (PeVI), combined with the PCA analysis, allows the important drivers of vulnerability to be revealed by using a quantitative method. Furthermore, that vulnerability in the context of SIDS can be studied using three dimensions: susceptibility, coping capacities and (lack of) adaptation capacities. The index also has significance for policymakers and disaster risk managers to support their decisions in a multidimensional vulnerability identification approach.

Index-based socioeconomic vulnerability assessments are the system's "snapshot" at the moment at which the study was performed (or when the multiple inputs were collected), but many (if not all) of the elements used to compute the index are dynamic and adjust in time as a response to the complex interactions and interconnection of the social, political and environmental systems. Hence, it is important to acknowledge this limitation and keep updating the system's current vulnerability status in terms of disaster risk.

Socioeconomic vulnerability is categorised as a non-linear phenomenon and driven by location-specific variables. Still, to a certain extent, our methodology can be used as a framework to assess socioeconomic vulnerability in the context of SIDS given the flexibility in its design to remove or incorporate variables that are relevant for a particular case.

Given the high societal component of vulnerability, such an assessment should be done by multidisciplinary teams that can bring different concepts, theories and preferences to the table. Also, vulnerability assessment should not rely exclusively on desk-based information; it is necessary to "feel" and "experience" the actual environment in which the vulnerability evolves. Surveying in a post-disaster environment presented some challenges to the research team, from the ethical dilemmas of interviewing an individual in distress to logistical issues such as the timing of the study and the number of question one may pose to an interviewee. Still, it offers invaluable lessons on what elements to include in our vulnerability assessment.

However, challenges remain regarding the implementation of the index-based approach. Many of the variables used to compute PeVI are qualitative and based on perceptions of the survey respondent; hence, the subjectivity and validity of such answers can increase uncertainty in the results. Many of the data needed are only possible to reveal after a disaster has occurred, such as evacuation or protective behaviour intentions, or how a disaster may have changed how someone values its risk. Besides, vulnerability assessments require a considerable amount of data at very detailed aggregation levels to be useful. However, such data is often not accessible to this type of study due to the protection of identity issues, or lack of willingness or time to cooperate.

10.3.2 On Exposure Assessment

Reflecting on the assessment of exposure, the first lesson for our modelling purposes and a better understanding of why (most) people do not evacuate on Sint Maarten is contextualising risk in SIDS settings. The size of SIDS compared to the size and intensities of major hurricanes means that no matter where on the island an individual is, the exposure cannot be avoided to a certain extent. Under certainty of being exposed, an individual (or household) will always prefer to face the hazard in their place of residence.

However, the above does not hold true if the perception of risk at their location is perceived to be high, or they do not trust the house's structural integrity. In this regard, it is imperative that disaster risk managers properly identify high-risk areas using a multi-hazard assessment, and most importantly communicate the level of risk to those located in those areas. Perception of being located in a risk area has been proved to be one of the variables that better promotes protective actions at the household level.

When communicating warning information, not only a warning but also the message content is relevant; a Category 5 hurricane warning will not be sufficient to promote protective actions (at least not on Sint Maarten). Warning and awareness of a threat are necessary to take protective actions but offer an incomplete picture of the underlying and complex process that may lead a person to take action. How the message is delivered, accepted and understood by individuals and the community plays a major role in warning communication strategies to lower the exposure. In this regard, trust in the source of information plays a significant role when an individual interprets and assesses his/her level of perceived risk.

The role of traditional media and social media in evacuation behaviour seems to have a greater impact on taking protective actions. However, most of the effort to increase the adoption of protective measures is put into official sources of information. There is also a clear need to understand better the effects that non-official sources of information have on promoting or limiting protective actions, and how new technologies can be used to achieve better responses to protect against a potential hazard.

Our method to assess exposure (using the ABM - Chapter 7) proves that to understand protective action behaviour, future research should not only focus on the characteristics of those who perform any type of protection, but also move forward to a better understanding of what other internal or external factors influence the decision-making process. How individuals (or households) perform decision-making during a possible threat is based on their risk assessment. This assessment is very complicated in a natural-hazard context due to the high uncertainty associated with natural phenomena such as floods and hurricane. Will it happen? Moreover, if so, how severe it will be? Is the household going to be affected? Is it better (safer) in my current location or shall we evacuate, are questions that every individual (or household) must answer before taking action or staying after the awareness of a possible threat.

One limitation on evacuation and exposure to natural hazards studies is the many variables involved in an individual's decision-making during an evacuation. Understanding which variables apply to a particular case study is vital; demographics, prior knowledge and evacuation experience, social network, perceived versus actual risk, warning content and how and by whom it was communicated, among others can all be good predictors of actual evacuation behaviour. Literature reviews on hurricane evacuation do not agree on which variables are good predictors or not, and often contradictory findings have been reported. Therefore, it is necessary to evaluate which particular variables can be used in a particular case study. Only by doing this can more realistic protective action models be achieved. This will help decision makers to take the necessary measures to mitigate the loss of lives, measures such as improvement of shelters, better warning communications (i.e. direct evacuation orders), and better risk awareness campaigns.

10.3.3 On Adaptive Risk Assessment

Disaster risk assessments are traditionally estimated using a static approach. However, disaster risk is dynamic and adaptive to changes in the components of vulnerability and exposure. Failure to incorporate such adaptive behaviour into DRA may give an incomplete picture or scale of the disaster risk. In this dissertation, we show how to possibly incorporate the adaptive nature of the exposure component by either using probabilistic maps or results from ABM simulations, and what can be gained from such an approach to have a different system understanding.

Based on the key drivers of decision-making, particularly for Sint Maarten, our implementation of adaptive exposure focuses on hazard-related information. However, implementations in other regions should consider which elements are more important to explore. The current implementation of the ABM can be adjusted to reflect those local needs, but a good system understanding should be provided beforehand, and some programming skills are required.

We learned from our modelling process that for disasters associated with extreme events, there is a low probability of occurrence but a high impact; the critical component in DRA is the exposure. There is a threshold on the magnitude of a hazard where everyone will be affected, and the only option will be to avoid (or limit) the exposure to the hazard, which is particularly true for SIDS given their size compared to a major hurricane. For SIDS, where the exposure component cannot be entirely removed (e.g. timely evacuation), the strategies to reduce exposure should be focused on proper land use planning, complemented with non-structural measures for preventing or mitigating risk. Early warning systems, proper identification and communication of risk, schemes for loss compensation, and adequate preparation and planning are all necessary for DRR in the SIDS context.

The ADRA framework and traditional disaster risk assessment methods cannot be seen as contradictory but as complementary methods. On the one hand, ADRA assessment can show which protective measures can be more useful to lower risk (to life) and show where those measures will have a more significant impact. For example, ADRA outputs can be used to detect those areas where more awareness campaigns need to be carried out or to identify areas where potentially extra resources will be needed in the aftermath of a disaster, allowing the resource allocation to be planned. On the other hand, traditional approaches are useful in assessing the overall impact of a hazard in a particular region of interest. Traditional approaches can be used for evaluating the impact on infrastructure and economy, as well as for zoning and regulations

10.3.4 On ABM for DRM

The lack of consensus in the definition or in the terminology used to describe the modelling of complex systems through the use of agents is a current challenge that researchers must face when dealing with a literature search on a specific topic. Different disciplines use different terminology referring to the same concepts and principles, making it difficult to keep track of the progress on ABM and more difficult (or confusing) to understand for those who are just starting to use ABM as a modelling technique. In addition, the use of different terminology makes multidisciplinary cooperation more challenging.

There is also a lack of agreement and poor definition of what ABM really means. ABM across the papers included in our review (Chapter 6) is referred to in multiple ways, as a toolkit, as a modelling paradigm or technique, or just as software. The lack of a unified definition adds another layer of complexity to the use of ABM not only in the WR-DRM domain but in general.

Based on the results in the different search engines used in our review (Chapter 6) and on the selection of relevant articles after the manual selection and exclusion of papers, we proposed that the set of words 'Agent-Based Models' should be the standard when referring to the use of modelling complex adaptive systems based on microsimulation of individual agents in DRM. We came up with this proposition as Agent-based models is the most used terminology nowadays, and therefore, this makes it easier to adopt it as the standard.

Finally, we suggest that a good ABM for WR-DRM should include the following eight elements: (i) a clearly stated research goal or objective, (ii) a review phase of existing ABMs that may serve as a starting point either for ideas or to reuse the code, (iii) explicit use of a standard to describe and document the design of the ABM (we suggest the ODD+D protocol) (iv) a clear definition and documentation of the protocol, ground theory and rules used in the architecture for agents' decision-making process, (v) a section to explain precisely how the verification and validation of the model were done, (vi) some type of model analysis and preferably sensitivity analysis on the most relevant parameters influencing the model, (vii) a reflection on how the research has contributed to the understanding of the processes being modelled beyond the case study application and offer a guide on transferability to other case studies and (viii) finally, share the source code; we suggest using the OpenABM initiative. These outlined steps should be met to help ABM become broadly used and accepted for WR-DRM.

10.3.5 On Sint Maarten DRR

Devastation like that observed on Sint Maarten after Hurricane Irma needs to be seen beyond the disaster. It can be an opportunity to rebuild a less vulnerable island. One key issue on Sint Maarten is the materials and methods used to construct houses/buildings. The government should improve the outdated building codes, increase inspections, and assist in rebuilding both financially and technically all over the island. The government should also review the land leasing model to implement more control over the quality of constructions in those areas identified as the most vulnerable during the household survey.

Residents perceive that one of the most critical factors of vulnerability is posed by the slow recovery pace after Hurricane Irma. The government can address this situation among others by implementing a hurricane fund and adding financial resources to the fund every year from taxation. A recovery fund will allow the independent island state to finance the reconstruction with less dependency on the Dutch government or other external financial organisations or donors.

In hindsight, closing most public shelters before the hurricane was the right decision because they are not designed to withstand Category 5 hurricanes, as demonstrated by the collapse of part of the roof of two of the shelters – Sister Marie-Laurence School and the New Testament Baptist Church. However, it is important to highlight that some residents needed the shelters to be open before the hurricane struck since their housing conditions were even less resistant, and their social links (family or friends) were not useful or did not exist at all, as for example in the case of some undocumented immigrants. The recommendation is that the government invests in improving the existing shelters or building new dedicated shelters. In addition, the provision of water, food, and beds must be guaranteed if the government wants people to make use of the shelters. Implementing awareness campaigns for the overall population of the island is necessary if the government decides to undergo improvements of the existing shelters.

Based on the interviews, the Sint Maarten government has lost much credibility among the island's residents regarding disaster management, warnings, and evacuation communication, which increases vulnerability for the next hurricane seasons on the island. It is advisable to run several awareness campaigns before every new hurricane season to regain the inhabitants' trust and adequately communicate the possible threats the island may face in the future. Awareness campaigns also help to maintain the residents' social memory to adapt and become better prepared prior to future hurricanes.

We have also detected that given the multiracial and multicultural environment of immigrants on Sint Maarten (undocumented or not) the fact that warning information is mainly disseminated in English and Dutch is excluding large sectors of the population with little or no knowledge of these languages, especially the Hispanic and French-speaking communities. Hence, we suggest that an effective measure to reduce vulnerability is through improving the communication of the warning messages, by including more languages, and by simplifying the content of the message so it can be easily understood for non-educated inhabitants.

Radio is the most important source of information on the island. However, the fact that social media platforms such as WhatsApp and Facebook were also broadly used during and after Hurricane Irma shows the shift in technologies and the important role of mobile applications in disaster management.

To minimise the impacts of a disaster such as the one caused by Hurricane Irma, the government of Sint Maarten needs to promote policies and strategies to diversify the economy of the island to not only depend on tourism; this could potentially decrease the level of vulnerability since the economic coverage was the predominant factor driving socio-economic vulnerability on the island.

10.3.6 On my PhD journey

I want to end this part of the dissertation with a personal note. Nothing can prepare you to see someone break down in tears in front of you because they lost everything they have to a natural hazard, from material objects to relatives. The experience of travelling to a devastated area after Hurricane Irma taught me a great lesson in humility. Following an engineering approach, (I) we mainly focus on the numbers in our models, we calibrate to the third decimal place, we polish our maps to be aligned correctly and with the best colour palette, and we forget that behind the simulations, the maps, and the reports we produce, there are people.

We shall never forget that working in DRM, we owe all of our work to people. We need to produce meaningful and usable science. This PhD journey has taught me to be more empathetic, and less rigid in carrying out my work. It will be very satisfactory from a personal and professional point of view if some of the practical recommendations we offer throughout this dissertation resonate in the people in charge of DRM on Sint Marten, and some of the recommendations are adopted to reduce risk on the island.

10.4 Future Directions

Index-based approaches for vulnerability assessments are built using large datasets that are often not available in developing nations. Not having all the base information in an index-based approach such as PeVI can create an unbalanced index with some dimensions fully represented by variables while other cannot be explored in detail, offering a partial picture of the vulnerability of a case study. We partially face this limitation; in PeVI, the number of factors used to measure the lack of adaptation component is lower than the other two vulnerability components. We could not include as many elements as desired due to the complexity of the component and the limitations with data access on the island. Future vulnerability analysis on Sint Maarten should include more elements of adaptation capacities to have a more balanced index. Examples of variables that could be included are climate change perception, adoption of green infrastructure and nature-based solutions, an income parity ratio, air quality data, enhancements of early warning systems and the implementation of a hurricane (or disaster) fund.

One limitation with our methodology is that as a design choice, ADRA only includes adaptability in the exposure component; changes and adaptability into the vulnerability are not accounted for in the model. Adaptation was not included in the vulnerability component because we were interested in evaluating the (temporary) protective measures undertaken in the face of a weather-related hazard, and we simulated our virtual city in time steps of days. In contrast, vulnerability reduction measures are typically planned for the long term, requiring longer time steps or total simulation time. We acknowledge that incorporating vulnerability and long-term measures gives a more robust tool for DRM, and that it will make it possible to include adaptation in the aftermath of a disaster and how the vulnerability is re-shaped after a disaster. Future implementations of ADRA methodology can include a dynamic and adaptive version of vulnerability as well.

In the same way, an operational ABM such as the one developed in this thesis will also benefit by linking it with institutional ABM (Abebe et al., 2019b; Haer et al., 2019); this aims to evaluate the impact of policies and regulations in flood risk management, and evaluate the impact of the coupled ABMs.

Another constraint of the current ABM implementation is that agents (households) do not move into the environment (urban city). Again, this was a modelling choice based on the research aim and the system's properties being evaluated (a hurricane on Sint Maarten). For other settings, such as evaluating large-scale evacuations, or a different type of hazard (e.g. flash floods), where location and time may be relevant to assess the impact of the hazard on the agents and the environment, it will be necessary to incorporate moving agents. The literature review in Chapter 6 presents several examples of ABM used for evacuation purposes, and we have already suggested ways to improve and implement such types of ABM.

Given the timing of this dissertation, an exciting scenario to use the ABM's potentialities would be to explore the effects of the Covid-19 pandemic in protective actions in times of emergencies, mainly how Covid-19 may have limited evacuation behaviour in the 2020 hurricane season, which was the most active on record (29 named storms). Collecting data of evacuation behaviours during the pandemic could be used to set up an ABM.

We believe that involving the stakeholders in different ABM modelling cycle phases is a relevant practice in ABMs for WR-DRM. Stakeholders can add value in the setup, validation or calibration of ABMs because this will allow modellers to define the rules and behaviours based on cultural and behavioural context. In this modelling approach, some precautions need to be kept in the mind of the modellers. Stakeholder selection is critical, the ABM modeller needs to carefully select a representative number of possible stakeholders to minimise bias or avoid conflicts of interest and take into account the effects of group observation on the participants' behaviour. A protocol or standard will be needed to select stakeholders and report and measure the different stakeholders' decision-making process if participatory modelling results are intended to parameterise the agent's rules.

In addition, the use of new data sources such as big data and near or real-time location from cell phone location data remains unexplored for validation or calibration of ABMs for WR-DRM. From our findings, only the work of Yin et al. (2019) presents a validation of an evacuation model using population distributions generated by mobile phone location data. This research can give us a glance into the potential of using these types of data model validation. We believe big data will play a fundamental role in model validation of ABMs for WR-DRM and can make model outputs more robust and widely accepted. In terms of human decision making, one can ask, what can be the potential use of artificial intelligence or neuronal networks in a more sophisticated implementation (or calibration) of the agent's decision making? Mustafa et al. (2018) give a preview of the potential of such technologies, as they calibrated their model parameters using genetic algorithms.

In terms of the predictor used to assess evacuation behaviour on Sint Maarten, this study also has some limitations; due to the restrictions in the post-disaster environment, we did not ask several potential predictors of evacuation in the survey. Amongst those are household income; high income increases the probability of people going to a hotel rather than to a public shelter (Lee et al., 2018; Smith and McCarty, 2009), or in the case of Sint Maarten, the possibility to evacuate the island before hurricane landfall. Prior evacuation behaviour has also been reported as a good predictor for future evacuation (Thompson et al., 2017). Unfortunately, we did not collect this information and exploring its statistical significance could add another helpful predictor to our logistic regression models. The role of faith and religious groups in the evacuation decision making would also be interesting to explore. We did not directly include such an element. Hence, no statistical significance on this can be drawn from our data, but in our data collection, the respondents repeatedly mentioned that they just put their lives in the hands of 'God' and expect the best outcome when faced with disasters triggered by natural hazards. These limitations offer opportunities for further research. Also, we did not ask those who did not evacuate if they performed protective actions in situ and to what extent, information that can be used to validate the findings of our ABM in Chapter 7.

Current warning messages are too technical; they might be useful for government and DRM officials but not for the general public. People need a more concrete and simplified version. In terms of winds, besides the category of the hurricane, the message should include a short note with a scale they can associate with (e.g. this category is expected to cause damage to well-built framed homes). Estimated time of impact was also valued as important in the fieldwork findings; not having an estimated time limited the protective action. In addition, the current forecast (based on NHC reports) does not account for other hazards associated with a hurricane, and not showing potential storm surges or potential inland flooding may create a false sense of security in those who believe their house is strong enough to sustain the winds of the forecasted hurricane but that may be prone to other hazards. For the 2020 hurricane season, the NHC started experimenting with a visualisation of the expected storm surge inundation values (NHC, 2020). The new content of the NHC report is a good way to increase awareness and risk perception and to promote adoption of protective actions. However, the experimental phase was only available for mainland USA and its territories. In addition, with the observed increase in hurricane intensity, one can ask if it is time to add an extra category to the Saffir-Simpson scale to accurately represent off-the-chart winds and storm surges.

DRA frameworks typically use economic losses to assess the impact of different DRA. The assessment of impacts in DRA should move further from direct physical and monetary losses and include some indirect socioeconomic impacts associated with disasters triggered by natural disasters, for example, post-traumatic stress syndrome, loss of job and depression. The socioeconomic impacts are rarely studied or integrated into DRA frameworks. DRA will benefit from such integration in their frameworks as they are critical for recovery after a disaster.

REFERENCES

(CRED-UNISDR), Centre for Research on the Epidemiology of Disasters - CRED and United Nations International Office for Disaster Risk Reduction-UNISDR, 2015. The Human Cost of Natural Disasters: A Global Perspective 2015. United Nations: Geneva.

Abar, S., Theodoropoulos, G.K., Lemarinier, P., O'Hare, G.M.P., 2017. Agent Based Modelling and Simulation tools: A review of the state-of-art software. Computer Science Review 24 13-33.

Abdi, H., Williams, L.J., 2010. Principal component analysis. Wiley Interdisciplinary Reviews: Computational Statistics 2(4) 433-459.

Abebe, Y.A., Ghorbani, A., Nikolic, I., Vojinovic, Z., Sanchez, A., 2019a. A coupled flood-agent-institution modelling (CLAIM) framework for urban flood risk management. Environmental Modelling & Software 111 483-492.

Abebe, Y.A., Ghorbani, A., Nikolic, I., Vojinovic, Z., Sanchez, A., 2019b. Flood risk management in Sint Maarten - A coupled agent-based and flood modelling method. J Environ Manage 248 Article 109317.

Aboagye, D., 2012. Living with Familiar Hazards: Flood Experiences and Human Vulnerability in Accra, Ghana. Journal of Urban Research Briefings 48(12).

Abson, D.J., Dougill, A.J., Stringer, L.C., 2012. Using Principal Component Analysis for information-rich socio-ecological vulnerability mapping in Southern Africa. Applied Geography 35(1-2) 515-524.

Adger, W.N., 2006. Vulnerability. Global Environmental Change 16(3) 268-281.

Aerts, J.C.J.H., Botzen, W.J., Clarke, K.C., Cutter, S.L., Hall, J.W., Merz, B., Michel-Kerjan, E., Mysiak, J., Surminski, S., Kunreuther, H., 2018. Integrating human behaviour dynamics into flood disaster risk assessment. Nature Climate Change 8(3) 193-199.

Alexander, D., 2015. Disaster and Emergency Planning for Preparedness, Response, and Recovery. Oxford University Press, USA.

American Red Cross, POSM - In the Field. Planning fieldwork in general. Retrieved on January 2018 from http://posm.io/in-the-field/.

Amin, S., Goldstein, M., 2008. Data Against Natural Disasters. Establishing Effective Systems for Relief, Recovery, and Reconstruction. World Bank: Washington, DC, p. 432.

Augusiak, J., Van den Brink, P.J., Grimm, V., 2014. Merging validation and evaluation of ecological models to 'evaludation': A review of terminology and a practical approach. Ecological Modelling 280 117-128.

Bach, C., Gupta, A.K., Nair, S.S., Birkmann, J., 2013. Critical Infrastructures and disaster risk reduction. National Institute of Disaster Management and Deutsche Gesellschaft für internationale Zusammenarbeit GmbH (GIZ): New Delhi, India.

Baeza, A., Bojorquez-Tapia, L.A., Janssen, M.A., Eakin, H., 2019. Operationalizing the feedback between institutional decision-making, socio-political infrastructure, and environmental risk in urban vulnerability analysis. J Environ Manage 241 407-417.

Bahk, C., Baptista, L., Winokur, C., Colodzin, R., Domdouzis, K., 2017. The ATHENA Mobile Application, Application of Social Media in Crisis Management, pp. 97-113.

Baker, E.J., 1991. Hurricane evacuation behavior. International Journal of Mass Emergencies and Disasters 9(2) 287-310.

Balica, S.F., Douben, N., Wright, N.G., 2009. Flood vulnerability indices at varying spatial scales. Water Sci Technol 60(10) 2571-2580.

Balica, S.F., Wright, N.G., van der Meulen, F., 2012. A flood vulnerability index for coastal cities and its use in assessing climate change impacts. Natural Hazards 64(1) 73-105.

Balke, T., Gilbert, N., 2014. How Do Agents Make Decisions? A Survey. Journal of Artificial Societies and Social Simulation 17(4) 13.

Baltar, F., Brunet, I., 2012. Social research 2.0: virtual snowball sampling method using Facebook. Internet Research 22(1) 57-74.

Bar-Ilan, J., 2007. Which h-index? — A comparison of WoS, Scopus and Google Scholar. Scientometrics 74(2) 257-271.

Barben, R., 2010. Vulnerability Assessment of Electric Power Supply under Extreme Weather Conditions, Laboratoire Des Systemes Energetiques. Ecole Pytechnique Federale De Laussane: Laussane.

Bateman, J.M., Edwards, B., 2002. Gender and Evacuation: A Closer Look at Why Women Are More Likely to Evacuate for Hurricanes. Natural Hazards Review 3(3) 107-117.

Becu, N., Amalric, M., Anselme, B., Beck, E., Bertin, X., Delay, E., Long, N., Marilleau, N., Pignon-Mussaud, C., Rousseaux, F., 2017. Participatory simulation to foster social learning on coastal flooding prevention. Environmental Modelling & Software 98 1-11.

Benight, C.C., McFarlane, A.C., 2007. Challenges for disaster research: Recommendations for planning and implementing disaster mental health studies. Journal of Loss and Trauma 12(5) 419-434.

Benjamin, E., Bassily-Marcus, A.M., Babu, E., Silver, L., Martin, M.L., 2011. Principles and practice of disaster relief: lessons from Haiti. Mt Sinai J Med 78(3) 306-318.

Berkoune, D., Renaud, J., Rekik, M., Ruiz, A., 2012. Transportation in disaster response operations. Socio-Economic Planning Sciences 46(1) 23-32.

Billari, F., Fent, T., Prskawetz, A., Scheffran, J., 2006. Agent-Based Computational Modelling. Applications in Demography, Social, Economic and Environmental Sciences. Physica-Verlag, Germany.

Bird, D.K., Chagué-Goff, C., Gero, A., 2011. Human Response to Extreme Events: a review of three post-tsunami disaster case studies. Australian Geographer 42(3) 225-239.

Birkmann, g., 2008. Assessing Vulnerability Before, During and After a Natural Disaster in Fragile Regions. Research Paper No. 2008/50. Institute for Environment and Human Security, United Nations University - UNU: Bonn, Germany.

Birkmann, J., Agboola, J.I., Welle, T., Ahove, M., Odunuga, S., von Streit, J., Pelling, M., 2016. Vulnerability, Resilience and Transformation of Urban Areas in the Coastal Megacity Lagos: Findings of Local Assessments and a Household Survey in Highly Exposed Areas. Journal of Extreme Events 3(3) 24.

Birkmann, J., Cardona, O.D., Carreño, M.L., Barbat, A.H., Pelling, M., Schneiderbauer, S., Kienberger, S., Keiler, M., Alexander, D., Zeil, P., Welle, T., 2013. Framing vulnerability, risk and societal responses: the MOVE framework. Natural Hazards 67(2) 193-211.

Birkmann, J., Fernando, N., 2008. Measuring revealed and emergent vulnerabilities of coastal communities to tsunami in Sri Lanka. Disasters 32 82-104.

Birkmann, J., 2006. Measuring vulnerability to natural hazards towards disaster resilient societies. United Nations University, Tokyo; New York.

Borshchev, A., Filipov, A., 2004. From System Dynamics and Discrete Event to Practical Agent Based Modeling: Reasons, Techniques, Tools, The 22nd International Conference of the System Dynamics Society.

Bosch, M., 2017. Institutional dimension of flood risk: Understanding institutional complexity in Flood Risk Management for the case of St Maarten, Technology, Policy and Management - TPM. Delft University of Technology: Delft, The Netherlands, p. 108.

Boulos, M.N.K., Resch, B., Crowley, D.N., Breslin, J.G., Sohn, G., Burtner, R., Pike, W.A., Jezierski, E., Chuang, K.-Y.S., 2011. Crowdsourcing, citizen sensing and sensor web technologies for public and environmental health surveillance and crisis management: trends, OGC standards and application examples. International Journal of Health Geographics 10(67).

Brown, S., Parton, H., Driver, C., Norman, C., 2016. Evacuation During Hurricane Sandy: Data from a Rapid Community Assessment. PLoS Curr 8.

Camps-Mur, D., Garcia-Saavedra, A., Serrano, P., 2013. Device to device communications with Wi-Fi direct: overview and experimentation. IEEE Wireless 20(3) 96-104.

Cangialosi, J.P., Latto, A.S., Berg, R., 2018. Hurricane Irma: 30 August – 12 September 2017, Tropical Cyclone Report, AL112027 ed. NOAA National Hurricane Center.

Cardona, O.D., Aalst, M.K.v., Birkmann, J., Fordham, M., McGregor, G., Perez, R., Pulwarty, R.S., Schipper, E.L.F., Sinh, B.T., 2012. Determinants of Risk: Exposure and Vulnerability, Managing the Risks of Extreme Events and Disasters to Advance Climate Change Adaptation. Cambridge University Press: New York, pp. pp.-65-108.

Castle, C.J.E., Crooks, A.T., 2006. Principles and concepts of agent-based modelling for developing geospatial simulations, CASA working paper series.

CNES, Centre national d'études spatiales, 2018. CNES Projects library - SPOT. Retrieved on November 2018 from: https://spot.cnes.fr/en/SPOT/index.htm

Chan, S., 2001. Complex adaptive systems, ESD.83 Research Seminar in Engineering Systems. 31 October/6 November, Massachusetts Institute of Technology.

Chandra-Putra, H., Zhang, H., Andrews, C.J., 2015. Modeling Real Estate Market Responses to Climate Change in the Coastal Zone. Journal of Artificial Societies and Social Simulation 18(2) 18.

Chandra-Putra, H., Andrews, C.J., 2019. An integrated model of real estate market responses to coastal flooding. Journal of Industrial Ecology(24) 424-435.

Cheff, I., Nistor, I., Palermo, D., 2019. Pedestrian evacuation modelling of a Canadian West Coast community from a near-field Tsunami event. Natural Hazards 98(1) 229-249.

Chen, X., 2011. Microsimulation of Hurricane Evacuation Strategies of Galveston Island. The Professional Geographer 60(2) 160-173.

Chen, X., 2012. Agent-based micro-simulation of staged evacuations. International Journal of Advanced Intelligence Paradigms 4(1) 22-35.

Chen, X., 2015. Activity-based Modeling and Microsimulation of Emergency Evacuations. International Journal of Applied Geospatial Research 6(3) 21-38.

Chen, X., Meaker, J.W., Zhan, F.B., 2006. Agent-Based Modeling and Analysis of Hurricane Evacuation Procedures for the Florida Keys. Natural Hazards 38(3) 321-338.

Chen, X., Zhan, F.B., 2008. Agent-based modelling and simulation of urban evacuation Relative effectiveness of simultaneous and staged evacuation strategies. Journal of the Operational Research Society 59(1) 25-33.

Ciurean, R.L., Schroter, D., Glade, T., 2013. Conceptual Frameworks of Vulnerability Assessments for Natural Disasters Reduction, In: InTech (Ed.), Approaches to Disaster Management - Examining the Implications of Hazards, Emergencies and Disasters: London, UK, pp. 3-32.

Coates, G., Hawe, G.I., Wright, N.G., Ahilan, S., 2014. Agent-based modelling and inundation prediction to enable the identification of businesses affected by flooding. WIT Transactions on Ecology and the Environment 184 13-22.

Coates, G., Li, C., Ahilan, S., Wright, N., Alharbi, M., 2019. Agent-based modeling and simulation to assess flood preparedness and recovery of manufacturing small and medium-sized enterprises. Engineering Applications of Artificial Intelligence 78 195-217.

Colten, C.E., 2006. Vulnerability and Place: Flat Land and Uneven risk in New orleans. American Anthropologist 108(4) 731-734.

Colten, C.E., Giancarlo, A., 2011. Losing Resilience on the Gulf Coast: Hurricanes and Social Memory. Environment: Science and Policy for Sustainable Development 53(4) 6-19.

Comfort, L., Wisner, B., Cutter, S., Pulwarty, R., Hewitt, K., Oliver-Smith, A., Wiener, J., Fordham, M., Peacock, W., Krimgold, F., 1999. Reframing disaster policy: the global evolution of vulnerable communities. Environmental Hazards 1(1) 39-44.

Connor, R.F., Hiroki, K., 2005. Development of a method for assessing flood vulnerability. Water Science & Technology 51(5) 61-67.

Crooks, A.T., Heppenstall, A.J., 2012. Introduction to Agent-Based Modelling, In: Heppenstall, A., Crooks, A., See, L.M., Batty, M. (Eds.), Agent-Based Models of Geographical Systems. Springer: Netherlands, pp. 85-105.

Cutter, S.L., Boruff, B.J., Shirley, W.L., 2003. Social Vulnerability to Environmental Hazards. Social Science Quarterly 84(2) 242-261.

Cutter, S.L., Osman-Elasha, B., Campbell, J., Cheong, S.-M., McCormick, S., Pulwarty, R., Supratid, S., Ziervogel, G., 2012. Managing the risks from climate extremes at the local level. In: Managing the Risks of Extreme Events and Disasters to Advance Climate Change Adaptation. Chapter 5., In: Field, C.B., V. Barros, T.F. Stocker, D. Qin, D.J. Dokken, K.L. Ebi, M.D. Mastrandrea, K.J. Mach, G.-K. Plattner, S.K. Allen, M. Tignor, and P.M. Midgley (Ed.), Special Report of Working Groups I and II of the Intergovernmental Panel on Climate Change (IPCC). IPCC: Cambridge, UK, and New York, NY, USA, pp. 291-338.

Dash, N., Gladwin, H., 2007. Evacuation Decision Making and Behavioral Responses: Individual and Household. Natural Hazards Review 8(3) 69-77.

Dash, N., Morrow, B.H., 2000. Return delays and evacuation order compliance: the case of Hurricane Georges and the Florida Keys. Global Environmental Change Part B: Environmental Hazards 2(3) 119-128.

David, N., 2006. Validation and Verification in Social Simulation: Patterns and Clarification of Terminology, In: Squazzoni, F. (Ed.), Epistemological Aspects of Computer Simulation in the Social Sciences. EPOS. Springer: Berlin, Heidelberg, pp. 117-129.

Dawson, R.J., Peppe, R., Wang, M., 2011. An agent-based model for risk-based flood incident management. Natural Hazards 59(1) 167-189.

de Hamer, J., 2019. Disaster governance on ST. Maarten. A study on how disaster governance in combination with St. Maarten's development affected the disaster response in the wake of Hurricane Irma, Development and Rural Innovation. Wageningen University: Wageningen, The netherlands, p. 101.

de Ruiter, M.C., Ward, P.J., Daniell, J.E., Aerts, J.C., 2017. A comparison of flood and earthquake vulnerability assessment indicators. Natural Hazards and Earth System Sciences 17(7) 1231-1251.

de Sherbinin, A., Bukvic, A., Rohat, G., Gall, M., McCusker, B., Preston, B., Apotsos, A., Fish, C., Kienberger, S., Muhonda, P., Wilhelmi, O., Macharia, D., Shubert, W., Sliuzas, R., Tomaszewski, B., Zhang, S., 2019. Climate vulnerability mapping: A systematic review and future prospects. Wiley Interdisciplinary Reviews: Climate Change 10(5) e600.

Demuth, J.L., Morss, R.E., Lazo, J.K., Trumbo, C., 2016. The Effects of Past Hurricane Experiences on Evacuation Intentions through Risk Perception and Efficacy Beliefs: A Mediation Analysis. Weather, Climate, and Society 8(4) 327-344.

Depietri, Y., Welle, T., Renaud, F.G., 2013. Social vulnerability assessment of the Cologne urban area (Germany) to heat waves: links to ecosystem services. International Journal of Disaster Risk Reduction 6 98-117.

Dieker, M., 2018. Keep Moving, Stay Tuned. Transfers 8(2) 67-86.

Dijkstra, E.W., 1959. A note on two problems in connexion with graphs. Numerische Mathematik 1(1) 269-271.

Dos Santos França, R., Steinberger, M.B., Maria das Graças, B.M., Omar, N., 2012. Simulating Collective Behavior in Natural Disaster Situations: A Multi-Agent Approach. INTECH Open Access Publisher 161-171.

Dressler, G., Müller, B., Frank, K., Kuhlicke, C., 2016. Towards thresholds of disaster management performance under demographic change: exploring functional relationships using agent-based modeling. Natural Hazards and Earth System Sciences 16(10) 2287-2301.

Du, E., Cai, X., Sun, Z., Minsker, B., 2017a. Exploring the Role of Social Media and Individual Behaviors in Flood Evacuation Processes: An Agent-Based Modeling Approach. Water Resources Research 53(11) 9164-9180.

Du, E., Rivera, S., Cai, X., Myers, L., Ernest, A., Minsker, B., 2017b. Impacts of Human Behavioral Heterogeneity on the Benefits of Probabilistic Flood Warnings: An Agent-Based Modeling Framework. JAWRA Journal of the American Water Resources Association 53(2) 316-332.

Dubbelboer, J., Nikolic, I., Jenkins, K., Hall, J., 2017. An Agent-Based Model of Flood Risk and Insurance. Journal of Artificial Societies and Social Simulation 20(1) 6.

Duffy, B., Smith, K., Terhanian, G., Bremer, J., 2010. Comparing data from online and face-to-face surveys. International Journal of Market Research 47(6) 615-639.

ECLAC, Economic Commission for Latin America and the Caribbean, 2017. Assessment of the Effects and Impacts of Hurricane Irma: Sint Maarten: Santiago, Chile.

Edmonds, B., Le Page, C., Bithell, M., Chattoe-Brown, E., Grimm, V., Meyer, R., Montañola-Sales, C., Ormerod, P., Root, H., Squazzoni, F., 2019. Different Modelling Purposes. Journal of Artificial Societies and Social Simulation 22(3) 6.

Eid, M.S., El-adaway, I.H., 2018. Decision-Making Framework for Holistic Sustainable Disaster Recovery: Agent-Based Approach for Decreasing Vulnerabilities of the Associated Communities. Journal of Infrastructure Systems 24(3) 04018009.

Eiser, J.R., Bostrom, A., Burton, I., Johnston, D.M., McClure, J., Paton, D., van der Pligt, J., White, M.P., 2012. Risk interpretation and action: A conceptual framework for responses to natural hazards. International Journal of Disaster Risk Reduction 1 5-16.

Erdlenbruch, K., Bonté, B., 2018. Simulating the dynamics of individual adaptation to floods. Environmental Science & Policy 84 134-148.

ERN Ingenieros Consultores, 2009. ERN-Hurricane, 1.0.0.0 ed. CAPRA project: https://ecapra.org/topics/ern-hurricane.

Eros, E., 2018. Technologies for Development. From Innovation to Social Impact, Putting 200 Million People "on the Map": Evolving Methods and Tools. Springer Open: Switzerland, pp. 177-186.

Etinay, N., Egbu, C., Murray, V., 2018. Building Urban Resilience for Disaster Risk Management and Disaster Risk Reduction. Procedia Engineering 212 575-582.

EU-JRC, European Commission - Joint Research Centre, 2017. Hurricane Irma in Sint Maarten (2017-09-07), In: Joint Research Centre (JRC) [Dataset] (Ed.): Ispra, Italy.

Fekete, A., 2009. Validation of a social vulnerability index in context to river floods in germany. Nat. Hazards Earth Syst. 9 393-403.

FEMA, 2006. Design Guidance for Shelters and Safe Rooms. FEMA.

FEMA, 2013. Surviving the Storm. A guide to hurricane Preparedness, In: U.S. Department of Homeland Security/Federal Emergency Management Agency (Ed.): USA.

Fernandez, P., Mourato, S., Moreira, M., Pereira, L., 2016. A new approach for computing a flood vulnerability index using cluster analysis. Physics and Chemistry of the Earth, Parts A/B/C 94 47-55.

Filatova, T., 2014. Market-based instruments for flood risk management: A review of theory, practice and perspectives for climate adaptation policy. Environmental Science & Policy 37 227-242.

Galán, J.M., Izquierdo, L.R., Izquierdo, S.S., Santos, J.I., Olmo, R.d., López-Paredes, A., Edmonds, B., 2009. Errors and Artefacts in Agent-Based Modelling. Journal of Artificial Societies and Social Simulation 12(1) 1.

Garschagen, M., Hagenlocher, M., Comes, M., Dubbert, M., Sabelfeld, R., Lee, Y.J., Grunewald, L., Lanzendörfer, M., Mucke, P., Neuschäfer, O., Pott, S., Post, J., Schramm, S., Schumann-Bölsche, D., Vandemeulebroecke, B., Welle, T., Birkmann, J., 2016. World Risk Report 2016. Entwicklung Hilft and United Nations University Institute of Environment and Human Security (UNU-EHS): Bonn.

Geerds, D., de With, C., 2011. The Transition of Undocumented Students to Foundation Based Education on Sint Maarten -Bottlenecks Concerning the Transition of Students in the Ages of Seven to Nine Years from the Perspective of Teachers, Parents, Students and Student Care Coordinators -, Faculty of Social and Behavioural Sciences. Utrecht University: Utrecht, The Netherlands.

Gehlot, H., Zhan, X., Qian, X., Thompson, C., Kulkarni, M., Ukkusuri, S.V., 2019. A-RESCUE 2.0: A High-Fidelity, Parallel, Agent-Based Evacuation Simulator. Journal of Computing in Civil Engineering 33(2) 04018059.

Ghavami, S.M., Maleki, J., Arentze, T., 2019. A multi-agent assisted approach for spatial Group Decision Support Systems: A case study of disaster management practice. International Journal of Disaster Risk Reduction 38 101223.

Ghorbani, A., 2013. Structuring Sociotechnical Complexity - Modelling Agent Systems Using Institutional Analysis. Delft University of Technology.

Gladwin, C.H., Gladwin, H., Peacock, W.G., 2001. Modeling Hurricane Evacuation Decisions with Ethnographic Methods. International Journal of Mass Emergencies and Disasters 19(2) 117-143.

Grimm, V., Berger, U., Bastiansen, F., Eliassen, S., Ginot, V., Giske, J., Goss-Custard, J., Grand, T., Heinz, S.K., Huse, G., Huth, A., Jepsen, J.U., Jørgensen, C., Mooij, W.M., Müller, B., Pe'er, G., Piou, C., Railsback, S.F., Robbins, A.M., Robbins, M.M., Rossmanith, E., Rüger, N., Strand, E., Souissi, S., Stillman, R.A., Vabø, R., Visser, U., DeAngelis, D.L., 2006. A standard protocol for describing individual-based and agent-based models. Ecological Modelling 198(1-2) 115-126.

Grimm, V., Berger, U., DeAngelis, D.L., Polhill, J.G., Giske, J., Railsback, S.F., 2010. The ODD protocol: A review and first update. Ecological Modelling 221(23) 2760-2768.

Grimm, V., Polhill, G., Touza, J., 2017. Documenting Social Simulation Models: The ODD Protocol as a Standard, Simulating Social Complexity. Springer International Publishing, pp. 349-365.

Gu, D., 2019. Exposure and vulnerability to natural disasters for world's cities. United Nations - Department of Economic and Social Affairs: New York.

Gunderson, L., 2010. Ecological and human community resilience in response to natural disasters. Ecology and Society 15(2).

Haer, T., Botzen, W.J.W., Aerts, J.C.J.H., 2019. Advancing disaster policies by integrating dynamic adaptive behaviour in risk assessments using an agent-based modelling approach. Environmental Research Letters 14(4) 044022.

Haer, T., Botzen, W.J.W., de Moel, H., Aerts, J., 2017. Integrating Household Risk Mitigation Behavior in Flood Risk Analysis: An Agent-Based Model Approach. Risk Anal 37(10) 1977-1992.

Handford, D., Rogers, A., 2012. An agent-based social forces model for driver evacuation behaviours. Progress in Artificial Intelligence 1(2) 173-181.

Haney, T.J., Elliott, J.R., 2013. The Sociological Determination: A Reflexive Look at Conducting Local Disaster Research after Hurricane Katrina. Sociology Mind 03(01) 7-15.

Harrison, C.G., Williams, P.R., 2016. A systems approach to natural disaster resilience. Simul Model Pract Theory 65 11-31.

Hasan, S., Ukkusuri, S., Gladwin, H., Murray-Tuite, P., 2011. Behavioral Model to Understand Household-Level Hurricane Evacuation Decision Making. Journal of Transportation Engineering 137(5) 341-348.

Hashim, M.S., Hassan, S., Bakar, A.A., 2018. Developing a Household Flood Vulnerability Index: A Case Study of Kelantan. Sch. J. Econ. Bus. Manag.

Heckbert, S., Baynes, T., Reeson, A., 2010 Agent-based modeling in ecological economics. Ann N Y Acad Sci. 1185 39-53.

Henderson, T.L., Sirois, M., Chen, A.C.-C., Airriess, C., Swanson, D.A., Banks, D., 2009. After a Disaster: Lessons in Survey Methodology from Hurricane Katrina. Population Research and Policy Review 28(1) 67-92.

Hendriks, E., Lichtenberg, J., Voorthuis, J., Quanjel, E., Gijsbers, R., 2015. Practical and ethical issues of Post-disaster research. Alam Cipta : International Journal of Sustainable Tropical Design Research and Practice 8(Special) 60-69.

Hoeppe, P., 2016. Trends in weather related disasters – Consequences for insurers and society. Weather and Climate Extremes 11 70-79.

Holbrook, A.L., Green, M.C., Krosnick, J.A., 2003. Telephone versus Face-to-Face Interviewing of National Probability Samples with Long Questionnaires. Public Opinion Quarterly 67(1) 79-125.

Homberg, M.v.d., Visser, J., Veen, M.v.d., 2017. Unpacking Data Preparedness from a humanitarian decision making perspective, 14th ISCRAM Conference: Albi, France.

Hong, Y., Kim, J.-S., Xiong, L., 2019. Media exposure and individuals' emergency preparedness behaviors for coping with natural and human-made disasters. Journal of Environmental Psychology 63 82-91.

Huang, S.-K., Lindell, M.K., Prater, C.S., 2016. Who Leaves and Who Stays? A Review and Statistical Meta-Analysis of Hurricane Evacuation Studies. Environment and Behavior 48(8) 991-1029.

Huang, S.-K., Lindell, M.K., Prater, C.S., Wu, H.-C., Siebeneck, L.K., 2012. Household Evacuation Decision Making in Response to Hurricane Ike. Natural Hazards Review 13(4) 283-296.

Husson, F., Josse, J., 2016. missMDA: a package for handling missing values in multivariate data analysis. Journal of Statistical Software 70(1) 1-31.

Husson, F., Lê, S., Pagès, J., 2010. Principal Component Analysis (PCA), Exploratory Multivariate Analysis by Example Using R. Chapman & Hall/CRC, pp. 16-60.

Husson, F., Lê, S., Pagès, J., 2017. Multiple Correspondence Analysis (MCA), Exploratory Multivariate Analysis by Example using R. CRC Press: Boca Raton, FL, pp. 131-172.

IGN, Institut National de L'Informa Géographique et Forestiére. 2017. Saint-Martin February 2017, IGN: France.

Imamura, F., Muhari, A., Mas, E., Pradono, M.H., Post, J., Sugimoto, M.A., 2012. Tsunami disaster mitigation by integrating comprehensive countermeasures in Padang city, Indonesia. Journal of Disaster Research 7(1) 48-64.

Insurance Information Institute, 2019. Facts and Statistics: Global catastrophes.

IPCC, 2012a. Glossary of terms. In: Managing the Risks of Extreme Events and Disasters to Advance Climate Change Adaptation, In: Field, C.B., V. Barros, T.F. Stocker, D. Qin, D.J. Dokken, K.L. Ebi, M.D. Mastrandrea, K.J. Mach, G.-K. Plattner, S.K. Allen, M. Tignor, and P.M. Midgley (Ed.), A Special Report of Working Groups I and II of the Intergovernmental Panel on Climate Change (IPCC). Intergovernmental Panel on Climate Change: Cambridge, UK, and New York, NY, USA,, pp. 555-564.

IPCC, 2012b. Managing the risks of extreme events and disasters to advance climate change adaptation: special report of the Intergovernmental Panel on Climate Change. Intergovernmental Panel on Climate Change: New York.

IPCC, 2014. Climate Change 2014: Impacts, Adaptation, and Vulnerability. Part A: Global and Sectoral Aspects. Contribution of Working Group II to the Fifth Assessment Report of the Intergovernmental Panel on Climate Change. Intergovernmental Panel on Climate Change: Cambridge, United Kingdom.

Irvine, A., Drew, P., Sainsbury, R., 2012. 'Am I not answering your questions properly?' Clarification, adequacy and responsiveness in semi-structured telephone and face-to-face interviews. Qualitative Research 13(1) 87-106.

Jenkins, K., Surminski, S., Hall, J., Crick, F., 2017. Assessing surface water flood risk and management strategies under future climate change: Insights from an Agent-Based Model. Sci Total Environ 595 159-168.

Jufri, F.H., Kim, J.-S., Jung, J., 2017. Analysis of Determinants of the Impact and the Grid Capability to Evaluate and Improve Grid Resilience from Extreme Weather Event. Energies 10(11).

Kaiser, R., Spiegel, P.B., Henderson, A.K., Gerber, M.L., 2003. The application of geographic information systems and global positioning systems in humanitarian emergencies: lessons learned, programme implications and future research. Disasters 27(2) 127-140.

Kappes, M.S., Papathoma-Köhle, M., Keiler, M., 2012. Assessing physical vulnerability for multi-hazards using an indicator-based methodology. Applied Geography 32(2) 577-590.

Karaye, I.M., Thompson, C., Horney, J.A., 2019. Evacuation Shelter Deficits for Socially Vulnerable Texas Residents During Hurricane Harvey. Health Serv Res Manag Epidemiol 6.

Kavanaugh, A., Sheetz, S.D., Quek, F., Kim, B.J., 2013. Cell Phone Use with Social Ties During Crises: The Case of the Virginia Tech Tragedy, In: Jennex, M.E. (Ed.), Using Social and Information Technologies for Disaster and Crisis Management. IGI Global: Hershey, PA, pp. 84-97.

Keller, S., Atzl, A., 2014. Mapping Natural Hazard Impacts on Road Infrastructure—The Extreme Precipitation in Baden-Württemberg, Germany, June 2013. International Journal of Disaster Risk Science 5(3) 227-241.

Kessler, R.C., Keane, T.M., Ursano, R.J., Mokdad, A., Zaslavsky, A.M., 2008. Sample and design considerations in post-disaster mental health needs assessment tracking surveys. International Journal of Methods in Psychiatric Research 17 Suppl 2 S6-S20.

Khunwishit, S., McEntire, D.A., 2012. Testing Social Vulnerability Theory: A Quantitative Study of Hurricane Katrina's Perceived Impact on Residents living in FEMA Designated Disaster Areas. Journal of Homeland Security and Emergency Management 9(1) 1-16.

Kitchenham, S., Charters, S., 2007. Guidelines for performing systematic literature reviews in software engineering, Technical report, Ver. 2.3 EBSE Technical Report. EBSE. Keele University: UK.

Kleinosky, L.R., Yarnal, B., Fisher, A., 2006. Vulnerability of Hampton Roads, Virginia to Storm-Surge Flooding and Sea-Level Rise. Natural Hazards 40(1) 43-70.

Koks, E.E., Rozenberg, J., Zorn, C., Tariverdi, M., Vousdoukas, M., Fraser, S.A., Hall, J.W., Hallegatte, S., 2019. A global multi-hazard risk analysis of road and railway infrastructure assets. Nat Commun 10(1) 2677.

Koning, K., Filatova, T., Bin, O., 2019. Capitalization of Flood Insurance and Risk Perceptions in Housing Prices: An Empirical Agent-Based Model Approach. Southern Economic Journal 85(4) 1159-1179.

Kremer, K., 2017. Anticipative interfaces for emergency situations. Information Design Journal 23(1) 32-38.

Kundzewicz, Z.W., Kanae, S., Seneviratne, S.I., Handmer, J., Nicholls, N., Peduzzi, P., Mechler, R., Bouwer, L.M., Arnell, N., Mach, K., Muir-Wood, R., Brakenridge, G.R., Kron, W., Benito, G., Honda, Y., Takahashi, K., Sherstyukov, B., 2013. Flood risk and climate change: global and regional perspectives. Hydrological Sciences Journal 59(1) 1-28.

Kuntiyawichai, K., Dau, Q.V., Sri-Amporn, W., Suryadi, F.X., 2016. An Assessment of Flood Hazard and Risk Zoning in the Lower Nam Phong River Basin, Thailand. International Journal of Technology 7(7).

Laird, J.E., 2012. The Soar Cognitive Architecture. The MIT Press.

Latif, S., Islam, K.M.R., Khan, M.M.I., Ahmed, S.I., 2011. OpenStreetMap for the Disaster Management in Bangladesh, 2011 IEEE Conference on Open Systems. IEEE Xplore: Langkawi, pp. 429-433.

Lavin, R.P., Schemmel-Rettenmeier, L., Frommelt-Kuhle, M., 2012. Conducting research during disasters. Annual Review of Nursing Research 30(1) 1-19.

Lavrakas, P.J., 2008. Face-to-Face Interviewing, Encyclopedia of Survey Research Methods. Sage Publications, Inc Thousand Oaks, CA, p. 1041.

Lazo, J.K., Bostrom, A., Morss, R.E., Demuth, J.L., Lazrus, H., 2015. Factors Affecting Hurricane Evacuation Intentions. Risk Anal 35(10) 1837-1857.

Lazo, J.K., Waldman, D.M., Morrow, B.H., Thacher, J.A., 2010. Household Evacuation Decision Making and the Benefits of Improved Hurricane Forecasting: Developing a Framework for Assessment. Weather and Forecasting 25(1) 207-219.

Lazrus, H., Demuth, J.L., Morss, R.E., Palen, L., Barton, C.M., Davis, C.A., Snyder, C., Wilhelmi, O.V., Anderson, K.M., Ahijevych, D.A., Anderson, J., Bica, M., Fossell, K.R., Henderson, J., Kogan, M., Stowe, K., Watts, J., 2017. Hazardous Weather Prediction and Communication in the Modern Information Environment. Bulletin of the American Meteorological Society 98(12) 2653-2674.

Lê, S., Josee, J., Husson, F., 2008. FactoMineR: An R Package for Multivariate Analysis. Journal of Statistical Software 25(1) 1-18.

Lechowska, E., 2018. What determines flood risk perception? A review of factors of flood risk perception and relations between its basic elements. Natural Hazards 94(3) 1341-1366.

Lee, D., Yoon, S., Park, E.-S., Kim, Y., Yoon, D.K., 2018. Factors Contributing to Disaster Evacuation: The Case of South Korea. Sustainability 10(10).

Lee, J.-S., Filatova, T., Ligmann-Zielinska, A., Hassani-Mahmooei, B., Stonedahl, F., Lorscheid, I., Voinov, A., Polhill, G., Sun, Z., Parker, D.C., 2015. The Complexities of Agent-Based Modeling Output Analysis. Journal of Artificial Societies and Social Simulation 18(4) 4.

Lefever, S., Dal, M., Matthíasdóttir, Á., 2007. Online data collection in academic research: advantages and limitations. British Journal of Educational Technology 38(4) 574-582.

Li, C., Coates, G., 2016. Design and development of an agent-based model for business operations faced with flood disruption. International Journal of Design & Nature and Ecodynamics 11(2) 97-106.

Li, K., Rollins, J., Yan, E., 2018. Web of Science use in published research and review papers 1997-2017: a selective, dynamic, cross-domain, content-based analysis. Scientometrics 115(1) 1-20.

Liang, D., Cong, L., Brown, T., Song, L., 2012. Comparison of Sampling Methods for Post-Hurricane Damage Survey. Journal of Homeland Security and Emergency Management 9(2).

Liang, W., Lam, N.S.N., Qin, X., Ju, W., 2015. A Two-level Agent-Based Model for Hurricane Evacuation in New Orleans. Journal of Homeland Security and Emergency Management 12(2) 407-435.

Lianxiao, Morimoto, T., 2019. Spatial Analysis of Social Vulnerability to Floods Based on the MOVE Framework and Information Entropy Method: Case Study of Katsushika Ward, Tokyo. Sustainability 11(2).

Lindell, M.K., Lu, J.-C., Prater, C.S., 2005. Household Decision Making and Evacuation in Response to Hurricane Lili. Natural Hazards Review 6(4).

Lindell, M.K., Perry, R.W., 2004. Communicating Environmental Risk in Multiethnic Communities. Sage, Thousands Oaks, CA.

Lindell, M.K., Perry, R.W., 2012. The protective action decision model: theoretical modifications and additional evidence. Risk Anal 32(4) 616-632.

Lipiec, E., Ruggiero, P., Mills, A., Serafin, K.A., Bolte, J., Corcoran, P., Stevenson, J., Zanocco, C., Lach, D., 2018. Mapping Out Climate Change: Assessing How Coastal Communities Adapt Using Alternative Future Scenarios. Journal of Coastal Research 34(5) 1196-1208.

Liu, X., Lim, S., 2018. An agent-based evacuation model for the 2011 Brisbane City-scale riverine flood. Natural Hazards 94(1) 53-70.

Lovari, A., Bowen, S.A., 2019. Social media in disaster communication: A case study of strategies, barriers, and ethical implications. Journal of Public Affairs 20(1).

Lovell, E., Mitchell, T., 2015. Disaster damage to critical infrastructure. Overseas Development Institute: London.

Löwe, R., Urich, C., Sto. Domingo, N., Mark, O., Deletic, A., Arnbjerg-Nielsen, K., 2017. Assessment of urban pluvial flood risk and efficiency of adaptation options through simulations – A new generation of urban planning tools. Journal of Hydrology 550 355-367.

Macal, C.M., North, M.J., 2009. Agent-based modeling and simulation. Winter Simulation Conference, pp. 86-98.

Macal, C.M., North, M.J., 2010. Tutorial on agent-based modelling and simulation. Journal of Simulation 4(3) 151-162.

Magliocca, N.R., Walls, M., 2018. The role of subjective risk perceptions in shaping coastal development dynamics. Computers, Environment and Urban Systems 71 1-13.

Maitland, C.F., Pogrebnyakov, N., Gorp, A.F.v., 2006. A Fragile Link: Disaster Relief, ICTs and Development, International Conference on Information and Communication Technology and Development: Berkeley, CA, USA, pp. 339-346.

Maldonado, A., Collins, T.W., Grineski, S.E., Chakraborty, J., 2016. Exposure to Flood Hazards in Miami and Houston: Are Hispanic Immigrants at Greater Risk than Other Social Groups? Int J Environ Res Public Health 13(8).

Manson, S., An, L., Clarke, K.C., Heppenstall, A., Koch, J., Krzyzanowski, B., Morgan, F., O'Sullivan, D., Runck, B.C., Shook, E., Tesfatsion, L., 2020. Methodological Issues of Spatial Agent-Based Models. Journal of Artificial Societies and Social Simulation 23(1) 3.

Manzoor, U., Zubair, M., Batool, K., Zafar, B., 2014. A multi-agent framework for efficient food distribution in disaster areas. International Journal of Internet Technology and Secured Transactions 5(4) 327-343.

Markolf, S.A., Hoehne, C., Fraser, A., Chester, M.V., Underwood, B.S., 2019. Transportation resilience to climate change and extreme weather events – Beyond risk and robustness. Transport Policy 74 174-186.

Maryam, H., Javaid, Q., Shah, M.A., Kamran, M., 2016. Survey on smartphones systems for emergency management. International Journal of Advanced Computer Science and Applications (IJACSA) 7(6) 301-311.

Mas, E., Adriano, B., Koshimura, S., 2013. An integrated simulation of tsunami hazard and human evacuation in La Punta, Peru. Journal of Disaster Research 8(2) 285-295.

McCarthy, J.J., Canziani, O.F., Leary, N.A., Dokken, D.J., White, K.S., 2001. Climate Change 2001: Impacts, Adaptation and Vulnerability. Cambridge University Press: Cambridge.

McNamara, D.E., Keeler, A., 2013. A coupled physical and economic model of the response of coastal real estate to climate risk. Nature Climate Change 3(6) 559-562.

MDS, Meteorological Department St. Maarten.-. 2018. Climatological Summary 2017. Meteorogical Department St. Maarten - MDS: Sint Maarten, p. 34.

Medina, N., 2017. EvacuAPP. Last Accessed on December 2020 from: https://ihe-delft.maps.arcgis.com/apps/webappviewer/index.html?id=39ac411859f94fe699f7e79937735be1.

Medina, N., Abebe, Y., Sanchez, A., Vojinović, Z., Nikolic, I., 2019. Surveying After a Disaster. Capturing Elements of Vulnerability, Risk and Lessons Learned from a Household Survey in the Case Study of Hurricane Irma in Sint Maarten. Journal of Extreme Events 6(2) 1950001.

Medina, N., Abebe, Y.A., Sanchez, A., Vojinović, Z., 2020. Assessing Socioeconomic Vulnerability after a Hurricane: A Combined Use of an Index-Based approach and Principal Components Analysis. Sustainability 12(4) 1452.

Medina, N., Sanchez, A., Vojinović, Z., Under Review. Evacuation Behaviour in a Small Island Developing State. The case study of Sint Maarten during Hurricane Irma: Manuscript in review.

Mesa-Arango, R., Hasan, S., Ukkusuri, S.V., Murray-Tuite, P., 2013. Household-Level Model for Hurricane Evacuation Destination Type Choice Using Hurricane Ivan Data. Natural Hazards Review 14(1) 11-20.

Messner, F., Meyer, V., 2006. Flood Damage, Vulnerability and Risk Perception – Challenges for Flood Damage Research. Springer Netherlands: Dordrecht, pp. 149-167.

MDC, Meteorological Department Curaçao, 2015. Hurricanes and Tropical Storms in the Dutch Caribbean: Willemstad, Curaçao.

Michael, C., 2014. Missing Maps: nothing less than a human genome project for cities. Retrieved on December 2017, https://www.theguardian.com/cities/2014/oct/06/missing-mapshuman-genome-project-unmapped-cities.

Milch, K., Broad, K., Orlove, B., Meyer, R., 2018. Decision Science Perspectives on Hurricane Vulnerability: Evidence from the 2010–2012 Atlantic Hurricane Seasons. Atmosphere 9(1).

Miller, J.H., Page, S.E., 2007. Complex adaptive systems an introduction to computational models of social life. Princeton University Press, Princeton, N.J.

Ministry of Public Housing Spatial Planning Environment and Infrastructure [VROMI], 2015. Ministry Plan 2015 - 2018, In: Ministry of Public Housing Spatial Planning Environment and Infrastructure (Ed.): Great Bay, Sint Maarten.

Mohagheghi, S., Javanbakht, P., 2015. Power Grid and Natural Disasters: A Framework for Vulnerability Assessment, 2015 Seventh Annual IEEE Green Technologies Conference, pp. 199-205.

Morton, M., Levy, J.L., 2011. Challenges in disaster data collection during recent disasters. Prehospital and Disaster Medicine 26(3) 196-201.

Mostafizi, A., Wang, H., Cox, D., Cramer, L.A., Dong, S., 2017. Agent-based tsunami evacuation modeling of unplanned network disruptions for evidence-driven resource allocation and retrofitting strategies. Natural Hazards 88(3) 1347-1372.

Müller, B., Bohn, F., Dreßler, G., Groeneveld, J., Klassert, C., Martin, R., Schlüter, M., Schulze, J., Weise, H., Schwarz, N., 2013. Describing human decisions in agent-based models – ODD + D, an extension of the ODD protocol. Environmental Modelling & Software 48 37-48.

Mustafa, A., Bruwier, M., Archambeau, P., Erpicum, S., Pirotton, M., Dewals, B., Teller, J., 2018. Effects of spatial planning on future flood risks in urban environments. J Environ Manage 225 193-204.

Mustapha, K., McHeick, H., Mellouli, S., 2013. Modeling and Simulation Agent-based of Natural Disaster Complex Systems. Procedia Computer Science 21 148-155.

Mythili, S., Shalini, E., 2016. A comparative study of Smart Phone Emergency Applications for Disaster Management. International Research Journal of Engineering and Technology 3(12) 392-395.

Nachappa, T.G., Ghorbanzadeh, O., Gholamnia, K., Blaschke, T., 2020. Multi-Hazard Exposure Mapping Using Machine Learning for the State of Salzburg, Austria. Remote Sensing 12(17).

Nagarajan, M., Shaw, D., Albores, P., 2012. Disseminating a warning message to evacuate: A simulation study of the behaviour of neighbours. European Journal of Operational Research 220(3) 810-819.

Naghawi, H., Wolshon, B., 2010. Transit-Based Emergency Evacuation Simulation Modeling. Journal of Transportation Safety & Security 2(2) 184-201.

Nakagawa, Y., Shaw, R., 2004. Social Capital: A Missing Link to Disaster Recovery. International Journal of Mass Emergencies and Disasters 22(1) 5-34.

Nandalal, H.K., Ratnayake, U.R., 2011. Flood risk analysis using fuzzy models. Journal of Flood Risk Management 4(2) 128-139.

Naqvi, A.A., Rehm, M., 2014. A multi-agent model of a low income economy: simulating the distributional effects of natural disasters. Journal of Economic Interaction and Coordination 9(2) 275-309.

NRC, National Research Council, 2006. Research on disaster response and recovery, Facing Hazards and Disasters: Understanding Human Dimensions. The National Academies Press: Washington, DC, pp. 124-179.

Neil Adger, W., Arnell, N.W., Tompkins, E.L., 2005. Successful adaptation to climate change across scales. Global Environmental Change 15(2) 77-86.

Netherlands Red Cross, 2017. First public report about the national campaign "The Netherlands helps St. Maarten". Netherlands Red Cross: The Netherlands.

Nguyen, T.T.X., Bonetti, J., Rogers, K., Woodroffe, C.D., 2016. Indicator-based assessment of climate-change impacts on coasts: A review of concepts, methodological approaches and vulnerability indices. Ocean & Coastal Management 123 18-43.

NHC, National Hurricane Center, 2020. Experimental Peak Storm Surge Forecast. National Hurricane Center, NHC and Central Pacific Hurricane Center. Last Accesed October 2020 from https://www.nhc.noaa.gov/productexamples/Peak_Storm_Surge_Forecast.shtml.

Nikolic, I., Dijkema, G.P.J., 2010. On the development of agent-based models for infrastructure evolution. International journal of critical infrastructures 6(2) 148-167.

NOOA, National Oceanic and Atmospheric Administration, 2017. Extremely active 2017 Atlantic hurricane season finally ends. NOAA, National Hurricane Center, Retrieved from https://www.noaa.gov/media-release/extremely-active-2017-atlantic-hurricane-season-finallyends

Norling, E.J., 2009. Modelling Human Behaviour with BDI Agents. University of Melbourne.

Norris, F.H., 2006. Disaster research methods: past progress and future directions. Journal of Traumatic Stress 19(2) 173-184.

Nurse, L.A., McLean, R.F., Agard, J., Briguglio, L.P., Duvat-Magnan, V., Pelesikoti, N., Tompkins, E., Webb, A., 2014. Small islands. In: Climate Change 2014: Impacts, Adaptation, and Vulnerability. Part B: Regional Aspects. Contribution of Working Group II to the Fifth Assessment Report of the Intergovernmental Panel on Climate Change. IPCC: United Kingdom and New York, pp. 1613-1654.

O'Mathúna, D., 2010. Conducting research in the aftermath of disasters: ethical considerations. Journal of evidence-based medicine 3(2) 65-75.

O'Sullivan, D., Evans, T., Manson, S., Metcalf, S., Ligmann-Zielinska, A., Bone, C., 2016. Strategic directions for agent-based modeling: avoiding the YAAWN syndrome. J Land Use Sci 11(2) 177-187.

OCHA, UN Office for the Coordination of Humanitarian Affairs, 2016. Humanitarian Data Centre in the Netherlands will increase Data Use and Impact in Humanitarian Sector. Retrieved on August 2018 from: https://reliefweb.int/report/world/humanitarian-data-centrenetherlands-will-increase-data-use-and-impact-humanitarian

Ogie, R.I., Pradhan, B., 2019. Natural Hazards and Social Vulnerability of Place: The Strength-Based Approach Applied to Wollongong, Australia. International Journal of Disaster Risk Science 10(3) 404-420.

Omaier, H.T., Alharbi, A.Z., Alotaibi, M.F., Ibrahim, D.M., 2019. Comparative Study between Emergency Response Mobile Applications. International Journal of Computer Science and Information Security (IJCSIS) 17(2) 87-91.

Ormerod, P., Rosewell, B., 2006. Validation and Verification of Agent-Based Models in the Social Sciences, In: Squazzoni, F. (Ed.), Epistemological Aspects of Computer Simulation in the Social Sciences. EPOS. Springer: Berlin, Heidelberg, pp. 130-140.

Oulahen, G., Mortsch, L., Tang, K., Harford, D., 2015. Unequal Vulnerability to Flood Hazards: "Ground Truthing" a Social Vulnerability Index of Five Municipalities in Metro Vancouver, Canada. Annals of the Association of American Geographers 105(3) 473-495.

Ouyang, M., Dueñas-Osorio, L., 2014. Multi-dimensional hurricane resilience assessment of electric power systems. Structural Safety 48 15-24.

Panteli, M., Mancarella, P., 2017. Modeling and Evaluating the Resilience of Critical Electrical Power Infrastructure to Extreme Weather Events. IEEE Systems Journal 11(3) 1733-1742.

Papathoma-Kohle, M., Schlogl, M., Fuchs, S., 2019. Vulnerability indicators for natural hazards: an innovative selection and weighting approach. Sci Rep 9(1) 15026.

Paton, D., 2008. Risk communication and natural hazard mitigation: how trust influences its effectiveness. International Journal of Global Environmental Issues 8(2).

Paul, S.K., 2013. Vulnerability Concepts and its Application in Various Fields: A Review on Geographical Perspective. Journal of Life and Earth Science 8 63-81.

PEARL, 2013. Preparing for Extreme And Rare events in coastaL regions, Description of Work.

PEARL, 2016. D3.1 Holistic and Multiple Risk Assessment Framework. UNESCO-IHE: Delft, The Netherlands.

PEARL, 2018. D3.5 Hurricane Irma Special Report. From Risk to Resilience: A fact finding and needs assessment report in the aftermath of Hurricane Irma on Sint Maarten. UNESCO-IHE: Delft, The Netherlands.

Percival, S., Teeuw, R., 2019. A methodology for urban micro-scale coastal flood vulnerability and risk assessment and mapping. Natural Hazards 97(1) 355-377.

Petak, W.J., 1985. Emergency Management: A Challenge for Public Administration. American Society for Public Administration 45(Special Issue: Emergency Management: AChallenge for Public Administration) 3-7.

Preston, B., Smith, T., Brooke, C., Gorddard, R., Measham, T., Withycombe, G., McInnes, K., Abbs, D., Beveridge, B., Morrison, C., 2008. Mapping climate change vulnerability in the Sydney Coastal Councils Group, In: CSIRO, S.C.C.G.a.t.A.G.O. (Ed.): Melbourne.

Priest, S., Tapsell, S., Penning-Rowsell, E., Viavattene, C., Wi, T., 2009. Building models to estimate loss of life for flood events. FLOODsite.

Rand, W., Herrmanna, J., Scheinb, B., Vodopivec, N., 2015. An agent-based model of urgent diffusion in social media. Journal of Artificial Societies and Social Simulation 18(2) 1.

Riad, J.K., Norris, F.H., Ruback, R.B., 1999. Predicting Evacuation in Two Major Disasters: Risk Perception, Social Influence, and Access to Resources. Journal of Applied Social Psychology 29 918-934.

Richardson, R.C., Plummer, C.A., Barthelemy, J.J., Cain, D.S., 2009. Research after Natural Disasters: Recommendations and Lessons Learned. Journal of Community Engagement and Scholarship 2(1) 3-11.

Robinson, S., 2017. Climate change adaptation trends in small island developing states. Mitigation and Adaptation Strategies for Global Change 22(4) 669-691.

Roper, W.L., Mays, G.P., 1999. GIS and Public Health Policy: A New Frontier for Improving Community Health. Journal of Public Health Management and Practice 5(2) vi-vii

Rufat, S., Tate, E., Burton, C.G., Maroof, A.S., 2015. Social vulnerability to floods: Review of case studies and implications for measurement. International Journal of Disaster Risk Reduction 14 470-486.

Sadri, A.M., Ukkusuri, S.V., Murray-Tuite, P., Gladwin, H., 2014. Analysis of hurricane evacuee mode choice behavior. Transportation Research Part C: Emerging Technologies 48 37-46.

SAMHSA, Substance Abuse and Mental Health Services Administration, 2016. Challenges and Considerations in Disaster Research. Retrieved on August 2018 from https://www.samhsa.gov/sites/default/files/dtac/supplemental-research-bulletin-jan-2016.pdf

Samuels, P., Huntington, S., Allsop, W., Harrop, J., 2009. Flood risk management research and practice: proceedings of the European conference on flood risk management research into practice (FLOODrisk 2008), Oxford, UK, 30 September-2 October 2008. CRC Press, Boca Raton (Fla.); London; New York [etc.].

Sayers, P., Yuanyuan, L., Galloway, G., Penning-Rowsell, E., Fuxin, S., Kang, W., Yiwei, C., Le Quesne, T., 2013. Flood Risk Management: A Strategic Approach, UNESCO ed. UNESCO, Paris.

Scandurra, G., Romano, A.A., Ronghi, M., Carfora, A., 2018. On the vulnerability of Small Island Developing States: A dynamic analysis. Ecological Indicators 84 382-392.

Scawthorn, C., Flores, P., Blais, N., Seligson, H., Tate, E., Chang, S., Mifflin, E., Thomas, W., Murphy, J., Jones, C., Lawrence, M., 2006. HAZUS-MH flood loss estimation methodology. II. Damage and loss assessment. Natural Hazards Review 7 72-81.

Scheuer, S., Haase, D., Meyer, V., 2010. Exploring multicriteria flood vulnerability by integrating economic, social and ecological dimensions of flood risk and coping capacity: from a starting point view towards an end point view of vulnerability. Natural Hazards 58(2) 731-751.

Schmolke, A., Thorbek, P., DeAngelis, D.L., Grimm, V., 2010. Ecological models supporting environmental decision making: a strategy for the future. Trends Ecol Evol 25(8) 479-486.

Schott, T., Landsea, C., Hafele, G., Lorens, J., Taylor, A., Thurm, H., Ward, B., Willis, M., Zaleski, W., 2019. The Saffir-Simpson Hurricane Wind Scale. NOAA - NHC: https://www.nhc.noaa.gov/aboutsshws.php.

Schulze, J., Müller, B., Groeneveld, J., Grimm, V., 2017. Agent-Based Modelling of Social-Ecological Systems: Achievements, Challenges, and a Way Forward. Journal of Artificial Societies and Social Simulation 20(2) 8.

Sebastien, O., Harivelo, F., 2015. Enhancing Disaster Management by Taking Advantage of General Public Mobile Devices: Trends and Possible Scenarios, Wireless Public Safety Networks 1, pp. 261-295.

Shah, A.A., Ye, J., Abid, M., Khan, J., Amir, S.M., 2018. Flood hazards: household vulnerability and resilience in disaster-prone districts of Khyber Pakhtunkhwa province, Pakistan. Natural Hazards 93(1) 147-165.

Singh, P., Sinha, V.S.P., Vijhani, A., Pahuja, N., 2018. Vulnerability assessment of urban road network from urban flood. International Journal of Disaster Risk Reduction 28 237-250.

Skinner, C.J., Rao, N.K., 1996. Estimation in dual frame surveys with complex designs. Journal of the American Statistical Association 91(433) 349-356.

Slovic, P., 1987. Perception of risk. Science 236(4799) 280-285.

Smith, B., Pilifosova, O., 2001. Adaptation to climate change in the context of sustainable development and equity, In: McCarthy, J.J., Canziani, O., Leary, N.A., Dokken, D.J., White, K.S. (Eds.) (Ed.), Climate Change 2001: Impacts, Adaptation and Vulnerability. IPCC Working Group II: Cambridge, pp. 877-912.

Smith, M., Goodchild, M.F., Longley, P.A., 2018. Geospatial Analysis. A Comprehensive Guide to Principles, Techniques and Software Tools- 6th edition. The Winchelsea Press, Leicester.

Smith, S.K., McCarty, C., 2009. Fleeing the storm(s): an examination of evacuation behavior during florida's 2004 hurricane season. Demography 46 127-145.

Sorg, L., Medina, N., Feldmeyer, D., Sanchez, A., Vojinovic, Z., Birkmann, J., Marchese, A., 2018. Capturing the multifaceted phenomena of socioeconomic vulnerability. Natural Hazards 92(1) 257-282.

STAT, Department of Statistics Sint Maarten, 2017. Statistical yearbook 2017, Retrieved from STAT: http://www.stat.gov.sx/downloads/YearBook/Statistical_Yearbook_2017.pdf, p. 74.

Steinführer, A., Kuhlicke, C., 2006. Social vulnerability and the 2002 flood: country report Germany (Mulde River). FLOODsite: Wallingford, UK.

Sterzel, T., Ludeke, M.K.B., Walther, C., Kok, M.T., Sietz, D., Lucas, P.L., 2020. Typology of coastal urban vulnerability under rapid urbanization. PLoS ONE 15(1) e0220936.

Stevens, E., 2018. What Is The Difference Between A Mobile App And A Web App?

Sumathipala, A, Jafarey A, De Castro LD, Ahmad A, Marcer D, Srinivasan S, Kumar N, Siribaddana S, Sutaryo S, Bhan A, Waidyaratne D, Beneragama S, Jayasekera C, Edirisingha S and Siriwardhana C, 2010. Ethical issues in post-disaster clinical interventions and research: A developing world perspective. key findings from a drafting and consensus generation meeting of the working group on disaster research and ethics (WGDRE) 2007. Asian Bioethics Review 2(2) 124-142.

Sun, Y., Yamori, K., 2018. Risk Management and Technology: Case Studies of Tsunami Evacuation Drills in Japan. Sustainability 10(9) 2982.

Taillandier, F., Adam, C., 2018. Games Ready to Use_ A Serious Game for Teaching Natural Risk Management. Simulation & Gaming 49(4) 441-470.

Takabatake, T., Shibayama, T., Esteban, M., Ishii, H., 2018. Advanced casualty estimation based on tsunami evacuation intended behavior: case study at Yuigahama Beach, Kamakura, Japan. Natural Hazards 92(3) 1763-1788.

Tan, C., Fang, W., 2018. Mapping the Wind Hazard of Global Tropical Cyclones with Parametric Wind Field Models by Considering the Effects of Local Factors. International Journal of Disaster Risk Science 9(1) 86-99.

Tan, M.L., Prasanna, R., Stock, K., Doyle, E.E.H., Leonard, G., Johnston, D., 2020. Understanding end-users' perspectives: Towards developing usability guidelines for disaster apps. Progress in Disaster Science 7.

Tan, N.T., 2013. Emergency management and social recovery from disasters in different countries. J Soc Work Disabil Rehabil 12(1-2) 8-18.

Tapsell, S.M., Penning-Rowsell, E.C., Tunstall, S.M., Wilson, T.L., 2002. Vulnerability to flooding: Health and social dimensions. Philosophical Transactions of the Royal Society 360 1511-1525.

Taramelli, A., Valentini, E., Sterlacchini, S., 2015. A GIS-based approach for hurricane hazard and vulnerability assessment in the Cayman Islands. Ocean & Coastal Management 108 116-130.

ten Broeke, G., van Voorn, G., Ligtenberg, A., 2016. Which Sensitivity Analysis Method Should I Use for My Agent-Based Model? Journal of Artificial Societies and Social Simulation 19(1) 5.

Thompson, R.R., Garfin, D.R., Silver, R.C., 2017. Evacuation from Natural Disasters: A Systematic Review of the Literature. Risk Anal 37(4) 812-839.

Tonn, G.L., Guikema, S.D., 2018. An Agent-Based Model of Evolving Community Flood Risk. Risk Anal 38(6) 1258-1278.

Turner, B., Kasperson, R., Matson, P., McCarthy, J., Corell, R., Christensen, L., EckleY, N., Kasperson, J., Luers, A., Martello, M., Polsky, C., Pulsipher, A., Schiller, A., 2003. A framework for vulnerability analysis in sustainability science. Proc Natl Acad Sci 100(14) 8074-8079.

Turvey, R., 2007. Vulnerability Assessment of Developing Countries: The Case of Small-island Developing States. Development Policy Review 25(2) 243-264.

U.S. Congress, 1990. Physical Vulnerability of Electric System to Natural Disasters and Sabotage, In: Office of Technology Assessment (Ed.). U.S. Government Printing Office: Washington, DC.

UNDP, 2010. Disaster Risk Assessment. United Nations Development Programme: New York, USA.

UNDRR, 2017. Terminology on Disaster Risk Reduction. United Nations Office for Disaster Risk Reduction: Geneva.

United Nations, 2019. World Population Prospects 2019. Department of Economic and Social Affairs - Population Division: New York, NY, USA.

United Nations Development Programme [UNDP], 2012. Innovation and technology in risk mitigation and development planning in SIDS: Towards flood risk reduction in Sint Maarten. United Nations Development Programme, Barbados and the OECS: New York, NY, USA.

United Nations Framework Convention on Climate Change [UNFCCC], 2005. Climate Change: Small Island Developing States. Climate Change Secretariat: Bonn.

Urban, C., Bernd, S., 2001. PECS–Agent-Based Modelling of Human Behaviour, AAAI Fall Symposium Series.

USAID, 2003. Shelters and Shelter Management: Reference Guide. USAID from the American people.

Vagias, W.M., 2006. Likert-type scale response anchors. Clemson International Institute for Tourism & Research Development - Department of Parks, Recreation and Tourism Management: Clemson University.

Valkering, P., Rotmans, J., Krywkow, J., van der Veen, A., 2016. Simulating Stakeholder Support in a Policy Process: An Application to River Management. SIMULATION 81(10) 701-718.

Van Dam, K.H., Nikolic, I., Lukszo, Z., 2012. Agent-based modelling of socio-technical systems. Springer Science & Business Media.

Vári, A., Ferencz, Z., Hochrainer-Stigler, S., 2013. Social Indicators of Vulnerability to Floods: An Empirical Case Study in Two Upper Tisza Flood Basins, In: Amendola, A., Ermolieva, T., Linnerooth-Bayer, J., Mechler, R. (Eds.), Integrated Catastrophe Risk Modeling: Supporting Policy Processes. Springer Netherlands: Dordrecht, pp. 181-198.

Vojinović, Z., 2015. Flood risk: the Holistic Perspective. From integrated to interactive Planning for Flood Resilience. IWA Publishing, London.

Vojinović, Z., 2015. Flood Risk: The Holistic Perspective. From Integrated to Interactive Planning for Flood Resilience. IWA Publishing, London, UK.

Vojinović, Z., Abbott, M.B., 2012. Flood risk and social justice: from quantitative to qualitative flood risk assessment and mitigation. IWA Publishing, London.

Vojinović, Z., Hammond, M., Golub, D., Hirunsalee, S., Weesakul, S., Meesuk, V., Medina, N., Sanchez, A., Kumura, S., Abbott, M.B., 2016. Holistic approach to flood risk assessment in areas with cultural heritage: a practical application in Ayutthaya, Thailand. Natural Hazards 81 589-616.

Vojinović, Z., Van Teeffelen, J., 2007. An integrated stormwater management approach for small islands in tropical climates. Urban Water Journal 4(3) 211-231.

Vorst, H.C.M., 2010. Evacuation models and disaster psychology. Procedia Engineering 3 15-21.

Waldrop, M., 1992. Complexity: The Emerging Science at Edge of Order and Chaos. Simon and Schuster, New York.

Walle, B.V.d., Comes, T., Brugghemans, B., Chan, J., Meesters, K., Homberg, M.v.d., 2013. A journey into the information Typhoon Haiyan Disaster : Resilience Lab Field Report findings and research insights : Part II - The Role of Information. The Disaster Resilience Lab: The Netherlands.

Walls, M., Magliocca, N., McConnell, V., 2018. Modeling coastal land and housing markets: Understanding the competing influences of amenities and storm risks. Ocean & Coastal Management 157 95-110.

Walter, L., 2010. Web Users Increasingly Rely on Social Media to Seek Help in a Disaster. Red Cross: EHS today.

Watts, J., 2019. CHIME ABM Hurricane Evacuation Model" (Version 1.4.0): CoMSES Computational Model Library. Retrieved from: https://doi.org/10.25937/hbnh-af93.

Watts, J., Morss, R.E., Barton, C.M., Demuth, J.L., 2019. Conceptualizing and implementing an agent-based model of information flow and decision making during hurricane threats. Environmental Modelling & Software 122 104524.

Wehn, U., Rusca, M., Evers, J., Lanfranchi, V., 2015. Participation in flood risk management and the potential of citizen observatories: A governance analysis. Environmental Science & Policy 48 225-236.

Wei, W., Shi, S., Zhang, X., Zhou, L., Xie, B., Zhou, J., Li, C., 2020. Regional-scale assessment of environmental vulnerability in an arid inland basin. Ecological Indicators 109.

Welle, T., Depietri, Y., Angignard, M., Birkmann, J., Renaud, F., Greiving, S., 2014. Vulnerability Assessment to Heat Waves, Floods, and Earthquakes Using the MOVE Framework, Assessment of Vulnerability to Natural Hazards, pp. 91-124.

Whitehead, J.C., Edwards, B., Van Willigen, M., Maiolo, J.R., Wilson, K., Smith, K.T., 2001. Heading for higher ground: factors affecting real and hypothetical hurricane evacuation behavior. Environmental Hazards 2(4) 133-142.

Wilensky, U., 1999. NetLogo. Center for Connected Learning and Computer-Based Modeling, Northwestern University: Evanston, IL.

Wilson, S.N., Tiefenbacher, J.P., 2012. The barriers impeding precautionary behaviours by undocumented immigrants in emergencies: The Hurricane Ike experience in Houston, Texas, USA. Environmental Hazards 11(3) 194-212.

Wirtz, A., Kron, W., Löw, P., Steuer, M., 2012. The need for data: natural disasters and the challenges of database management. Natural Hazards 70(1) 135-157.

Wu, H.-C., Lindell, M.K., Prater, C.S., 2012. Logistics of hurricane evacuation in Hurricanes Katrina and Rita. Transportation Research Part F: Traffic Psychology and Behaviour 15(4) 445-461.

Yang, L.E., Scheffran, J., Süsser, D., Dawson, R., Chen, Y.D., 2018. Assessment of Flood Losses with Household Responses: Agent-Based Simulation in an Urban Catchment Area. Environmental Modeling & Assessment 23(4) 369-388.

Yang, Y., Mao, L., Metcalf, S.S., 2019. Diffusion of hurricane evacuation behavior through a home-workplace social network: A spatially explicit agent-based simulation model. Computers, Environment and Urban Systems 74 13-22.

Yin, L., Chen, J., Zhang, H., Yang, Z., Wan, Q., Ning, L., Hu, J., Yu, Q., 2019. Improving emergency evacuation planning with mobile phone location data. Environment and Planning B: Urban Analytics and City Science 47(6) 964-980.

Zlatanova, S., Fabbri, A.G., 2009. Geo-ICT for Risk and Disaster Management, In: Scholten, H.J., van de Velde, R., van Manen, N. (Eds.), Geospatial Technology and the Role of Location in Science. Springer Netherlands: Dordrecht, pp. 239-266.

APPENDIX A. HOUSEHOLD SURVEY – FACE-TO-FACE INTERVIEW

Questionnaire Post- Hurricane IRMA.	pearl IHE DELFT

To citizens of Sint Maarten,

Thank you for supporting this household survey.

This questionnaire is part of an international research project. IHE Delft – Institute for water education is performing a study aiming to gather valuable information on the perception, preparedness, and risk for hurricane evacuation, as well other specific insights on hurricane and flood related hazards potentially affecting the population of Sint Maarten.

Your participation in this survey will be treated **anonymously** and all the information will be kept confidential and individual records won't be shared with any officials of the island or any other administrative entity.

Your answers will help the research to gain a deeper understanding of flood risks and hurricane evacuation in the island which will enhance future decision making processes for land use and spatial planning, risk reduction and infrastructure improvement as well as evacuation plans if they are found necessary.

The questionnaire consists of three parts: General information, preparedness and reaction and risk perception/awareness. It is estimated to take you around 10 to 15 minutes to fill it in completely. Please read every question carefully and in case of doubts do not hesitate to contact the responsible team that handed the survey to you.

Thank you again for your support. With your help we can build a safer Sint Maarten for all!!!

Questionnaire Post- Hurricane IRMA.			
Survey ID:_____	Map #:_______	Field Paper:________	 

Part 1. General / Household Information

#	Question	Answer
1.1	Were you born in Sint Maarten?	❑ Yes ❑ NO where: ________________
1.2	Year of birth (interviewee)	__________
1.3	Gender	❑ Female ❑ Male ❑ Other
1.4	Which year did you move to the part of the island you are currently living?	__________
1.5	Total number of inhabitants in the household (**including yourself**)	__________
1.6	Are you tenant or owner of the house / apartment?	❑ Tenant ❑ Owner
1.7	Do you know the year of construction of the house?	❑ Yes Specify: __________ ❑ No
1.8	If working. What describe best your job location:	❑ Permanent or Fixed location ❑ Changing location
1.9	How many cars are within your household?	__________
1.10	How many smartphones/tablets are within your household?	__________
1.11	Do you have any pets or animals?	❑ Yes ❑ No
1.12	Is your home insured for natural disasters?	❑ Yes ❑ No
1.13	Is your home insured against lootings or riots?	❑ Yes ❑ No

Questionnaire Post- Hurricane IRMA.

Survey ID:_____ Map #:_____ Field Paper:______

pearl IHE DELFT

Part 2. Events, Preparedness and reaction

#	Question	Answer
2.1	How many hurricanes and tropical storms, if any, do you remember to have hit Sint Maarten during the time you have lived here?	________
2.2	Do you know where to get up-to-date information on early warnings and actual evacuation news/instructions?	☐ To a great extent ☐ To a moderate extent ☐ To some extent ☐ To a small extent ☐ Not at all ☐ Prefer not to answer / Not Applicable
2.3	From where do you get the latest updates on warnings or evacuation information? Mark all that apply.	☐ TV ☐ Radio ☐ Sint Maarten government channel ☐ Sirens / Speakers / Megaphone ☐ Sign posted in my neighbourhood ☐ Mobile App. Name:______ ☐ Friend or relative ☐ Internet ☐ Other services. Please specify ______
2.4	Regarding Hurricane IRMA. Did you receive any warning information before the hurricane hit the island?	☐ Yes → Please Specify # days: ____ ☐ No
2.4.a	If warning information was received. From whom did you receive it? Mark all that apply	☐ Sint Maarten government official ☐ Army or Police department ☐ Fire department ☐ Weather Broadcast (TV or internet) ☐ Family member or Friend ☐ Red Cross / Civil defence ☐ Other: ______ ☐ Prefer not to answer / Not Applicable
2.5	Did you and your family evacuate for Hurricane IRMA?	☐ Yes → Please answer from **2.6.a to 2.6.e** ☐ No → Please answer from **2.7.a to 2.7.m**
If you answered **YES** to question 2.5 please answer questions 2.6.a to 2.6.e		
2.6.a	When did you evacuate (leave your home or work) to go someplace safe during hurricane IRMA?	☐ Before the hurricane arrived ☐ Just as the hurricane arrived ☐ During the hurricane ☐ After the hurricane has passed ☐ Prefer not to answer / Not Applicable

Questionnaire Post- Hurricane IRMA.

Survey ID:_____ Map #:_____ Field Paper:______

pearl IHE DELFT

#	Question	Answer
2.6.b	Was the warning/evacuation information given with sufficient time to take actions?	☐ Yes ☐ No → Specify # of days you would need to complete a safe evacuation. ______
2.6.b.1	If no sufficient time, and in order to complete a safe evacuation. Could you estimate how many days in advance would need to receive the evacuation information?	☐ 1 -2 days ☐ 3 - 4 days ☐ 4 - 6 days ☐ More than 6 days
2.6.c	What type of information did you receive? Mark all that apply	☐ General evacuation order ☐ Shelter location ☐ Expected date/time of hurricane ☐ Driving/walking direction ☐ What to bring to shelter ☐ Other. Please specify ______
2.6.d	Did you follow the given instructions?	☐ Yes ☐ No ☐ Prefer not to answer / Not Applicable
2.6.e	If you evacuated from hurricane IRMA before the hurricane hit the island, where did you choose to go?	☐ Public shelter ☐ Special needs Shelter ☐ Pet-Friendly shelter ☐ Home of a relative or friend ☐ Hotel ☐ I left the island ☐ Other. Please specify ______

IF QUESTION 2.5 was Negatively answer. This is you did **NOT** evacuate please answer questions **2.7.a** to **2.7.m**

2.7 How strongly did each of the following factors influence your decision to remain at home and **NOT** evacuate during hurricane IRMA?
For each factor, please use the scale, ranging from "Not at all influential" to "Extremely Influential".

		Not at all influential	Slightly influential	Somewhat Influential	Very influential	Extremely influential	Prefer not to Answer
2.7.b	I felt Hurricane IRMA would not be a threat						
2.7.c	My home is strong enough to resist a hurricane						
2.7.d	Someone I know said there was no need for me to evacuate						
2.7.e	I did not know where to evacuate						

Questionnaire Post- Hurricane IRMA.

Survey ID:______ Map #:______ Field Paper:________

pearl IHE DELFT

#	Question	Not at all Influential	Slightly Influential	Somewhat Influential	Very Influential	Extremely Influential	Prefer not to Answer
2.7.f	I did not trust the official warning						
2.7.g	I needed assistance to evacuate myself or a relative						
2.7.i	I did not receive an official warning						
2.7.j	In my experience, it is better to stay at home						
2.7.k	I was waiting to reunite with family and/or friends						
2.7.l	I did not want to leave my property alone for fear of looters						
2.7.m	I consider the facilities of the shelters are not adequate to evacuate						

Continue questions to all respondents

#	Question	Answer
2.8	Based on your experiences during previous evacuations. Do you trust official sources of warning or evacuation in the island?	☐ To a great extent ☐ To a moderate extent ☐ To a some extent ☐ To a small extent ☐ Not at all ☐ Prefer not to answer / Not Applicable

2.9 To what extent do you agree with the following statements: In Sint Maarten...

#	Statement	Strongly Disagree	Disagree	Slightly Disagree	Slightly Agree	Agree	Strongly Agree	Prefer not to answer
2.9.a	*"... the number of available shelters is adequate."*							
2.9.b	*"...the location of the shelters are adequate"*							
2.9.c	*"...the road infrastructure to evacuate is adequate""*							

#	Question	Answer
2.10	Did your household experience a shortage of critical infrastructure services due to Hurricane Irma? Please choose all that apply	☐ Medical services / Access to hospitals ☐ Electricity ☐ Water supply ☐ Sanitation ☐ Transport ☐ No affected ☐ Other ________ ☐ Prefer not to answer / Not Applicable

Page 4 of 7

Questionnaire Post- Hurricane IRMA.

Survey ID:______ Map #:______ Field Paper:________

pearl IHE DELFT

Part 3. Risk Perception / Awareness

#	Question	Answer
3.1	With regard to Hurricanes, floods and natural disasters who do you think is responsible for taking action in Sint Maarten? Please choose and rank the three most relevant to you. Where 1 is the most important and 3 the least important	___ Sint Maarten Government ___ Dutch Government ___ Police / Fire department ___ Citizens ___ Red Cross / Civil Defence ___ Others. Please Specify ________

3.2 How likely is it that you would evacuate if the following events are forecasted to hit your local area? ...

#	Event	Definitely would not	Probably would not	About 50/50	Probably would	Definitely would	Prefer not to answer
3.2.a	Tropical depression (winds less than 68 km/h)						
3.2.b	Tropical depression (winds 63–118 km/h)						
3.2.c	Hurricane Category 1 (winds 19–153 km/h)						
3.2.d	Hurricane Category 2 (winds 154–177 km/h)						
3.2.e	Hurricane Category 3 (winds 178–208 km/h)						
3.2.f	Hurricane Category 4 (winds 209–251 km/h)						
3.2.g	Hurricane Category 5 (winds more than 252 km/h)						

3.3. To what extent do you agree with the following statements:

#	Statement	Strongly Disagree	Disagree	Slightly Disagree	Slightly Agree	Agree	Strongly Agree	Prefer not to answer
3.3.a	*"If the early warnings that I receive would be more precise and would reach me more directly. I would follow them more than I do now"*							
3.3.b	*"In Sint Maarten, the losses due to the recent hurricane could have been prevented by more appropriate planning and management from the City authorities"*							

Page 5 of 7

Questionnaire Post- Hurricane IRMA.			pearl IHE DELFT
Survey ID:____	Map #:____	Field Paper:____	

#	Question	Answer
3.4	When a hurricane or tropical storm approaches your local area, how frequently, do you check the forecasts on TV, mobile, radio, and/or on the internet?	☐ Less than once a day ☐ About once a day ☐ Several times a day ☐ Every couple of hours ☐ Throughout the whole day ☐ No Answer / Not Applicable
3.5	Have you previously received any official training/ or had community meetings, regarding procedures for hurricane or disaster evacuation?	☐ Yes ☐ No ☐ No Answer / Not Applicable
3.6	To the best of your knowledge, which one of the following is the most likely cause of injury or death during a hurricane? Please choose all that apply	___ Flying or falling objects from high winds ___ Rising water levels and high waves (storm surge) ___ Flooding from heavy rains ___ Accidents during evacuation ___ Not really sure ___ Others Specify ________

3.7. To what extent do you agree with the following statements.

		Strongly Disagree	Disagree	Slightly Disagree	Slightly Agree	Agree	Strongly Agree	Prefer not to answer
3.7.a	*"After hurricane Irma I would definitely acquire a home insurance for natural disasters"*							
3.7.b	*"After hurricane Irma I would definitely acquire a home insurance for Riots and Looting"*							
3.7.c	*"After hurricane Irma I would definitely acquire a life insurance for myself or my relatives"*							

Final Remarks.

Please share any additional comments, suggestions or information related with Hurricane evacuation, floods, natural disasters and institutions in Sint Maarten.

____

____

____

____

Page **6** of **7**

Questionnaire Post- Hurricane IRMA.			pearl IHE DELFT
Survey ID:____	Map #:____	Field Paper:____	

Thank you for your time completing this questionnaire. If you want to be contacted once we have the results from this survey, please provide us with your personal information.

Name: ____________________

Telephone: ____________________

e-mail: ____________________

Page **7** of **7**

APPENDIX B. VULNERABILITY INDEX FOR EACH NEIGHBOURHOOD

Table D.1 Assessment of Vulnerability. Components and PeVI values.

ID	Neighbourhood	Susceptibility	Lack of Coping Capacities	Lack of Adaptation Capacities	PeVI – Vulnerability Index
1	Low Lands	25.16	32.51	32.25	29.97
2	Point Pirouette	24.83	34.13	2.19	20.38
3	Maho	23.82	21.60	**54.06**	33.16
4	Beacon Hill	29.56	41.68	13.48	28.24
5	The Airport	No Data	No Data	No Data	No Data
6	Simpson Bay Village	**34.46**	39.06	24.41	32.64
7	Cole Bay Lagoon	20.39	40.44	17.00	25.95
8	Billy Folly	27.39	35.65	14.95	26.00
9	Cay Bay	29.69	40.93	43.02	37.88
10	Diamond	30.18	34.71	6.36	23.75
11	Cockpit	31.58	40.42	22.91	31.64
12	Cole Bay Village	27.85	27.51	15.11	23.49
13	Orange Grove	29.31	38.14	25.66	31.04
14	Wind Sor	**34.42**	33.70	32.49	33.54
15	Reward	24.50	36.68	32.65	31.28
16	Ebenezer	22.95	29.43	16.00	22.79
17	St Peters	31.81	38.31	28.67	32.93
18	Betty's Estate	19.89	21.98	18.41	20.09
19	Retreat Estate	22.97	27.64	17.17	22.60
20	St John Estate	24.89	19.75	4.43	16.36

Table D-1 (Continuation)

ID	Neighbourhood	Susceptibility	Lack of Coping Capacities	Lack of Adaptation Capacities	PeVI – Vulnerability Index
21	Saunders	25.93	36.92	22.25	28.37
22	Mary's Estate	31.74	37.35	24.45	31.18
23	Sentry Hill	30.82	42.13	41.15	38.03
24	Cay Hill	30.59	42.00	32.23	34.94
25	Belair	32.64	36.00	15.66	28.10
26	Fort Hill	32.88	38.24	26.94	32.69
27	Bethlehem	24.81	23.21	24.16	24.06
28	Nazareth	30.65	36.63	8.44	25.24
29	Union Farm	25.27	39.43	25.10	29.93
30	Zorg En Rust	29.83	43.21	24.33	32.46
31	Mount William	29.45	**55.17**	31.60	**38.74**
32	Madame's Estate	32.56	38.02	27.76	32.78
33	Over the Pond	**34.36**	**44.94**	37.73	**39.01**
34	Belvedere	26.23	29.68	29.08	28.33
35	Bishop Hill	26.97	**63.55**	27.54	**39.35**
36	Dutch Quarter	**41.27**	36.22	**57.34**	**44.94**
37	Middle Region	**34.01**	41.06	31.54	35.54
38	Easter Fresh Pond	20.00	30.45	36.29	28.91
39	Philipsburg	30.09	**46.68**	27.36	34.71
40	Pond Island	21.25	33.33	28.81	27.80
41	Salt Pans	No Data	No Data	No Data	No Data
42	The Harbour	No Data	No Data	No Data	No Data

Table D-1 (Continuation)

ID	**Neighbourhood**	**Susceptibility**	**Lack of Coping Capacities**	**Lack of Adaptation Capacities**	**PeVI – Vulnerability Index**
43	Oyster Pond	26.20	34.90	24.18	28.43
44	Defiance	26.71	29.22	26.02	27.32
45	Ocean Terrace	17.19	35.03	**50.00**	34.08
46	Dawn Beach	22.13	15.17	22.85	20.05
47	Sucker Garden	33.52	38.53	41.38	37.81
48	Guana Bay	23.87	27.75	37.92	29.85
49	Hope State	30.46	25.50	30.54	28.84
50	Geneva Bay	No Data	No Data	No Data	No Data
51	Back Bay	No Data	No Data	No Data	No Data
52	Over the Bank	27.74	**44.09**	**51.32**	**41.05**
53	Vineyard	25.81	19.23	**52.56**	32.54
54	Pointe Blanche	27.36	24.38	26.39	26.04

APPENDIX C. Review of evacuation predictors

Table C.1 *Principal variables affecting evacuation behaviour found in previous research, extracted from [1] Baker (1991), [2] Thompson et al. (2017), [3] Dash and Gladwin (2007) and [4] Huang et al. (2016). Positively (+): predictor of evacuation. Negatively (-): predictor of non-evacuation. Not conclusive: (+) in some studies and (-) in others. No effect: no statistical significance have been found either to promote evacuation or not.*

Group	Variable / predictor	Contribute to evacuate	Reference	Note
Demographic characteristics	Gender	Positively	[2] – [3]	Females are more likely to follow an evacuation order.
		Not conclusive	[1] – [4]	Variable that is not typically associated with actual evacuation rates, or non-significant results have been found.
	Age	Negatively	[2] – [3]	The older segment of the population is associated with limited mobility, hence less likelihood of evacuation.
		Positively	[1] – [2] [3]	Families with children tend to seek refuge more. Elderly residents in retirement areas have a higher tendency to evacuate (assisted).
		No effect	[4]	Non-significant statistical correlation is reported.
	Race / ethnicity	Positively	[2] – [3]	White – Caucasians were found to evacuate more than other races
		Negatively	[2]	Black – Hispanic are reported to evacuate less; it might be a cofounder of income.
		No effect	[4]	Non-significant statistical correlation is reported
	Car ownership	Not conclusive	[1]	Variable that it is not typically associated or non-significant results have been found with actual evacuation rates
	Disabled population	Not conclusive	[2] – [3]	Limited mobility may produce less evacuation behaviour. Or due to the limited mobility, this segment of the population may start the evacuation early
	Pets	Negatively	[2]	Households with pets tend to evacuate less than those without one. Difficulty to accommodate pets in shelters or hotels may explain this behaviour
		Not conclusive	[1]	Variable that it is not typically associated or non-significant results have been found with actual evacuation rates s

Table C.1 (Continuation)

Group	Variable / predictor	Contribute to evacuate	Reference	Note
Socio-economic characteristics	Level of education	Not conclusive	[1] – [2]	Variable that is not typically associated with actual evacuation rates, or non-significant results have been found. Some studies found a high correlation, others no correlation at all.
		No effect	[4]	Non-significant statistical correlation is reported.
	Household income	Not conclusive	[2]	Some studies found a high correlation, others no correlation at all between the income of a household and actual evacuation behaviour.
		Positively	[3]	Higher incomes were associated with higher evacuation rates.
		No effect	[4]	Non-significant statistical correlation is reported.
	Home ownership	Negatively	[2] – [4]	Owning a house has often been found to affect evacuation behaviour. Residents feel safer at home or prefer to stay to do repairs.
		Not conclusive	[1]	Variable is not typically associated with evacuation.
	Household size	Negatively	[3]	Single families have a lower tendency to evacuate.
		Positively	[2]	Houses with children have a higher tendency to evacuate.
		No effect	[4]	Non-significant statistical correlation is reported.
Housing characteristics	Type of house	Positively	[1] – [2] [4]	Those with fragile houses, such as mobile homes or boats, have a higher tendency to evacuate.
		Negatively	[1] – [2]	Perception of having a strong house may lead to low evacuation rates.
	Protection of the house	Negatively	[1] – [3]	Stay home to protect from looters or to do some repairs during the storm.
		No effect	[4]	Non-significant statistical correlation is reported.
	Property damage	Positively	[1] – [2] [3] – [4]	The bigger the loss (past or expected), the more likely to evacuate.

Table C.1 (Continuation)

Group	Variable / predictor	Contribute to evacuate	Reference	Note
Information	Government evacuation order	Positively	[1] – [2] [3] – [4]	If a mandatory evacuation is communicated. In some cultures obeying authority figures or being afraid of receiving a fine may lead to higher evacuation rates.
	Message content	Positively	[2] – [3]	The more specific and personalised the message, and the more urgency to evacuate, the higher the evacuation rates.
	Information from neighbours, friends or family	Not conclusive	[1]	Social cohesion may lead to higher evacuation rate to follow or reunite with family or close friends. Neighbours of relatives not evacuating may lead to lower evacuation rates due to peer pressure.
		Positively	[2] – [3]	Peers, friends or family members acting as a warning information source have resulted in evacuation behaviour, especially in communities where the family is the centre of society (i.e. Hispanic). Faith groups also have a role in disseminating evacuation orders and a higher number of evacuees.
		No effect	[4]	Non-significant statistical correlation is reported.
	False alarms, 'crying wolf' phenomenon	Negatively	[2]	Near miss experiences lead to failure to evacuate in future warnings
	False alarms, 'crying wolf' phenomenon	No effect	[1]	No significance between evacuation and the source of information. (i.e. official, TV, radio, friends).
	Frequency of gathering information	No effect	[1]	No significance between evacuation and frequency of media attention, keeping a tracking chart.
	Source of information	Positively	[2] – [4]	Perceived trustworthiness of the source has been found as a good predictor of accepting an evacuation order (or advice).
		No effect	[1]	No significance between evacuation and the source of information. (i.e. official, TV, radio, friends).

Table C.1 (Continuation)

Group	Variable / predictor	Contribute to evacuate	Reference	Note
Place, geography and storm characteristics	Length of residence in a place	Not conclusive	[1] – [2]	Newcomers do not know about potential risks. Long-term residents have a better knowledge of the risk in their residence area.
		Negatively	[3]	Number of years in a place influences evacuation behaviour; the longer the resident has lived in an area, the lower the evacuation rates.
	Hazard awareness	Positively	[3]	Being aware of living in a high-risk area increases the evacuation probability.
		No effect	[1]	Weak correlation between evacuation and belief that the storm will hit.
		Not conclusive	[4]	Actual evacuation studies have reported positive correlation and non-significant correlation.
	High risk areas	Positively	[1] – [3] [4]	Households located in low-lying/flood-prone areas have a higher tendency to evacuate.
	Previous disaster experience	Positively	[3]	Previous experience plays a role in evacuation behaviour.
		Not conclusive	[1] – [2] [4]	Self-reported past experiences have no predictive power of what households did in subsequent hurricanes.
	Prior evacuation behaviour	Positively	[1] – [2]	People that have already evacuated under previous evacuations orders are more likely to evacuate again.
	Disaster (perceived) intensity	Positively	[1] – [2] [3] – [4]	More intention to evacuate is found under threat of a larger or more intense disaster, such as a higher category hurricane.
	Number of storeys	Positively	[2]	Those living on ground floors have a higher tendency to evacuate in flood-prone areas.
	Risk perception	Positively	[1] – [2] [3] – [4]	The higher the perceived risk, the higher the tendency to evacuate.
	Discomfort of evacuation	Negatively	[1] – [3]	Forecast traffic jams, shelter conditions, not having anywhere to go, or the impossibility to return home in the aftermath of the disaster.

APPENDIX D. LIST OF PAPERS USED IN THE LITERATURE REVIEW OF ABM FOR WR-DRM

1) Abebe, Y.A., Ghorbani, A., Nikolic, I., Vojinovic, Z. and Sanchez, A. (2019) A coupled flood-agent-institution modelling (CLAIM) framework for urban flood risk management. Environmental Modelling & Software 111, 483-492 doi:10.1016/j.envsoft.2018.10.015.
2) Abebe, Y.A., Ghorbani, A., Nikolic, I., Vojinovic, Z. and Sanchez, A. (2019) Flood risk management in Sint Maarten – A coupled agent-based and flood modelling method. J Environ Manage 248, Article 109317 doi:10.1016/j.jenvman.2019.109317.
3) Aerts, J.C.J.H., Botzen, W.J., Clarke, K.C., Cutter, S.L., Hall, J.W., Merz, B., Michel-Kerjan, E., Mysiak, J., Surminski, S. and Kunreuther, H. (2018) Integrating human behaviour dynamics into flood disaster risk assessment. Nature Climate Change 8(3), 193-199 doi:10.1038/s41558-018-0085-1.
4) Assaf, H. (2011) Framework for Modeling Mass Disasters. Natural Hazards Review 12(2), 47-61 doi:10.1061/(asce)nh.1527-6996.0000033.
5) Baeza, A., Bojorquez-Tapia, L.A., Janssen, M.A. and Eakin, H. (2019) Operationalizing the feedback between institutional decision-making, socio-political infrastructure, and environmental risk in urban vulnerability analysis. J Environ Manage 241, 407-417 doi:10.1016/j.jenvman.2019.03.138.
6) Becu, N., Amalric, M., Anselme, B., Beck, E., Bertin, X., Delay, E., Long, N., Marilleau, N., Pignon-Mussaud, C. and Rousseaux, F. (2017) Participatory simulation to foster social learning on coastal flooding prevention. Environmental Modelling & Software 98, 1-11 doi:10.1016/j.envsoft.2017.09.003.
7) Burstein, M.H. and Diller, D.E. (2004) A framework for dynamic information flow in mixed-initiative human_agent organizations. Applied Intelligence 20(3), 283-298.
8) Busby, J.S., Onggo, B.S.S. and Liu, Y. (2016) Agent-based computational modelling of social risk responses. European Journal of Operational Research 251(3), 1029-1042 doi:10.1016/j.ejor.2015.12.034.
9) Castro, S., Poulos, A., Herrera, J.C. and de la Llera, J.C. (2019) Modeling the Impact of Earthquake-Induced Debris on Tsunami Evacuation Times of Coastal Cities. Earthquake Spectra 35(1), 137-158 doi:10.1193/101917eqs218m.
10) Chandra-Putra, H. and Andrews, C.J. (2019) An integrated model of real estate market responses to coastal flooding. Journal of Industrial Ecology (24), 424-435 doi:10.1111/jiec.12957.
11) Chandra-Putra, H., Zhang, H. and Andrews, C.J. (2015) Modeling Real Estate Market Responses to Climate Change in the Coastal Zone. Journal of Artificial Societies and Social Simulation 18(2), 18 doi:10.18564/jasss.2577.
12) Chang, S., Ichikawa, M., Deguchi, H. and Kanatani, Y. (2017) Optimizing the Arrangement of Post-Disaster Rescue Activities: An Agent-Based Simulation approach. Journal of Advanced Computational Intelligence and Intelligent Informatics 21(7), 1202-1210 doi:10.20965/jaciii.2017.p1202.
13) Cheff, I., Nistor, I. and Palermo, D. (2019) Pedestrian evacuation modelling of a Canadian West Coast community from a near-field Tsunami event. Natural Hazards 98(1), 229-249 doi:10.1007/s11069-018-3487-5.
14) Chen, X. (2011) Microsimulation of Hurricane Evacuation Strategies of Galveston Island. The Professional Geographer 60(2), 160-173 doi:10.1080/00330120701873645.
15) Chen, X. (2012) Agent-based micro-simulation of staged evacuations. Int. J. Advanced Intelligence Paradigms 4(1), 22-35.
16) Chen, X. (2015) Activity-based Modeling and Microsimulation of Emergency Evacuations. International Journal of Applied Geospatial Research 6(3), 21-38 doi:10.4018/ijagr.2015070102.
17) Chen, X., Meaker, J.W. and Zhan, F.B. (2006) Agent-Based Modeling and Analysis of Hurricane Evacuation Procedures for the Florida Keys. Natural Hazards 38(3), 321-338 doi:10.1007/s11069-005-0263-0.
18) Chen, X. and Zhan, F.B. (2008) Agent-based modelling and simulation of urban evacuation Relative effectiveness of simultaneous and staged evacuation strategies. Journal of the Operational Research Society 59(1), 25-33.

19) Coates, G., Hawe, G.I., Wright, N.G. and Ahilan, S. (2014) Agent-based modelling and inundation prediction to enable the identification of businesses affected by flooding. WIT Transactions on Ecology and the Environment 184, 13-22
20) Coates, G., Li, C., Ahilan, S., Wright, N. and Alharbi, M. (2019) Agent-based e-entry and simulation to assess flood preparedness and recovery of manufacturing small and medium-sized enterprises. Engineering Applications of Artificial Intelligence 78, 195-217 doi:10.1016/j.engappai.2018.11.010.
21) Coates, G., Li, C., Wright, N.G. and Ahilan, S. (2016) Investigating the flood responsiveness of small and medium enterprises using agent-based modelling and simulation. International Journal of Safety and Security Engineering 6(3), 627-635 doi:10.2495/SAFE-V0-N0-1-9.
22) Connell, P.E. and Donnell, G. (2014) Towards modelling flood protection investment as a coupled human and natural system. Hydrology and Earth System Sciences 18(1), 155-171 doi:10.5194/hess-18-155-2014.
23) Crick, F., Jenkins, K. and Surminski, S. (2018) Strengthening insurance partnerships in the face of climate change – Insights from an agent-based model of flood insurance in the UK. Sci Total Environ 636, 192-204 doi:10.1016/j.scitotenv.2018.04.239.
24) Dawson, R.J., Peppe, R. and Wang, M. (2011) An agent-based model for risk-based flood incident management. Natural Hazards 59(1), 167-189 doi:10.1007/s11069-011-9745-4.
25) Di Mauro, M., Megawati, K., Cedillos, V. and Tucker, B. (2013) Tsunami risk reduction for densely populated Southeast Asian cities: analysis of vehicular and pedestrian evacuation for the city of Padang, Indonesia, and assessment of interventions. Natural Hazards 68(2), 373-404 doi:10.1007/s11069-013-0632-z.
26) Dixon, D.S., Mozumder, P., Vásquez, W.F. and Gladwin, H. (2017) Heterogeneity Within and Across Households in Hurricane Evacuation Response. Networks and Spatial Economics 17(2), 645-680 doi:10.1007/s11067-017-9339-0.
27) Dressler, G., Müller, B., Frank, K. and Kuhlicke, C. (2016) Towards thresholds of disaster management performance under demographic change: exploring functional relationships using agent-based e-entry. Natural Hazards and Earth System Sciences 16(10), 2287-2301 doi:10.5194/nhess-16-2287-2016.
28) Du, E., Cai, X., Sun, Z. and Minsker, B. (2017) Exploring the Role of Social Media and Individual Behaviors in Flood Evacuation Processes: An Agent-Based Modeling Approach. Water Resources Research 53(11), 9164-9180 doi:10.1002/2017wr021192.
29) Du, E., Rivera, S., Cai, X., Myers, L., Ernest, A. and Minsker, B. (2017) Impacts of Human Behavioral Heterogeneity on the Benefits of Probabilistic Flood Warnings: An Agent-Based Modeling Framework. JAWRA Journal of the American Water Resources Association 53(2), 316-332 doi:10.1111/1752-1688.12475.
30) Dubbelboer, J., Nikolic, I., Jenkins, K. and Hall, J. (2017) An Agent-Based Model of Flood Risk and Insurance. Journal of Artificial Societies and Social Simulation 20(1), 6 doi:10.18564/jasss.3135.
31) Eid, M.S. and El-adaway, I.H. (2017) Integrating the Social Vulnerability of Host Communities and the Objective Functions of Associated Stakeholders during Disaster Recovery Processes Using Agent-Based Modeling. Journal of Computing in Civil Engineering 31(5), 04017030 doi:10.1061/(asce)cp.1943-5487.0000680.
32) Eid, M.S. and El-adaway, I.H. (2018) Decision-Making Framework for Holistic Sustainable Disaster Recovery: Agent-Based Approach for Decreasing Vulnerabilities of the Associated Communities. Journal of Infrastructure Systems 24(3), 04018009 doi:10.1061/(asce)is.1943-555x.0000427.
33) Eivazy, H. and Malek, M.R. (2019) Flood Management in Aqala through an Agent-Based Solution and Crowdsourcing Services in an Enterprise Geospatial Information System. ISPRS International Journal of Geo-Information 8(9), 420 doi:10.3390/ijgi8090420.
34) Erdlenbruch, K. and Bonté, B. (2018) Simulating the dynamics of individual adaptation to floods. Environmental Science & Policy 84, 134-148 doi:10.1016/j.envsci.2018.03.005.
35) Fiedrich, F. and Burghardt, P. (2007) Agent-based systems for disaster management. Communications of the ACM 50(3), 41-42 doi:10.1145/1226736.1226763.
36) Gehlot, H., Zhan, X., Qian, X., Thompson, C., Kulkarni, M. and Ukkusuri, S.V. (2019) A-RESCUE 2.0: A High-Fidelity, Parallel, Agent-Based Evacuation Simulator. Journal of Computing in Civil Engineering 33(2), 04018059 doi:10.1061/(asce)cp.1943-5487.0000802.

37) Ghavami, S.M., Maleki, J. and Arentze, T. (2019) A multi-agent assisted approach for spatial Group Decision Support Systems: A case study of disaster management practice. International Journal of Disaster Risk Reduction 38, 101223 doi:10.1016/j.ijdrr.2019.101223.
38) Goto, Y., Affan, M., Agussabti, Nurdin, Y. and Yuliana, D.K. (2012) Tsunami evacuation simulation for disaster education and city planning. Journal of Disaster Research 7(1), 92-101.
39) Haer, T., Botzen, W.J.W. and Aerts, J.C.J.H. (2016) The effectiveness of flood risk communication strategies and the influence of social networks—Insights from an agent-based model. Environmental Science & Policy 60, 44-52 doi:10.1016/j.envsci.2016.03.006.
40) Haer, T., Botzen, W.J.W. and Aerts, J.C.J.H. (2019) Advancing disaster policies by integrating dynamic adaptive behaviour in risk assessments using an agent-based modelling approach. Environmental Research Letters 14(4), 044022 doi:10.1088/1748-9326/ab0770.
41) Haer, T., Botzen, W.J.W., de Moel, H. and Aerts, J. (2017) Integrating Household Risk Mitigation Behavior in Flood Risk Analysis: An Agent-Based Model Approach. Risk Anal 37(10), 1977-1992 doi:10.1111/risa.12740.
42) Hajhashemi, E., Murray-Tuite, P.M., Hotle, S.L. and Wernstedt, K. (2019) Using agent-based e-entry to evaluate the effects of Hurricane Sandy's recovery timeline on the ability to work. Transportation Research Part D: Transport and Environment 77, 506-524 doi:10.1016/j.trd.2019.08.011.
43) Han, Y. and Peng, Z.-r. (2019) The integration of local government, residents, and insurance in coastal adaptation: An agent-based e-entry approach. Computers, Environment and Urban Systems 76, 69-79 doi:10.1016/j.compenvurbsys.2019.04.001.
44) Handford, D. and Rogers, A. (2012) An agent-based social forces model for driver evacuation behaviours. Progress in Artificial Intelligence 1(2), 173-181 doi:10.1007/s13748-012-0015-9.
45) Husby, T.G. and Koks, E.E. (2017) Household migration in disaster impact analysis: incorporating behavioural responses to risk. Natural Hazards 87(1), 287-305 doi:10.1007/s11069-017-2763-0.
46) Imamura, F., Muhari, A., Mas, E., Pradono, M.H., Post, J. and Sugimoto, M.A. (2012) Tsunami disaster mitigation by integrating comprehensive countermeasures in Padang city, Indonesia. Journal of Disaster Research 7(1), 48-64.
47) Iwanaga, S. and Namatame, A. (2016) Contagion of Evacuation Decision Making on Real Map. Mobile Networks and Applications 21(1), 206-214 doi:10.1007/s11036-016-0704-x.
48) Jenkins, K., Surminski, S., Hall, J. and Crick, F. (2017) Assessing surface water flood risk and management strategies under future climate change: Insights from an Agent-Based Model. Sci Total Environ 595, 159-168 doi:10.1016/j.scitotenv.2017.03.242.
49) Katada, T., Kuwasawa, N., Shida, S. and Kojima, M. (2013) Scenario Analysis for evacuation strategies for residents in big cities during large-scale flooding. Journal of Japan Society of Civil Engineers 69(1), 71-82.
50) Kim, J., Lee, S. and Lee, S. (2016) An evacuation route choice model based on multi-agent simulation in order to prepare Tsunami disasters. Transportmetrica B: Transport Dynamics 5(4), 385-401 doi:10.1080/21680566.2016.1147002.
51) Koning, K., Filatova, T. and Bin, O. (2019) Capitalization of Flood Insurance and Risk Perceptions in Housing Prices: An Empirical Agent-Based Model Approach. Southern Economic Journal 85(4), 1159-1179 doi:10.1002/soej.12328.
52) Kunwar, B., Simini, F. and Johansson, A. (2016) Evacuation time estimate for total pedestrian evacuation using a queuing network model and volunteered geographic information. Phys Rev E 93(3), 032311 doi:10.1103/PhysRevE.93.032311.
53) Lämmel, G., Grether, D. and Nagel, K. (2010) The representation and implementation of time-dependent inundation in large-scale microscopic evacuation simulations. Transportation Research Part C: Emerging Technologies 18(1), 84-98 doi:10.1016/j.trc.2009.04.020.
54) León, J. and March, A. (2014) Urban morphology as a tool for supporting tsunami rapid resilience: A case study of Talcahuano, Chile. Habitat International 43, 250-262 doi:10.1016/j.habitatint.2014.04.006.
55) León, J. and March, A. (2016) An urban form response to disaster vulnerability: Improving tsunami evacuation in Iquique, Chile. Environment and Planning B: Planning and Design 43(5), 826-847 doi:10.1177/0265813515597229.

56) Li, C. and Coates, G. (2016) Design and development of an agent-based model for business operations faced with flood disruption. International Journal of Design & Nature and Ecodynamics 11(2), 97-106 doi:10.2495/dne-v11-n2-97-106.
57) Li, Y., Hu, B., Zhang, D., Gong, J., Song, Y. and Sun, J. (2019) Flood evacuation simulations using cellular automata and e-entry systems –a human-environment relationship perspective. International Journal of Geographical Information Science 33(11), 2241-2258 doi:10.1080/13658816.2019.1622015.
58) Liang, W., Lam, N.S.N., Qin, X. and Ju, W. (2015) A Two-level Agent-Based Model for Hurricane Evacuation in New Orleans. Journal of Homeland Security and Emergency Management 12(2), 407-435. Doi:10.1515/jhsem-2014-0057
59) Lipiec, E., Ruggiero, P., Mills, A., Serafin, K.A., Bolte, J., Corcoran, P., Stevenson, J., Zanocco, C. and Lach, D. (2018) Mapping Out Climate Change: Assessing How Coastal Communities Adapt Using Alternative Future Scenarios. Journal of Coastal Research 34(5), 1196-1208 doi:10.2112/jcoastres-d-17-00115.1.
60) Liu, X. and Lim, S. (2016) Integration of spatial analysis and an agent-based model into evacuation management for shelter assignment and routing. Journal of Spatial Science 61(2), 283-298 doi:10.1080/14498596.2016.1147393.
61) Liu, X. and Lim, S. (2018) An agent-based evacuation model for the 2011 Brisbane City-scale riverine flood. Natural Hazards 94(1), 53-70 doi:10.1007/s11069-018-3373-1.
62) Löwe, R., Urich, C., Sto. Domingo, N., Mark, O., Deletic, A. and Arnbjerg-Nielsen, K. (2017) Assessment of urban pluvial flood risk and efficiency of adaptation options through simulations – A new generation of urban planning tools. Journal of Hydrology 550, 355-367 doi:10.1016/j.jhydrol.2017.05.009.
63) Lumbroso, D. and Davison, M. (2018) Use of an agent-based model and Monte Carlo analysis to estimate the effectiveness of emergency management interventions to reduce loss of life during extreme floods. Journal of Flood Risk Management 11, S419-S433 doi:10.1111/jfr3.12230.
64) Lumbroso, D. and Di Mauro, M. (2008) Recent developments in loss of life and evacuation modelling for flood event management in the UK. WIT Transactions on Ecology and the Environment 118, 263-272 doi:10.2495/FRIAR080251.
65) Magliocca, N.R. and Walls, M. (2018) The role of subjective risk perceptions in shaping coastal development dynamics. Computers, Environment and Urban Systems 71, 1-13 doi:10.1016/j.compenvurbsys.2018.03.009.
66) Makinoshima, F., Imamura, F. and Abe, Y. (2018) Enhancing a tsunami evacuation simulation for a multi-scenario analysis using parallel computing. Simulation Modelling Practice and Theory 83, 36-50 doi:10.1016/j.simpat.2017.12.016.
67) Manzoor, U., Zubair, M., Batool, K. and Zafar, B. (2014) A multi-agent framework for efficient food distribution in disaster areas. International Journal of Internet Technology and Secured Transactions 5(4), 327-343 doi:10.1504/IJITST.2014.068711.
68) Mas, E., Adriano, B. and Koshimura, S. (2013) An integrated simulation of tsunami hazard and human evacuation in La Punta, Peru. Journal of Disaster Research 8(2), 285-295.
69) Mas, E., Felsenstein, D., Moya, L., Grinberger, A.Y., Das, R. and Koshimura, S. (2018) Dynamic Integrated Model for Disaster Management and Socioeconomic Analysis (DIM2SEA). Journal of Disaster Research 13(7), 1257-1271 doi:10.20965/jdr.2018.p1257.
70) Mas, E., Koshimura, S., Imamura, F., Suppasri, A., Muhari, A. and Adriano, B. (2015) Recent Advances in Agent-Based Tsunami Evacuation Simulations: Case Studies in Indonesia, Thailand, Japan and Peru. Pure and Applied Geophysics 172(12), 3409-3424 doi:10.1007/s00024-015-1105-y.
71) McNamara, D.E. and Keeler, A. (2013) A coupled physical and economic model of the response of coastal real estate to climate risk. Nature Climate Change 3(6), 559-562 doi:10.1038/nclimate1826.
72) Mostafizi, A., Wang, H., Cox, D., Cramer, L.A. and Dong, S. (2017) Agent-based tsunami evacuation e-entry of unplanned network disruptions for evidence-driven resource allocation and retrofitting strategies. Natural Hazards 88(3), 1347-1372 doi:10.1007/s11069-017-2927-y.
73) Mostafizi, A., Wang, H., Cox, D. and Dong, S. (2019) An agent-based vertical evacuation model for a near-field tsunami: Choice e-entry, logical shelter locations, and life safety. International Journal of Disaster Risk Reduction 34, 467-479 doi:10.1016/j.ijdrr.2018.12.018.

74) Mostafizi, A., Wang, H. and Dong, S. (2019) Understanding the Multimodal Evacuation Behavior for a Near-Field Tsunami. Transportation Research Record: Journal of the Transportation Research Board 2673(11), 480-492 doi:10.1177/0361198119837511.
75) Mustafa, A., Bruwier, M., Archambeau, P., Erpicum, S., Pirotton, M., Dewals, B. and Teller, J. (2018) Effects of spatial planning on future flood risks in urban environments. J Environ Manage 225, 193-204 doi:10.1016/j.jenvman.2018.07.090.
76) Nagarajan, M., Shaw, D. and Albores, P. (2012) Disseminating a warning message to evacuate: A simulation study of the behaviour of neighbours. European Journal of Operational Research 220(3), 810-819 doi:10.1016/j.ejor.2012.02.026.
77) Naghawi, H. and Wolshon, B. (2010) Transit-Based Emergency Evacuation Simulation Modeling. Journal of Transportation Safety & Security 2(2), 184-201 doi:10.1080/19439962.2010.488316.
78) Naghawi, H. and Wolshon, B. (2012) Performance of Traffic Networks during Multimodal Evacuations: Simulation-Based Assessment. Natural Hazards Review 13(3), 196-204 doi:10.1061/(asce)nh.1527-6996.0000065.
79) Nakanishi, H., Black, J. and Suenaga, Y. (2019) Investigating the flood evacuation behaviour of older people: A case study of a rural town in Japan. Research in Transportation Business & Management 30, Article 100376 doi:10.1016/j.rtbm.2019.100376.
80) Naqvi, A.A. and Rehm, M. (2014) A multi-agent model of a low income economy: simulating the distributional effects of natural disasters. Journal of Economic Interaction and Coordination 9(2), 275-309 doi:10.1007/s11403-014-0137-1.
81) Nejat, A. and Damnjanovic, I. (2012) Agent-Based Modeling of Behavioral Housing Recovery Following Disasters. Computer-Aided Civil and Infrastructure Engineering 27(10), 748-763 doi:10.1111/j.1467-8667.2012.00787.x.
82) Oh, B.H., Kim, K., Choi, H.-L. and Hwang, I. (2018) Cooperative Multiple Agent-Based Algorithm for Evacuation Planning for Victims with Different Urgencies. Journal of Aerospace Information Systems 15(6), 382-395 doi:10.2514/1.I010589.
83) Rand, W., Herrmanna, J., Scheinb, B. and Vodopivec, N. (2015) An agent-based model of urgent diffusion in social media. Journal of Artificial Societies and Social Simulation 18(2), 1 doi:10.18564/jasss.2616.
84) Saadi, I., Mustafa, A., Teller, J. and Cools, M. (2018) Investigating the impact of river floods on travel demand based on an agent-based e-entry approach: The case of Liège, Belgium. Transport Policy 67, 102-110 doi:10.1016/j.tranpol.2017.09.009.
85) Safarzyńska, K., Brouwer, R. and Hofkes, M. (2013) Evolutionary modelling of the macro-economic impacts of catastrophic flood events. Ecological Economics 88, 108-118 doi:10.1016/j.ecolecon.2013.01.016.
86) Sahal, A., Leone, F. and Péroche, M. (2013) Complementary methods to plan pedestrian evacuation of the French Riviera's beaches in case of tsunami threat: graph- and multi-agent-based modelling. Natural Hazards and Earth System Sciences 13(7), 1735-1743 doi:10.5194/nhess-13-1735-2013.
87) Simmonds, J., Gómez, J.A. and Ledezma, A. (2019) The role of agent-based e-entry and multi-agent systems in flood-based hydrological problems: a brief review. Journal of Water and Climate Change. Doi: 10.2166/wcc.2019.108
88) Solís, I.A. and Gazmuri, P. (2017) Evaluation of the risk and the evacuation policy in the case of a tsunami in the city of Iquique, Chile. Natural Hazards 88(1), 503-532 doi:10.1007/s11069-017-2876-5.
89) Sun, Y. and Yamori, K. (2018) Risk Management and Technology: Case Studies of Tsunami Evacuation Drills in Japan. Sustainability 10(9), 2982 doi:10.3390/su10092982.
90) Taillandier, F. and Adam, C. (2018) Games Ready to Use_ A Serious Game for Teaching Natural Risk Management. Simulation & Gaming 49(4), 441-470.
91) Takabatake, T., Shibayama, T., Esteban, M. and Ishii, H. (2018) Advanced casualty estimation based on tsunami evacuation intended e-entry: case study at Yuigahama Beach, Kamakura, Japan. Natural Hazards 92(3), 1763-1788 doi:10.1007/s11069-018-3277-0.
92) Takabatake, T., Shibayama, T., Esteban, M., Ishii, H. and Hamano, G. (2017) Simulated tsunami evacuation e-entry of local residents and visitors in Kamakura, Japan. International Journal of Disaster Risk Reduction 23, 1-14 doi:10.1016/j.ijdrr.2017.04.003.
93) Tonn, G.L. and Guikema, S.D. (2018) An Agent-Based Model of Evolving Community Flood Risk. Risk Anal 38(6), 1258-1278 doi:10.1111/risa.12939.

94) Triatmadja, R. (2015) Numerical simulations of an evacuation from a tsunami at Parangtritis beach in Indonesia. Journal of Tsunami Society International 34(1), 50-66.
95) Ukkusuri, S.V., Hasan, S., Luong, B., Doan, K., Zhan, X., Murray-Tuite, P. and Yin, W. (2016) A-RESCUE: An Agent based Regional Evacuation Simulator Coupled with User Enriched Behavior. Networks and Spatial Economics 17(1), 197-223 doi:10.1007/s11067-016-9323-0.
96) Uno, K. and Kashiyama, K. (2008) Development of Simulation System for the Disaster Evacuation Based on Multi-Agent Model Using GIS. Tsinghua Science and Technology 13(S1), 348-353.
97) Valkering, P., Rotmans, J., Krywkow, J. and van der Veen, A. (2016) Simulating Stakeholder Support in a Policy Process: An Application to River Management. Simulation 81(10), 701-718 doi:10.1177/0037549705060793.
98) Walls, M., Magliocca, N. and McConnell, V. (2018) Modeling coastal land and housing markets: Understanding the competing influences of amenities and storm risks. Ocean & Coastal Management 157, 95-110 doi:10.1016/j.ocecoaman.2018.01.021.
99) Wang, H., Mostafizi, A., Cramer, L.A., Cox, D. and Park, H. (2016) An agent-based model of a multimodal near-field tsunami evacuation: Decision-making and life safety. Transportation Research Part C: Emerging Technologies 64, 86-100 doi:10.1016/j.trc.2015.11.010.
100) Watts, J., Morss, R.E., Barton, C.M. and Demuth, J.L. (2019) Conceptualizing and implementing an agent-based model of information flow and decision making during hurricane threats. Environmental Modelling & Software 122, 104524 doi:10.1016/j.envsoft.2019.104524.
101) Widener, M.J., Horner, M.W. and Ma, K. (2015) Positioning Disaster Relief Teams Given Dynamic Service Demand: A Hybrid Agent-Based and Spatial Optimization Approach. Transactions in GIS 19(2), 279-295 doi:10.1111/tgis.12092.
102) Widener, M.J., Horner, M.W. and Metcalf, S.S. (2012) Simulating the effects of social networks on a population's hurricane evacuation participation. Journal of Geographical Systems 15(2), 193-209 doi:10.1007/s10109-012-0170-3.
103) Wolshon, B. and Dixit, V.V. (2012) Traffic modelling and simulation for regional multimodal evacuation analysis. Int. J. Advanced Intelligence Paradigms 4(1), 71-82.
104) Yang, B. and Ren, B. (2015) A parallel spatio-temporal model for emergency evacuation simulation. Int. J. Simulation and Process Modelling 10(1), 10-18.
105) Yang, L.E., Scheffran, J., Süsser, D., Dawson, R. and Chen, Y.D. (2018) Assessment of Flood Losses with Household Responses: Agent-Based Simulation in an Urban Catchment Area. Environmental Modeling & Assessment 23(4), 369-388 doi:10.1007/s10666-018-9597-3.
106) Yang, Y., Mao, L. and Metcalf, S.S. (2019) Diffusion of hurricane evacuation e-entry through a home-workplace social network: A spatially explicit agent-based simulation model. Computers, Environment and Urban Systems 74, 13-22 doi:10.1016/j.compenvurbsys.2018.11.010.
107) Yin, L. (2006) Agent-Based Simulations for Disaster Decision Support. Journal of Security Education 1(4), 169-175 doi:10.1300/J460v01n04_15.
108) Yin, L., Chen, J., Zhang, H., Yang, Z., Wan, Q., Ning, L., Hu, J. and Yu, Q. (2019) Improving emergency evacuation planning with mobile phone location data. Environment and Planning B: Urban Analytics and City Science 47(6), 964-980 doi:10.1177/2399808319874805.
109) Yin, W., Murray-Tuite, P., Ukkusuri, S.V. and Gladwin, H. (2014) An agent-based e-entry system for travel demand simulation for hurricane evacuation. Transportation Research Part C: Emerging Technologies 42, 44-59 doi:10.1016/j.trc.2014.02.015.
110) Yoshida, Y., Kimura, T., Minegishi, Y. and Sano, T. (2014) Tsunami safe town planning with evacuation simulation. Journal of Disaster Research 9(sp), 719-729.
111) Yu, J., Zhang, C., Wen, J., Li, W., Liu, R. and Xu, H. (2018) Integrating multi-agent evacuation simulation and multi-criteria evaluation for spatial allocation of urban emergency shelters. International Journal of Geographical Information Science 32(9), 1884-1910 doi:10.1080/13658816.2018.1463442.
112) Zhang, Z., Wolshon, B., Herrera, N. and Parr, S. (2019) Assessment of post-disaster e-entry traffic in megaregions using agent-based simulation. Transportation Research Part D: Transport and Environment 73, 307-317 doi:10.1016/j.trd.2019.06.010.

113) Zhu, J., Dai, Q., Deng, Y., Zhang, A., Zhang, Y. and Zhang, S. (2018) Indirect Damage of Urban Flooding: Investigation of Flood-Induced Traffic Congestion Using Dynamic Modeling. Water 10(5), 622 doi:10.3390/w10050622.
114) Zhu, X., Dai, Q., Han, D., Zhuo, L., Zhu, S. and Zhang, S. (2019) Modeling the high-resolution dynamic exposure to flooding in a city region. Hydrology and Earth System Sciences 23(8), 3353-3372 doi:10.5194/hess-23-3353-2019.
115) Zia, K., Farrahi, K., Riener, A. and Ferscha, A. (2013) An agent-based parallel geo-simulation of urban mobility during city-scale evacuation. Simulation 89(10), 1184-1214 doi:10.1177/0037549713485468.

APPENDIX E. ID NEIGHBOURHOODS - VROMI

ID	Neighbourhood	ID	Neighbourhood	ID	Neighbourhood
1	Mullet Bay	23	Sentry Hill	45	Ocean Terrace
2	Point Pirouette	24	Cay Hill	46	Dawn Beach
3	Maho	25	Belair	47	Sucker Garden
4	Beacon Hill	26	Fort Hill	48	Guana Bay
5	The Airport	27	Bethlehem	49	Hope State
6	Simpson Bay Village	28	Nazareth	50	Geneva Bay
7	Cole Bay Lagoon	29	Union Farm	51	Back Bay
8	Billy Folly	30	Zorg En Rust	52	Over the Bank
9	Cay Bay	31	Mount William	53	Vineyard
10	Diamond	32	Madame's Estate	54	Pointe Blanche
11	Cockpit	33	Over the Pond	55	Rockland
12	Cole Bay Village	34	Belvedere	56	Western Fresh Pond
13	Orange Grove	35	Bishop Hill	57	Little Bay Village
14	Wind Sor	36	Dutch Quarter	58	Welegelegen
15	Reward	37	Middle Region	59	Cay Hill Village
16	Ebenezer	38	Easter Fresh Pond	60	Little Cape Bay
17	St Peters	39	Philipsburg	61	Zaeger Gut
18	Betty's Estate	40	Pond Island	62	Cupecoy
19	Retreat Estate	41	Salt Pans	63	Bloomingdale
20	St John Estate	42	The Harbour	64	Foga
21	Saunders	43	Oyster Pond	65	Red Pond Estate
22	Mary's Estate	44	Defiance		

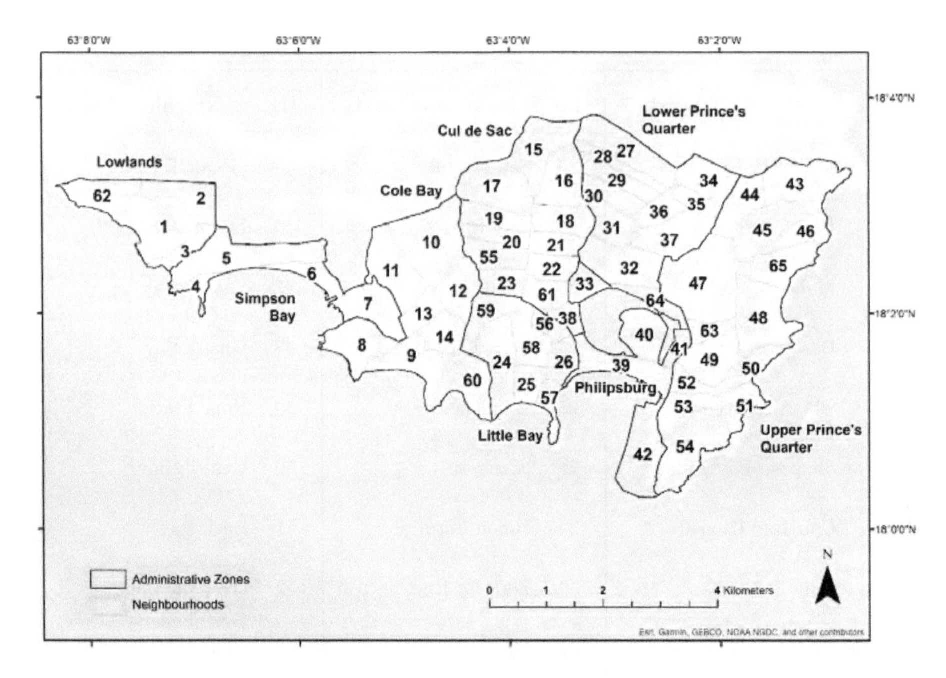

Figure E.1. Neighbourhood's ID – VROMI

APPENDIX F. ATTRIBUTES TABLE FOR SHELTER MANAGEMENT

Table F.1. Full list of attributes available for administrators on Shelters.

Attribute	Description
FAC_ID	Corresponds to the shelter Facility ID assigned to the building
Name	Name of the building
Address	Address of the main entrance of the building
Shelter	The official (or NOT) status of the building as a shelter
STATUS_OPEN	Whether the building is currently accepting people in the facility, to be assigned by the shelter manager
DateCollection	The latest update on the status as a shelter
Inspector	Name of the latest administrator to update information on the shelter
Category	Normal use of the building (i.e. airport, banking, education, government, healthcare, religious, other)
Shelter Telephone	Mobile phone or landline to contact the shelter
Shelter Owner	Name of person or organisation that owns the shelter
Owner Consent	Whether the shelter's owner has agreed to use the facilities for a specific evacuation
Ground_Elev	Elevation above sea level of the building
Flood_Risk_level	Risk level according to the domain: low, intermediate or high
Physical_Cond	Physical condition of the building in terms of vulnerability: weak, intermediate, or solid
Vulnerability	Vulnerability assessment for the building according to the risk and physical condition: low, medium, high, very high or extreme
Capacity	Total number of people the shelter can accommodate
Parking	Whether the shelter is equipped with parking units
NumPark	Number of parking spaces available at the shelter
Nbeds	Total number of beds available at the facility
Ntoilets	Total number of toilet units available at the facility

Table F.1 (Continuation)

Attribute	Description
RunningWater	Whether the shelter can provide water from the public water system at a given moment
WaterStorage	Whether the shelter is equipped with tanks to store drinking water
WS_Capac	Storage capacity (if any) of drinking water (m^3)
BottledWater	Current storage of bottled drinking water (m^3)
Storage_Capacity	Space available in m^3 for storage of supplies to cope with an evacuation
Roof_Structure	Structure of the building's roof
Roof_MAT	Type of material used in the building's roof
WindowsProt	Whether the windows or glass doors are protected by shutters
PowerGenerator	Whether the shelter is equipped with an alternative energy power source
FoodPrep	Whether the shelter is equipped with an inside kitchen facility
Photo	Photo of the shelter to identify the location

LIST OF ACRONYMS

ABM	Agent-Based Modelling
ADRA	Adaptive Disaster Risk Assessment
AIC	Akaike Information Criterion
AUC	Area Under the Curve
BDI	Belief-Desires-Intention
CAS	Complex Adaptive Systems
CNES	Centre national d'études spatiales
CoMSES	Computational Modeling in Social and Ecological Sciences
CRED	Centre for Research on the Epidemiology of Disasters
df	degrees of freedom
DRA	Disaster Risk Assessment
DRM	Disaster Risk Management
DRR	Disaster Risk Reduction
ECLAC	Economic Commission for Latin America and the Caribbean
ESF	Emergency Support Function
EU-JRC	European Commission – Joint Research Centre
FEMA	Federal Emergency Management Agency
GIS	Geographic Information Systems
GPS	Global Positioning System
GSA	Global Sensitivity Analysis
h	hour
HPC	High-Performance Computing
ICT	Information and Communication Technologies
IDE	Integrated Development Environment
IGN	Institut National de L'Informa Géographique et Forestiére
INSEE	Institut National de la Statistique et des Études Économiques
IPCC	Intergovernmental Panel on Climate Change
km	Kilometer
mb	Milibares
MCA	Multiple Correspondence Analysis
MDC	Meteorological Department Curaçao
MDS	Meteorogical Department St. Maarten
NASA	National Aeronautics and Space Administration
NGO	Non-Governmental organizations

NHC	National Hurricane Center
NOAA	National Oceanic and Atmospheric Administration
OCHA	Office for the Coordination of Humanitarian Affairs
ODD	Overview, Design concepts, Details protocol
ODD+D	Overview, Design concepts, Details, + Decision protocol
OFAT	One-Factor-At-A-Time
OSM	Open Street Maps
PC	Parallel computing
PCA	Principal Component Analysis
PEARL	Preparing for Extreme And Rare events in coastaL regions
PECS	Physical Conditions, Emotional State, Cognitive Capabilities and Social Status
PeVI	PEARL vulnerability index
PTSD	Post-Traumatic Stress Disorder
ROC curve	Receiver Operating Characteristic curve
SA	Sensitivity Analysis
SAMHSA	Substance Abuse and Mental Health Services Administration
SE	Standard Error
SIDS	Small Island Developing States
SOAR	State, Operator, And Result. Cognitive Architecture model
SPOT	Satellite Pour l'Observation de la Terre
STAT	Department of Statistics Sint Maarten
SXM	Abbreviation for Sint Maarten
TST	Total Simulation Time
UNDP	United Nations Development Programme
UNESCO-IHE	Educational institute for water education
UNFCCC	United Nations Framework Convention on Climate Change
UNISDR	United Nations International Office for Disaster Risk Reduction
USAID	United States Agency for International Development
VROMI	In Dutch: Ministerie van Volkshuisvesting, Ruimtelijke Ordening, Milieu & Infrastructuur In English: Ministry of Public Housing Spatial Planning Environment and Infrastructure
WEB-APP	web-based application
WR-DRM	Water-Related DRM

LIST OF TABLES

LIST OF FIGURES

About the Author

Neiler obtained his BSc, in Sanitary Engineering in 2005 from the faculty of Engineering in University of Antioquia, in Medellin, Colombia. He later obtained a post-graduate certificate as specialist in Geographic Information Systems (GIS), from the University of San Buenaventura, also in Medellin, Colombia in the year 2010. After graduation Neiler stay in the university working for one year as a research engineer in several environmental projects. After that he joined a water utility company where he worked for over 3 years in charge of the design of water distribution and sewer network systems. He later was the founder and CEO of a consultancy firm, where he further expand his expertise as a modeller of water and sewer systems as well as a GIS developer.

In 2010, looking for strengthen his knowledge and skills to fill some gaps he identified during his professional career, he moved to The Netherlands to study his Master of Science at UNESCO-IHE thanks to the support of Nuffic through the NFP scholarship. Neiler obtained his MSc with distinction in 2012 in water science and engineering with a specialization in Hydroinformatics. His master's topic was the optimal design of sewer network systems considering the effects of urbanization and land use change.

After graduating in 2012, Neiler stayed in UNESCO-IHE for six months as a research assistant. During this time he collaborate in the setup of 1D-2D models for flood control and disaster risk management in Sint Maarten and Virgin British Islands. Also, he contributed to the preparation of lecturing material for the course Urban Flood Management and Disaster Risk Mitigation. In 2013 he returned to his company in Colombia where he work in the design and construction of several projects related to water distribution and sewer systems as a lead designer in different municipalities across the country.

In March 2014, Neiler started his PhD research at IHE Delft, and the Faculty of Applied Sciences at TU Delft. His research was part of a larger project PEARL (Preparing for extreme and rare events in coastal regions), funded under the European Union FP7. This project ended in June 2018, since then her research has been partially funded by the project RECONECT (Regenerating ecosystems with nature-based solutions for hydro-meteorological risk reduction), from the European Union's Horizon 2020 Research and Innovation Programme.

LIST OF PUBLICATIONS

Journal Publications

Medina N., Abebe Y.A., Sanchez A., Vojinović Z., Nikolic I. (2021) Agent-Based Models for Water-Related Disaster Risk Management: A state-of-the-art review. Manuscript under review

Medina N., Sanchez A., Vojinović Z. (2021) Emergency Evacuation Behaviour in a Small Island Setting. The case study of Sint Maarten during Hurricane Irma. Manuscript under review

Medina N., Sanchez A., Vojinović Z. (2021) EvacuAPP: A web-based application for emergency management for hurricane disasters. Manuscript under review

Medina N., Abebe Y.A., Sanchez A., Vojinović Z., Nikolic I. (2021) Adaptive Disaster Risk Assessment. An Agent-Based Modeling Approach. Manuscript in preparation

Medina N., Abebe Y.A., Sanchez A., Vojinović Z. (2020) Assessing Socioeconomic Vulnerability after a Hurricane: A Combined Use of an Index-Based approach and Principal Components Analysis Sustainability 12:1452 doi:10.3390/su12041452

Medina N, Abebe Y.A., Sanchez A., Vojinović Z., Nikolic I. (2019) Surveying After a Disaster. Capturing Elements of Vulnerability, Risk and Lessons Learned from a Household Survey in the Case Study of Hurricane Irma in Sint Maarten Journal of Extreme Events 6:1950001 doi:10.1142/S2345737619500015

Sorg L., **Medina N.**, Feldmeyer D., Sanchez A., Vojinovic Z., Birkmann J., Marchese A. (2018) Capturing the multifaceted phenomena of socioeconomic vulnerability Natural Hazards 92:257-282 doi:10.1007/s11069-018-3207-1

Vojinovic Z., Keerakamolchai W., Weesakul S., Pudar R., **Medina N.**, Alves A. (2017) Combining Ecosystem Services with Cost-Benefit Analysis for Selection of Green and Grey Infrastructure for Flood Protection in a Cultural Setting Environments 4:3

Vojinović, Z., Hammond, M., Golub, D., Hirunsalee, S., Weesakul, S., Meesuk, V., **Medina, N.**, Sanchez, A., Kumura, S., Abbott, M.B. (2016) Holistic approach to flood risk assessment in areas with cultural heritage: a practical application in Ayutthaya, Thailand Natural Hazards 81:589-616 doi:10.1007/s11069-015-2098-7

Sanchez A., **Medina N.**, Vojinović Z., Price R. (2014) An integrated cellular automata evolutionary-based approach for evaluating future scenarios and the expansion of urban drainage networks Journal of Hydroinformatics 16:319 doi:10.2166/hydro.2013.302

Conference Proceedings and Book chapters

Medina, N., Sanchez, A., Vojinovic, Z. (2016) The Potential of Agent Based Models for Testing City Evacuation Strategies Under a Flood Event. Procedia Engineering, 154, 765-772

Medina, N., Sanchez, A., Nikolic, I; Vojinovic, Z. (2016) Agent based models for testing city evacuation strategies under a flood event as strategy to reduce flood risk. Presented at EGU- general assembly, Vienna, Austria

Medina N., Sanchez A., Vojinović Z. (2015) Automated runoff coefficient computation in urban drainage systems using Google satellite images and fuzzy classification. In: IAHR World Congress. The Hague, The Netherlands.

Sanchez A., Vojinović Z., **Medina N.**, Mynett A. (2015) Assessing the implications for urban drainage infrastructure of future scenarios of urban growth with cellular automata. In: IAHR World Congress. The Hague, The Netherlands.

Vojinović Z., Abbott, M., Makropoulos, C., Nikolic, I., Sanchez, A., Abebe, Y.A., Manojlovic, N., Pelling, M., **Medina, N.** (2014) Holistic Flood Risk Assessment In Coastal Areas-The PEARL Approach. In: 11th International Conference on Hydroinformatics. New York City, USA.

Newsletter Article

Abebe, Y.A., **Medina N.**, and Vojinovic, Z. (2018). Strengthening Sint Maarten: Lessons Learned after Hurricane Irma. Research Counts, 2(12). Boulder, CO: Natural Hazards Center, University of Colorado Boulder. Available at: http://bit.ly/RC_SXM_Irma

Netherlands Research School for the
Socio-Economic and Natural Sciences of the Environment

DIPLOMA

for specialised PhD training

The Netherlands research school for the
Socio-Economic and Natural Sciences of the Environment
(SENSE) declares that

Neiler de Jesús Medina Peña

born on 28 April 1981 in Medellín, Colombia

has successfully fulfilled all requirements of the
educational PhD programme of SENSE.

Delft, 21 June 2021

Chair of the SENSE board

Prof. dr. Martin Wassen

The SENSE Director

Prof. Philipp Pattberg

The SENSE Research School has been accredited by the Royal Netherlands Academy of Arts and Sciences (KNAW)

KONINKLIJKE NEDERLANDSE
AKADEMIE VAN WETENSCHAPPEN

The SENSE Research School declares that Neiler de Jesús Medina Peña has successfully fulfilled all requirements of the educational PhD programme of SENSE with a work load of 54.1 EC, including the following activities:

SENSE PhD Courses

- Environmental research in context (2014)
- Research in context activity: 'Preparing and presenting Summer course on 'Sensores Remotos y Sistemas de Información Geográfica para manejo y gestión de riesgos en Inundaciones' (GIS and Remote Sensing for Flood Risk and Disaster Management) on 23-26 April 2018 at University of Antioquia (Medellin, Colombia)'

Selection of Other PhD and Advanced MSc Courses

- Academic writing for PhD fellows, IHE Delft (2019)
- English for Academic Purposes-3, TU Delft (2016)
- PhD Start-Up modules A, B and C (including Scientific integrity), TU Delft (2015-2018)
- Self-Presentation: Presenting yourself and your work, TU Delft (2015)
- How to make a questionnaire and conduct an interview, TU Delft (2017)
- Popular Scientific Writing, TU Delft (2017)
- Brain Management, TU Delft (2017)
- Foundations of teaching, learning and assessment, TU Delft (2017)
- Model Thinking, University of Michigan/MOOC Coursera (2014)
- Methods and Statistics in Social Sciences. Quantitative Methods, Qualitative Research Methods, Basic Statistics and Inferential Statistics, University of Amsterdam/MOOC Coursera (2017)
- Urban Flood Management and Disaster Risk Mitigation, IHE Delft (2020)

Management and Didactic Skills Training

- Teaching in the MSc courses 'Urban Flood Management and Disaster Risk Mitigation', 'Urban Water Systems', 'Advanced Computer Modelling of Water Distribution Networks' and 'GIS and Remote Sensing Applications for the Water Sector' (2015-2018)

Oral Presentations

- *Agent based models for testing city evacuation strategies under a flood event as strategy to reduce flood risk.* European Geosciences Union - General Assembly, 17-22 April 2016, Vienna, Austria
- *The potential of agent based models for testing city evacuation strategies under a flood event.* 12th International Conference on Hydroinformatics, 21-26 August 2016, Incheon, South Korea
- *Planning strategies for flood disaster risk prevention with human behaviour models.* 37th IAHR World Congress, 13-18 August 2017, Kuala Lumpur , Malaysia

SENSE coordinator PhD education

Dr. ir. Peter Vermeulen

For Product Safety Concerns and Information please contact our EU representative GPSR@taylorandfrancis.com Taylor & Francis Verlag GmbH, Kaufingerstraße 24, 80331 München, Germany

T - #0137 - 160425 - C328 - 240/170/18 - PB - 9781032116174 - Gloss Lamination